Nonlinearity in Energy Harvesting Systems

Elena Blokhina · Abdelali El Aroudi
Eduard Alarcon · Dimitri Galayko
Editors

Nonlinearity in Energy Harvesting Systems

Micro- and Nanoscale Applications

Springer

Editors
Elena Blokhina
School of Electrical and Electronic Engineering
University College Dublin
Dublin
Ireland

Abdelali El Aroudi
Department of Electronic, Electrical and Automatic Engineering
University Rovira I Virgili
Tarragona
Spain

Eduard Alarcon
Technical University of Catalonia
Barcelona
Spain

Dimitri Galayko
Laboratory of Computer Sciences
UPMC—Sorbonne Universities
Paris
France

ISBN 978-3-319-20354-6 ISBN 978-3-319-20355-3 (eBook)
DOI 10.1007/978-3-319-20355-3

Library of Congress Control Number: 2016948783

Printed on acid-free paper

This Springer imprint is published by Springer Nature
The registered company is Springer International Publishing AG
The registered company address is: Gewerbestrasse 11, 6330 Cham, Switzerland

Preface

As circuits and systems become more miniature and less power-consuming, we move towards a complete synthesis of systems and renewable power sources. This is an essential step towards fully autonomous, micro- and nano-scale smart systems that will become a part of the Internet of Things (IoT) trend. The area of engineering that develops autonomous energy sources, able to derive energy from environment, is called energy harvesting. It is a versatile, multidisciplinary and rapidly developing area that combines electronic engineering, materials science and physics with circuit design and system-level integration.

There are many energy sources available for harvesting in the environment. Perhaps, one of the most common examples of an energy harvester known to every reader is a solar cell. But this is not the only example though. In this book, we discuss vibration energy harvesting, a particular type of energy conversion from mechanical motion. Due to the ubiquitous nature of mechanical motion and vibrations, they seem as a very promising approach to supply microsystems with energy. Moreover, since we would like to focus on micro- and nano-systems, we will consider only electrostatic and electromagnetic energy conversion mechanisms, which are particularly suitable for implementation through microtechnologies and can be scaled down.

Since the idea of vibration energy harvesting has been introduced, it has become clear that there are certain challenges that cannot be addressed applying a conventional approach based on linear mechanical resonators. In particular, modern energy harvesters are required to be robust, to respond to a large range of external frequencies, and in some cases, to operate with noise-like vibrations. Often perceived as parasitic, nonlinearity in energy harvesting systems can address these particular challenges. As we shall show throughout the book, nonlinearity is inevitable and arises due to the nature of energy harvesters. Since energy harvesters are 'mixed' systems combining the electrical and the mechanical domains, the presence of electromechanical coupling causes nonlinearity that should not be neglected in analysis and, actually, can be utilised and improve the performance of a harvester.

This book covers a range of different topics related to nonlinearity in vibration energy harvesters. We start, however, with some fundamentals and explain what is microtechnology and how it is employed to build a harvester. We also present introductory discussions on oscillators and transducers for energy harvesters. With these fundamentals covered, we proceed and discuss each particular case of nonlinearity due to nonlinear mechanical resonators (nonlinearity in the mechanical domain) and conditioning circuits in electrostatic and piezoelectric harvesters (nonlinearity due to electromechanical coupling).

We sincerely thank all the authors for their research and contribution to the field and to this book: Manuel Domínguez-Pumar, Joan Pons-Nin, Juan Chávez-Domínguez, Hassen M. Ouakad, Miguel López-Suárez, Riccardo Rurali Gabriel Abadal, Sukhdeep Kaur, Cuong Phu Le, Peter Harte, Orla Feely, Adrien Badel and Elie Lefeuvre.

Dublin, Ireland — Elena Blokhina
Tarragona, Spain — Abdelali El Aroudi
Barcelona, Spain — Eduard Alarcon
Paris, France — Dimitri Galayko

Contents

Chapter 1
Introduction to Vibration Energy Harvesting

Elena Blokhina, Abdelali El Aroudi, Eduard Alarcon and Dimitri Galayko

1.1 Historical Background

Energy harvesting has received increased attention in the recent years from different disciplines, mainly due to its potential impact as a key technology enabling self-powered ultra-low-power electronic systems.

In a general case, energy harvesting is the utilisation of ambient energy from the environment surrounding a system, converting this energy to a different form (usually electricity) and then using this energy to power the aforementioned system. This differs from powering a system using finite energy sources such as coal, oil, batteries, fuel cells, etc. As a concept, energy harvesting is not new: in fact, it has been in operation for millennia. Water wheels that harvest energy from the currents of streams to power mills have been in existence since at least the first century BC [1] while wind powered machines are known from the first century AD [2].

In the roughly two thousand years that have passed since those first recorded energy harvesting systems the world has changed radically, with electricity now being one of the main sources of energy for many of the systems used on the planet. In that time, the concept of energy harvesting has not changed but the sources for, and forms of energy that are outputted from energy harvesting systems have. Modern day

E. Blokhina (✉)
University College Dublin, Dublin, Ireland
e-mail: elena.blokhina@ucd.ie

A. El Aroudi
University Rovira I Virgili, Tarragona, Spain
e-mail: abdelali.elaroudi@urv.cat

E. Alarcon
Universitat Politécnica de Catalunya, Barcelona, Spain
e-mail: eduard.alarcon@upc.edu

D. Galayko
UPMC — Sorbonne Universities, Paris, France
e-mail: dimitri.galayko@lip6.fr

E. Blokhina et al. (eds.), *Nonlinearity in Energy Harvesting Systems*,
DOI 10.1007/978-3-319-20355-3_1

systems for large-scale renewable power generation such as solar arrays, wind farms and ocean wave generators follow the concept of energy harvesting. By downscaling the energy levels to those used in portable applications, a historic example of energy harvesting are the kinetic wristwatches, which harness the human body kinetic energy and store it mechanically in springs to actuate mechanical clocks. Nowadays, the term energy harvesting is usually understood to mean the generation of electrical energy for low powered, miniature, wireless devices [3]. It is this definition of energy harvesting that this book follows.

In recent years, energy harvesting has become a huge area of research and has attracted much interest from both academia [3–6] and industry [7–9]. This interest has been boosted even more since we enter the age of the Internet of Things (IoT), aiming at wearable technologies and micro-electronics for bio-applications. In many instances, the main objective of these IoT technology will be to sense information about the environment and communicate this information wirelessly. In this case, the communicating objects connected into networks are referred to as wireless sensor nodes (WSNs).

In numerous applications of WSNs, the sensors are located in remote locations that reduce their accessibility. This means that in a lot of cases, these devices will require their own power supply. A solution to this problem is to use a battery or a fuel cell. Such power supplies though, have finite lifetimes and depending on the power usage of the WSN in question, may not last very long. This creates a different problem: replacing these power sources. This is particularly true when the objects being powered are in difficulty to reach locations. Thus, it is desirable that WSNs are completely autonomous from a power point of view and this is the major motivating factor for energy harvesting research.

1.2 Ambient Energy Sources for Harvesting

There are many different ambient sources in a variety of environments from which energy can be harvested. The choice of which source to harvest from depends not only on its availability but also on the application in question. Below, some of the more common energy harvesting sources are outlined, along with some of their advantages, disadvantages and applications.

Ambient **solar energy** is the most popular source for energy harvesting, and as a result it is the most mature and commercially established field in energy harvesting research [10]. Solar energy harvesting is based on the principle of the photovoltaic effect. The photovoltaic effect was discovered by Alexandre-Edmond Becquerel and using this knowledge, he designed the first solar cell in 1839 [5, 11]. For nearly forty years, solar energy harvesting has been used in consumer electronics with the most ubiquitous example being the handheld calculator [12].

Standard solar cells have energy conversion efficiencies of between ten and twenty percent, and can produce roughly 100 mW/cm^2 in sunlight. This reduces by an order of magnitude to approximately 100 μW/cm^2 in a standard illuminated office [4]. This is a major disadvantage for solar energy harvesting. If the system being powered is located in a poorly lit environment, e.g. a bio-sensor monitoring a person's health from within their body, then the amount of energy converted reduces dramatically and thus solar energy scavenging is not a viable option.

Temperature variations can be converted to electrical energy through thermoelectric generators. These thermal energy harvesters generally consist of two different conductors joined in two places. When there is a temperature difference between these junctions, an open circuit voltage develops between them. This is the basis of the Seebeck effect, named after Thomas Johann Seebeck who discovered the phenomenon in 1821 [5, 6, 13]. The effectiveness of these harvesters hinges on the availability of large thermal gradients since the voltage produced by the generator is proportional to the temperature difference between the two junctions [14]. There are many heat sources where thermoelectric generators can be used including human body warmth [15] and the waste heat of car engines [16]. Harvesting thermal energy from the human body, in theory, is very attractive for wearable technologies but due to the small temperature gradient between the human body and the ambient temperature, these harvesters can only output less than 30 $\mu W/cm^2$ [15]. This leaves thermal energy harvesters somewhat ineffective for applications like bio-sensors that require very compact device geometries.

Ambient **radio frequency** energy is another possible source for energy harvesting. Ambient RF energy is available through the vast number of radio communications transmitters scattered throughout the modern world. These transmitters generate strong electromagnetic fields and RF energy harvesters can convert this into electrical energy. A major limitation of RF energy harvesting is the very low power density levels associated with it. Power density levels less than $1\,mW/m^2$ are expected around 25 m from a GSM base station and this is at least an order of magnitude less again from wireless LAN networks [6, 17, 18]. The power received by an RF harvester decreases dramatically as the distance from the source transmitter increases. Thus, this type of harvester requires either a very large collection antenna, which increases its size beyond practical use in a microsystem, or to be very close in proximity to the radiating source, thereby reducing its number of viable applications [4].

Microbial fuel cells operate on a similar principle to standard fuel cells. Their difference lies in the fact that unlike normal fuel cells, which have a finite lifetime, albeit with a very large energy capacity, microbial fuel cells source their energy from the potential of anaerobic bacteria to respire, oxidise and transport electrons efficiently [5]. They convert chemical energy into electrical energy through these 'electro-active' bacteria. Their potential is certainly not fully realised yet. In fact, their main applications appear not only to be in energy harvesting but also in the treatment of wastewater in a very environmentally friendly way [19, 20].

The final ambient energy source covered here is **kinetic energy**. Much research has gone into kinetic energy harvesting in the 21st Century [21–28]. This is due to certain advantages it has over other harvesting technologies. Kinetic motion energy sources are available in most environments and due to microelectomechanical systems (MEMS) technology, kinetic energy harvesters can be microscaled with great

Table 1.1 A sample of recorded power densities for ambient energy sources according to a variety of references. Directly comparing different energy harvesting methods is difficult due to the multitude of variables affecting energy harvester performance. This table serves only to give an idea of the order of magnitude of power density expected from current designs

Energy source	Power density	References
Solar	100 mW/cm^2 (direct sun)	[30]
	0.1 mW/cm^2 (office desk)	
Temperature gradients	0.015 mW/cm^2 (10 °C gradient)	[31]
RF waves	1 mW/m^2 (25 m from GSM base station)	[5]
Microbial fuel cell	6.86 W/m^2	[5]
Vibrations	0.01–0.1 mW/m^2	[14]

success. This means they can be used in a vast array of applications from bio-tech devices inside the human body to the structural monitoring of buildings during construction [29]. Vibration energy harvesters (VEHs) that are powered by ambient mechanical vibrations are a subset of a more general class of kinetic energy harvesters (KEHs) that are powered by any kind of motion (fluidic, air, mechanical and, etc.).

These are currently the main areas of energy harvesting research. A table comparing the presently expected power densities for harvesting devices with respect to the various sources can be seen in Table 1.1. Unfortunately, this table serves only to give an idea of the expected order of magnitude of power density given the multitude of variables affecting energy harvesting performance.

Some energy harvesting methods are more well established than others. The availability of the ambient source affects all of the methods, but certainly some sources are more readily available than others, e.g. vibrations are in abundance within the human body but sunlight is not. Even though an ambient energy may be available in a certain environment, the harvester may not be able to scavenge it effectively if it is not located in close proximity to the source of this energy, e.g. RF harvesters.

1.3 Vibration Energy Harvesters

In a broad sense, kinetic energy harvesters can convert any mechanical motion energy (like fluid flows, pressure variations and ambient vibrations) into electrical energy to power systems located in the environs of this 'free' mechanical energy [29]. Vibration energy harvesting is a subset of this harvesting method dedicated to the conversion of vibration energy into electrical energy. There are a large number of sources from which vibrations can be harvested and it is for this reason that this is a very active and growing area of research. These sources include the shaking of a car engine, the

Table 1.2 The acceleration and frequency characteristics of a variety of different sources of vibration energy [14, 33]

Vibration source	Acceleration [ms^{-2}]	Frequency [Hz]
Car engine	12	200
Base of 3-axis machine tool	10	70
Food blender casing	6.4	121
Clothes dryer	3.5	121
Door frame just after door closes	3	125
Microwave oven	2.25	121
Air conditioning vents in office building	0.2–1.5	60
External windows next to a busy street	0.7	100
CD drive on notebook computer	0.6	75
Washing machine	0.5	109
Refrigerator	0.1	240
Human motion	0.35	1–10

motion of a beating heart and the low level vibrations of a microwave oven amongst others [32]. The acceleration and frequency characteristics of a variety of vibration sources found in [14, 33] can be seen in Table 1.2.

A flow chart highlighting a generic process for vibration energy harvesting can be seen in Fig. 1.1. Mechanical vibration energy in an environment can be 'captured' by a mechanical resonator. Such a resonator is normally composed of a mass coupled to a spring. When this mechanical energy has been captured, it must then be converted to electrical energy. A transducer converts energy from one form to another. This electromechanical conversion is controlled by conditioning circuitry, which sometimes take the form of smart electronics. This conditioning circuitry also manages the transfer of the converted energy to the system or load it exists to supply for example, a WSN or a pacemaker.

Vibration energy harvesters are generally classified by their transduction method. The three most common transduction methods for VEHs are: electromagnetic, piezoelectric and electrostatic. These methods are discussed below.

Electromagnetic vibration energy harvesters (emVEHs) convert mechanical energy in the form of vibrations to electrical energy using electromagnetic transduction methods similar in theory to what most macroscale electrical generators are based upon [10]. Their basic design consists of a magnetic core and a coil, one of which is coupled to the mechanical resonator and the other remains static. Following Faraday's law of induction, when the resonator vibrates, the relative movement of

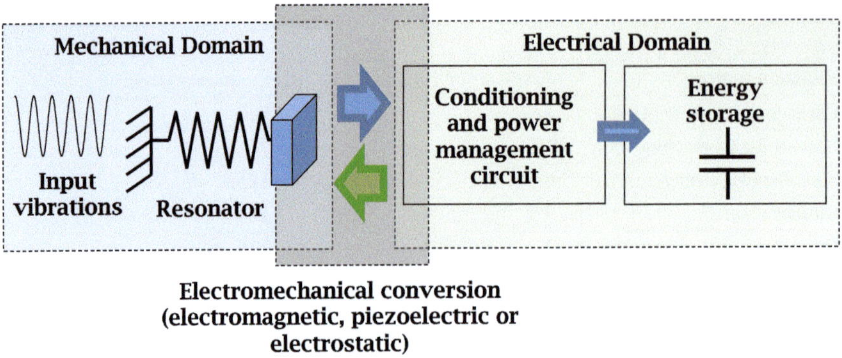

Fig. 1.1 Generic flow chart describing the vibration-to-electrical energy conversion. Mechanical vibrations in an environment are 'captured' by a mechanical resonator. This resonator will be coupled to a transducer which performs an electromechanical conversion of the energy, resulting in electricity. This process is controlled by some conditioning circuity which also manages the transfer of the harvested energy to the load/circuit that needs to be supplied, e.g. a pacemaker or a WSN

the magnet to the coil induces an AC voltage on the coil. It is in this way that the mechanical energy is converted to electrical energy. An example of the electromagnetic transduction mechanism can be seen in Fig. 1.2a.

Many emVEH designs have been fabricated both with single magnets [34, 35] and multiple magnets [36, 37]. Electromagnetic transduction is a very effective way of harnessing mechanical vibrations. EmVEHs have good power densities and do not require complex conditioning electronics [38]. Their major drawback lies in their inability to be microscaled due to their bulky magnets and coils. As a result, emVEHs are not suited to applications with strict size constraints.

Piezoelectrical vibration energy harvesters (pVEHs) are based upon the piezoelectric effect whereby a strain in a certain type of material produces an electric field within that material [10]. The piezoelectric transduction mechanism can be seen in Fig. 1.2c. In a pVEH, the transducer/resonator generally consists of a cantilever beam

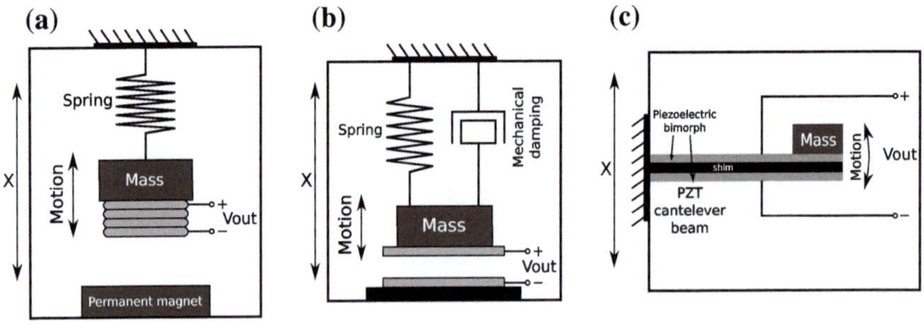

Fig. 1.2 Graphical representations of VEH transduction mechanisms from [14]. **a** Electromagnetic, **b** Electrostatic, **c** Piezoelectric

fixed on one end, with an oscillating mass at the opposite end. The cantilever contains a piezoelectric material layer and when the resonator vibrates, this layer is deformed. This deformation causes charge to be displaced across the piezoelectric element. The resulting potential difference can be used to power devices [33, 39–42].

The main advantages of pVEHs are their high output power, particularly for their size. That having been said, their cantilever beams suffer from material degradation including fatigue, depolarisation and micro-'fissure' [43]. This considerably reduces their longevity. This degradation is accelerated when pVEHs are microscaled using MEMS technology. More discussion about piezoelectric energy harvester will be provided in Chaps. 4 and 10.

Electrostatic vibration energy harvesters (eVEHs) employ high-quality mechanical resonators and conditioning electronics, coupled together through a variable capacitor (the transducer). Usually, the electromechanical coupling is implemented by employing the movable plate of the variable capacitor as the mass of the mechanical micro-resonator. The energy conversion occurs when there is a constant charge or voltage on the capacitor whilst its plates are separating. As the capacitance decreases, electricity is generated. The management of this conversion is carried out by control/conditioning electronics. The mechanism of an eVEH can be seen in Fig. 1.2b. Over the past fifteen years, many eVEHs have been reported on in [10, 22, 24, 25, 27, 44].

One of the major benefits of eVEHs is their ability to be microscaled using mature MEMS technology. Although their output power levels are not high (in the order of a few tens of microwatts), due to their diminutive size they have huge potential in a variety of applications that have strict constraints on package dimensions. These include wireless sensor nodes and bio-sensors. They also display highly nonlinear behaviour, which requires detailed analysis and study. The electrostatic transduction mechanism is described in detail in Chap. 3. A comparison of the characteristics of the three VEH classes can be found in Table 1.3.

Table 1.3 Some advantages and disadvantages of the three main VEH transduction mechanisms

Transducer	Advantages	Disadvantages
Electromagnetic	No voltage source needed	Low output voltage
	Robust	Difficult to microscale
Piezoelectric	No voltage source needed	Requires voltage rectification
	High output voltage and power	Fatigue of PZT beams
Electrostatic	High output voltage	Voltage source needed
	Suited to microscaling	Requires complex control electronics

1.4 Components of Vibration Energy Harvesters

In this section, we will give more details on the components of vibration energy harvesters and explain their role in the process of vibration-to-electricity conversion. A functional energy harvesting system requires (a) a mechanical subsystem that captures mechanical motion of the environment, (a) an energy conversion device, which is known as a transducer, and (c) a power management circuit that will control the storage and usage of converted energy (see Fig. 1.1).

In general, we know that energy has many different forms. For example, kinetic energy of vibrations cannot be used directly to power an electronic circuit. It must be **converted** to the electrical domain first. Thus, we should have a device or a mechanism that will transfer one form of energy (mechanical) to another (electrical). Such a device is called transducer. The transducer may be electromagnetic, piezoelectric or electrostatic (capacitive).

Electromechanical transducers are never attached directly to moving mechanical objects—this would be ineffective (and in some cases, technologically impossible). There must be an intermediate system that transfers external motion due to the environment to the motion of a transducer. This system must be purely mechanical since it involves the transformation of motion entirely in the mechanical domain.

Let us consider the example of a electrostatic (capacitive) transducer. How could we arrange a proper connection of a variable capacitor to a vibrating environment? Consider a system shown schematically in Fig. 1.3. The electrostatic transducer of this example has one fixed electrode and one movable electrode. The movable electrode is suspended on a spring. The spring is attached to an enclosing frame. Now the frame is placed in a vibrating environment and experiences acceleration due to external

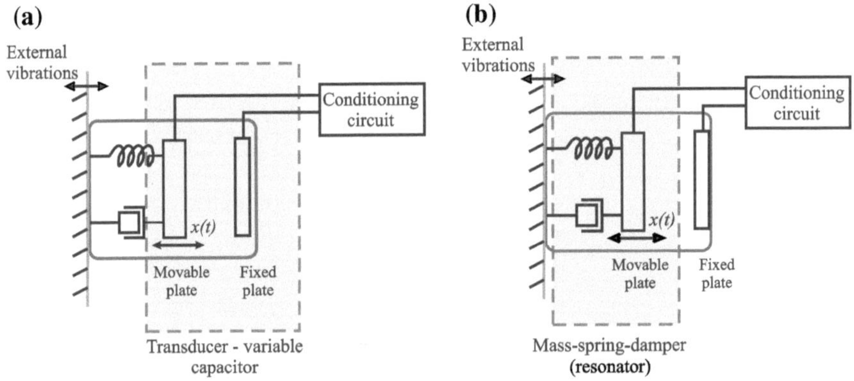

Fig. 1.3 Schematic view of a resonator, a transducer and a conditioning circuit of an electrostatic VEH. Figure **a** emphasises the transducer while figure **b** emphasises the resonator in this scheme. Note that the movable electrode is simultaneously a part of the transducer and a part of the resonator

vibrations. The relative displacement of the movable electrode with respect to the frame cases a change in the capacitance of the transducer.

This is shown in Fig. 1.3: case (a) highlights the variable capacitor (transducer) while case (b) emphasises the mechanical mass-spring system that controls the capacitance of the transducer. As is seen from the figure, the movable electrode is simultaneously a part of the mechanical system and a part of the transducer. This mass-spring system forms a resonator. Since there are always some losses due to dissipation (for example, air damping) in a realistic mechanical system, we add a damper to the system to accommodate these losses. Hence, we usually say that this is a mass-spring-damper system or a damped resonator. The ideal model of the mass-spring-damper system is shown in Fig. 1.4. The aim of the resonator is to constrain the motion of the movable electrode and capture external vibrations in the most efficient way, usually through the phenomenon of resonance.

Actual resonators do not look like the simplified structure shown in Figs. 1.3 and 1.4 with a proof mass attached on a spring. Realistic VEH resonators are distributed mechanical structures suspended on elastic arms, sometimes of a very complex form. Two examples of an electrostatic and piezoelectric VEHs are shown in Figs. 1.5 and 1.6. We note, however, that the 1D model shown in Fig. 1.4 is a very good approximation, simple and convenient for the use in analytical and semi-analytic studies of VEHs. We will discuss this model in Chap. 3.

Every VEHs also have electrical components. The role of this component is to condition the state of the transducer and to manage the converted energy. The circuit implementing these functions is called the conditioning circuit (see Figs. 1.1 and 1.3). The conditioning circuit operates entirely in the electrical domain. In the case of electrostatic transducers, the conditioning circuit implements dynamical biasing. Without biasing, a capacitive transducer cannot operate and convert energy. Biasing of electrostatic transducers is described in Chap. 4. In the case of piezoelectric and electromagnetic transducers, conditioning circuits help to maximise the converted energy. Conditioning circuits for piezoelectric transducers are discussed in details in Chap. 10.

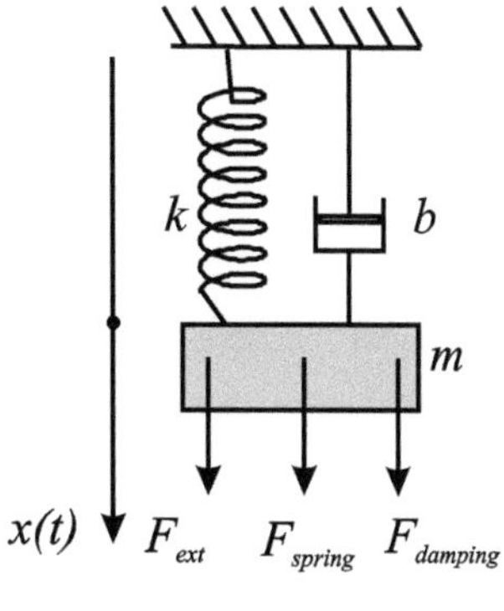

Fig. 1.4 Schematic view of the ideal model of a mass-spring-damper system (the resonator) driven by an external (mechanical) force F_{ext}

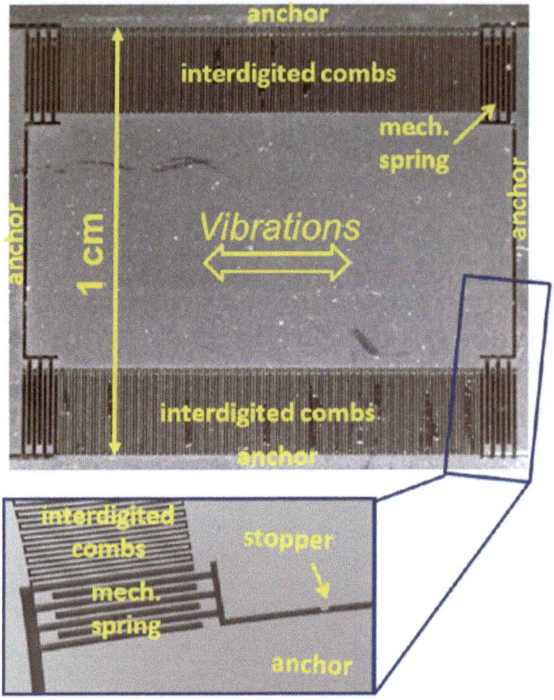

Fig. 1.5 A photo of a MEMS resonator and a transducer for a VEH described in [45]. Note the proof mass, the direction of vibrations and the shape of the springs that support the proof mass

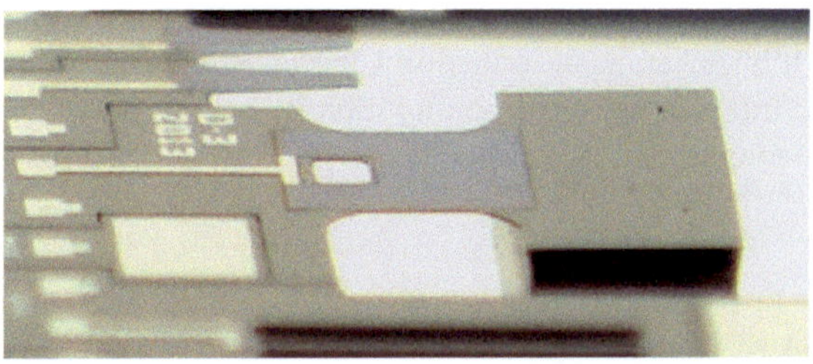

Fig. 1.6 Another example of a MEMS resonator: a photo of a resonator with a piezoelectric transducer. Note that the role of the spring is played by a thin and light elastic clamped-free beam while the proof mass is represented by a large and massive 'tip' attached to the beam [46]

Finally, electrical energy (voltage) generated by the transducer needs to be converted into a form suitable to supply to a load. The circuit that implements dc–dc conversion is called the power management circuit. The operation of this circuit is also performed entirely in the electrical domain. In many cases, the power management block is tightly integrated with the conditioning circuit.

Usually, very low levels of the converted power (above one milliwatt in the majority of cases) makes the design of conditioning and power management circuits extremely challenging. For this reason, electrical conditioning of transducers for vibration energy harvesters deserves particular attention and is a large area of research itself.

Thus, we have established that in order to capture the motion of the environment, we use a mass-spring-damper system (resonator). The resonator drives the movable plate of a variable capacitor (transducer) changing its capacitance and, therefore, changing its electrical energy. A conditioning circuit controls the voltage (or charge) applied to the transducer in order to maximise energy conversion. Converted energy is further managed by the power management circuit and stored. The resonator operates in the mechanical domain while the conditioning circuit, the power management circuit and energy storage operate in the electrical domain. The variable capacitor (transducer) couples the mechanical and electrical domains.

1.5 Nonlinearity in Vibration Energy Harvesters

1.5.1 Role of Nonlinearities

For thorough understanding and optimal design of energy harvesting systems, it is necessary to take account of the presence of nonlinearity, both desired and undesired, within these systems. This nonlinearity arises through a number of mechanisms. The fundamental methods of energy conversion are typically nonlinear. Nonlinearities appear both in the electrical domain, through elements such as diodes, and the mechanical, for example, through nonlinearity in the spring force and impact nonlinearities. A number of recent contributions seek to utilise nonlinearity in novel way to improve performance. High-Q mechanical resonators that commonly appear in vibration energy harvesters can efficiently harvest energy only when excited by vibrations very close to the resonant frequency. Designers are exploring the use of nonlinearity to widen the bandwidth and reduce the ambient frequency sensitivity of such harvesters.

The modelling, analysis and design of energy harvesters are already made complex by the fact that the systems combine elements from different domains. Nonlinearity adds to this complexity. The behaviour of linear systems is broadly well understood, and the tools available for their analysis are readily applied. With the introduction of nonlinearity, even apparently simple systems can give rise to highly complex behaviour. This can include the presence of coexisting solutions, sudden qualitative changes in behaviour with parameter variation, various forms of instability and the intricate patterns of behaviour known as chaos. The combination of nonlinearity with stochastic excitation brings still further complexity.

The need for understanding of nonlinearity is not unique to energy harvesting systems. For around a century now, the analysis of a variety of electrical systems has drawn on, and has also inspired, advances in the underpinning mathematical science of nonlinear systems. The theory and application of this science saw huge advances throughout the twentieth century, beginning with the work of Lyapunov [47] and Poincare [48] around the start of that century and proceeding through the work of van der Pol, Andronov, Lorenz and Smale to the popularisation and broad application of this work since the 1980s. The name of Leon Chua is synonymous with the application of nonlinear methods in electrical engineering, with seminal contributions in memristive circuits, chaotic circuits, neural networks and the development of a theoretical foundation for nonlinear circuits more generally.

The theory of nonlinear oscillators [49–52] was central to this development, and is also of particular relevance to the study of energy harvesting. The behaviour of electronic oscillators cannot be fully explained by application of linear theory. A number of authors delivered successful analyses through various forms of harmonic balance. Others applied a variety of methods of nonlinear dynamics to study oscillations by examining the existence, stability and bifurcation of limit cycles of related nonlinear differential equations. The long-observed entrainment of oscillators was successfully studied by application of techniques of nonlinear dynamics.

Nonlinearity, often considered as a parasitic effect by electrical engineers and system designers, is of great importance in the design of VEHs. There are a number of issues related to modern harvesters, and these issues cannot be addressed by employing linear 'building blocks' for VEHs. Since we just discussed the role of VEH components, we can take, for instance, the resonator and its role in a VEH and briefly summarise the advantages and disadvantages of linear and nonlinear resonators.

One of the most sought-after effects that designers aim to achieve is broadening of the frequency response of a VEH with respect to the frequency of external vibrations ω_{ext}. It is clearly a useful effect since it allows the resonator of a VEH to sustain mechanical motion even in the case when the external driving frequency is far from the natural frequency of the resonator. This is virtually impossible to achieve with a linear resonator: in order to widen the frequency response of linear resonator one would have to decrease its quality factor. This, in turn, will result in additional mechanical losses and worsen electromechanical coupling and energy conversion. However, a wider frequency response can be achieved using nonlinearities. We are going to overview three common sources of nonlinearities in VEHs: mechanical nonlinearity associated with nonlinear springs of the resonator, impact nonlinearity and nonlinearity that arises from electromechanical coupling.

1.5.2 Examples of Useful Nonlinearities

From Sect. 1.4 we know that a VEH must employ a mechanical resonator. Let us compare two energy harvesting systems, one employing a high-Q linear resonator and the other—a high-Q nonlinear resonator with mechanical nonlinearity arising from a nonlinear spring. (Here we imply that a reader has a basic knowledge of a resonator and resonance. A detailed discussion on these will be presented in Chap. 3.)

First, consider a case of a linear resonator having a natural resonance frequency ω_0. The resonance response (i.e. the displacement of the resonator as a function of the frequency of the driving vibrations) of a linear resonator is shown in Fig. 1.7, the line marked 'linear'. If we drive a harvester that utilised a linear resonator by a harmonic vibration (or vibrations) at a frequency (or frequencies) $\omega_{ext} \approx \omega_0$, we will obtain large amplitude vibrations. This phenomenon is called **resonance**. The characteristics of these vibrations are completely predictable. Now, if for some reason the frequency of the external vibrations ω_{ext} changes and shifts from ω_0, the linear resonator will be quickly 'detuned' from resonance, and the amplitude of vibrations will drop significantly.

For this reason, researchers and VEH designers seek to expand the resonance response of the mechanical part of a harvester (keeping the Q-factor of the resonator large) [53–59]. This is only possible by employing some sort of nonlinearity. For example, one can introduce mechanically induced nonlinearities to the resonator's springs. These nonlinearities lead to some fundamental nonlinear effects (the spring hardening or the spring softening effects). The spring hardening effect is observed when the effective resonance frequency increases with the amplitude of external vibrations and with the amplitude of the resonator vibrations. The spring softening effect is observed when the effective frequency, vice versa, decreases. These two fundamental cases are shown in Fig. 1.7 and are manifested through the 'tilt' of the resonance curve to the right (hardening) or to the left (softening).

Now, assume that we drive a harvester with a nonlinear resonator by a harmonic vibration at a frequency ω_{ext}. We see that the nonlinear resonator will continue to be 'in resonance' even for frequencies ω_{ext} that are not close to the natural frequency ω_0.

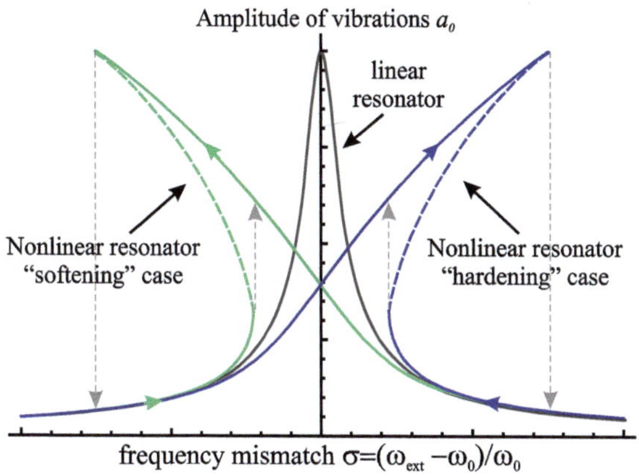

Fig. 1.7 Resonance in linear and nonlinear resonators. The amplitude of steady-state forced oscillations is shown as a function of the normalised frequency mismatch $\sigma = \omega_{\text{ext}}/\omega_0 - 1$. The *grey line* corresponds to the linear case while the *blue* and *green lines* correspond to nonlinear cases. For the nonlinear cases one can see multi-modality and hysteresis: there is a segment of the resonance curve where two stable solutions of a_0 and one unstable a_0 (shown by the *dashed line*) coexist

Therefore, for linear resonators

- Disadvantage: narrow bandwidth and high frequency selectivity;
- Advantage: unique amplitude and stability;

while for nonlinear resonators

- Disadvantage: multi-stability and hysteresis may lead to jumps between two or more possible solutions;
- Advantage: Wider bandwidth and less selective frequency response.

Nonlinear resonators seem to be particularly suitable for wideband energy harvesting and noise harvesting (i.e. when external vibrations represent noise) due to their wider frequency response. This property has been exploit in vibration energy harvesters by designing and fabricating MEMS resonators with nonlinear springs. For example, study [60] shows that MEMS energy harvesters with nonlinear springs showing strong softening effect (similar to the one shown in Fig. 1.7) respond effectively to noise actuation.

It is interesting that similar nonlinear effects (hardening or softening) can be introduced into the resonator using mechanisms other than the nonlinear design of springs. For instance, impacts (with stoppers or another mass) and electric nonlinearity due to the transducer force will bring similar nonlinear effects and can be utilised for the widening of the frequency response of the resonator.

Study [61] introduces a multiple mass system and demonstrate how the dynamics of multiple mass systems can be used to significantly enhance the power output of VEHs compared to single-mass VEH designs. The particular case of such a system with two-mass resonator is shown in Fig. 1.8. The reason why this structure was proposed for energy harvesting is that it allows us to explore the velocity amplification principle through the impact of the proof mass with another mass. Indeed, note that the second mass is detached from the lower (heavier) mass and experiences an impact with the lower mass if the system is driven by large amplitude vibrations. To explain the principle of velocity amplification, consider an elastic collision the two travelling masses. The conservation of total momentum dictates that the final velocity of the smaller mass v_{2f} is equal to

$$v_{2f} = \frac{(e+1)m_1 v_{1i} + (m_2 - e\,m_1)v_{2i}}{m_1 + m_2} \tag{1.1}$$

where e is the coefficient of restitution and v_{1i} and v_{2i} are the velocities of the first and second masses before the impact. By assuming that $e = 1$ and $m_1/m_2 \to \infty$, we can obtain the upper limit to the velocity of the second mass $v_{2f} = 2v_{1i} - v_{2i}$. In the case of equal but opposite velocities $v_{1i} = -v_{2i}$, one can obtain a substantial gain in the velocity of the second mass $v_{2f} = 3v_{2i}$. This is particularly important for VEHs whose transduction mechanism is proportional to the velocity. This is the case for electromagnetic harvesters where the induced e.m.f in the coil connected to the resonator is directly proportional to the velocity of the second mass v_2.

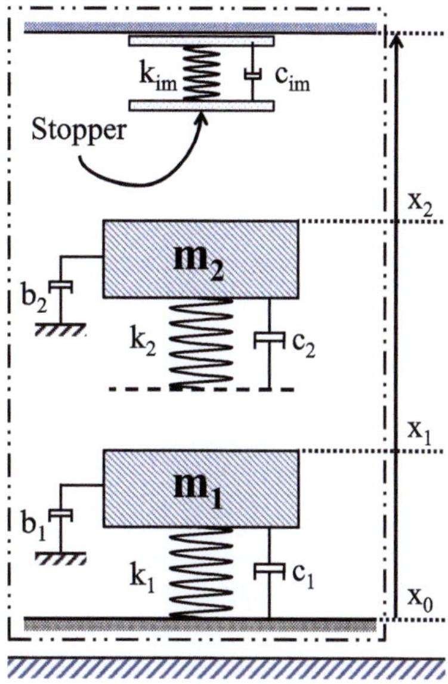

Fig. 1.8 Schematic structure of a two-mass resonator for wideband energy harvesting employing the velocity amplification principle from [61, 62]

However, the impact dynamics of the two-mass systems introduce nonlinearity and causes an additional effect—widening of the frequency response of the system. In study [62], it has been shown that although the dynamics each of the masses are subjected to linear equations, the impact instances between the two masses and the impact with the stopper lead to a set of piecewise equations. Therefore this system as a whole cannot be considered as a linear system and over a very broad range of parameters it behaves nonlinearly and chaotically. Figure 1.9 illustrates a typical response of such a system to a frequency sweep (the parameters of the device can be found in [61, 62]). Note that since the displacement of the first and the second mass is irregular, the frequency response is given in terms of the root mean square (RMS) of the power generated by the harvester. Also note that the bandwidth of the response is wide compared to a linear oscillator case and is increasing with A_{ext}.

Study [63] also employs impact to widen the frequency response of electrostatic energy harvesters. In this paper, the author suggests a new structure of the MEMS resonators for an electrostatic VEH that exploits mechanical frequency amplification by multiple mass impacts in combination with elastic stoppers, see Fig. 1.10. When the system is shaken at low frequency in a range of 10–60 Hz, a micro-ball, placed in the cavity of the MEMS proof mass, impacts with it and transfer kinetic energy to the gap-closing comb transducer. The transducer, however, resonates at its natural frequency, which is higher than the frequency of external vibrations. In addition, elastic stoppers amplify the proof mass and ball velocity throughout collision with the fixed frame. The systems combines the advantages of impact nonlinearity (wider frequency response) with the frequency-up conversion mechanism. Such a system can be driven by vibrations at a lower frequency (more common for mechanical

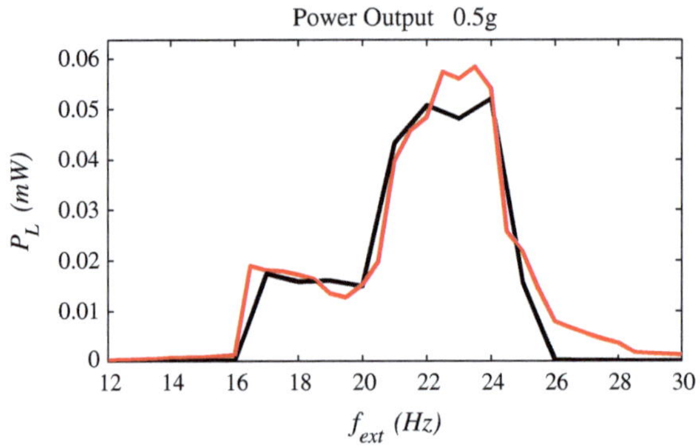

Fig. 1.9 Average converted power P_L versus the frequency of the input vibrations ω_{ext} of an electromagnetic harvesters utilising the two-mass resonator with velocity amplification. The details of the system can be found in [62]. The black line is the experimental measurement while the red line is the model proposed in the latter study. Note the wideband response of the system—i.e. the absence of clear and narrow peak around the resonance frequency

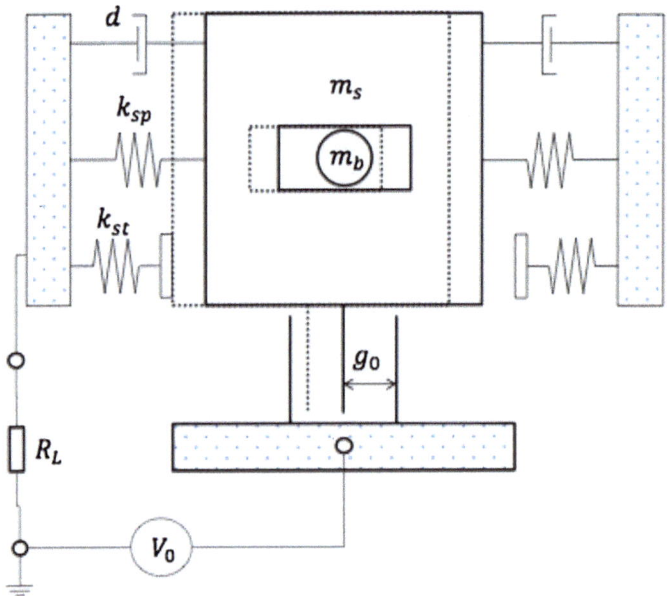

Fig. 1.10 Mechanical structure of the MEMS resonator employing impacts from study [63]

vibration and much lower than the natural frequency of the MEMS resonator) and it will respond to them as shown in that work.

The effect of stoppers on the performance of electrostatic harvesters has been exploited in studies [64, 65]. In these works, the authors show that through simulation and measurement results show that the impact with stoppers has positive effects. It

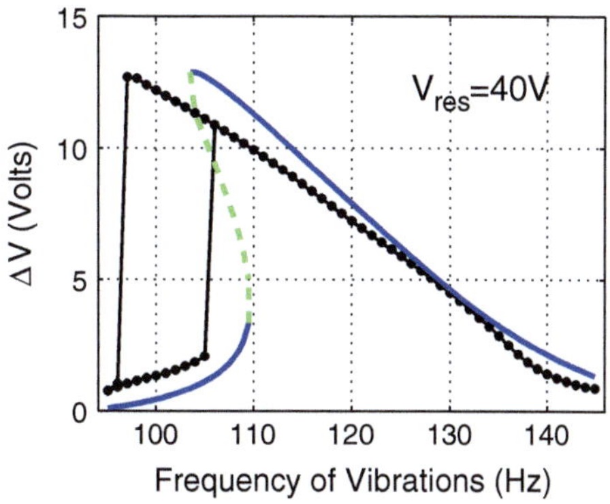

Fig. 1.11 Nonlinear effects that are caused by electrostatic transducer force F_t are similar to those caused by mechanical nonlinear forces. This figure shows the effect of hysteresis and frequency shift due to F_t (compare with Fig. 1.7). The two *curves* correspond to a theory and measurements whose details can be found in study [67]

increases the bandwidth of the resonator and moderately increases the converted power in saturation for large acceleration power spectral densities.

Finally, electromechanical coupling and the presence of a transducer force also causes nonlinear effects. These effects are similar to those produced by nonlinear mechanical forces. For instance, the transducer force causes hysteresis, multistability, and so-called electrostatic softening or hardening effects even in the absence of mechanical nonlinearity (see Fig. 1.11). In the context of electrostatic harvesters, these effects are quantified, described and measured in studies [45, 66, 67]. In particular, in study [67] the authors treat the transducer force in a way similar to nonlinear mechanical forces and provide a theory to explain and quantify the electromechanical coupling due to this force. The nonlinear effects due to electromechanical coupling are discussed in Chap. 9.

Therefore, mechanical and nonmechanical nonlinearities cause similar effects to the resonator of a VEH. The most 'obvious' and useful effect is the widening of the frequency response of the resonator with respect to the frequency of external vibrations ω_{ext}. This is clearly seen from the frequency characteristic given in Fig. 1.11 and is demonstrated in many studies of nonlinearity in energy harvesters. Additionally, nonlinearities shifts (modifies) the resonance frequency and may cause hysteresis (in the case of strong nonlinearities).

1.6 Scope and Structure of the Book

This book is aimed at vibration energy harvesters at micro- and nanoscale. Due to scaling issues, only electrostatic and piezoelectric energy conversion mechanisms are efficient at these scales and, therefore, we will focus on these type of energy harvesters. The main aim of the book is to highlight the positive role on nonlinearities in the design of such systems, regardless of the origin of this nonlinearity.

We start with the overview of MEMS technologies that are employed to fabricate micro- and nanoscale harvesters (Chap. 2). Chapters 3 and 4 provide a detailed explanation into the concepts of resonators and transducers and underline the energy conversion principle. Chapter 5 deals with advanced nonlinear mechanics that appears in the analysis of mechanical nonlinearities. These chapters are fundamental for understanding energy harvesting process. Particular nonlinearities that arise due to stoppers (impact), the use of graphene or the use of conditioning electronics in electrostatics and piezoelectric harvesters are discussed in Chaps. 6 through 10.

References

1. Wikander, O. (2000). Handbook of ancient water technology (741p). Brill: Leiden. ISBN 414176647.
2. Drachmann, A. (1961). Heron's windmill. *Centaurus*, *7*, 145–151.
3. Briand, D., Yeatman, E., Roundy, S., Brand, O., Fedder, G. K., Hierold, C., Korvink, J. G., & Tabata, O. (2015). Micro energy harvesting (Vol. 12). John Wiley & Sons.
4. Paradiso, J., Starner, T., et al. (2005). Energy scavenging for mobile and wireless electronics. *Pervasive Computing, IEEE*, *4*(1), 18–27.
5. Selvan, K. V., & Ali, M. S. M. (2016). Micro-scale energy harvesting devices: Review of methodological performances in the last decade. *Renewable and Sustainable Energy Reviews*, *54*, 1035–1047.
6. Vullers, R., van Schaijk, R., Doms, I., Van Hoof, C., & Mertens, R. (2009). Micropower energy harvesting. *Solid-State Electronics*, *53*(7), 684–693.
7. MicroGen Systems, Inc. https://www.microgensystems.com
8. Perpetuumm ltd. http://www.perpetuum.com.
9. Hayakawa, M. (1991). Electronic wristwatch with generator. https://www.google.ie/patents/US5001685 US Patent 5,001,685.
10. Mitcheson, P., Yeatman, E., Rao, G., Holmes, A., & Green, T. (2008). Energy harvesting from human and machine motion for wireless electronic devices. *Proceedings of the IEEE*, *96*(9), 1457–1486.
11. Becquerel, A. E. (1839). Mémoire sur les effets électriques produits sous l'influence des rayons solaires. *Comptes Rendus*, *9*, 561–567.
12. Sangani, K. (2007). Power solar-the sun in your pocket. *Engineering & Technology*, *2*(8), 36–38.
13. Rowe, D. M. (2005). Thermoelectrics handbook: Macro to nano. CRC press.
14. Dudka, A. (2014). Study, optimization and silicon implementation of a smart high-voltage conditioning circuit for electrostatic vibration energy harvesting system. Ph.D. thesis, Université Pierre et Marie Curie-Paris VI.
15. Settaluri, K. T., Lo, H., & Ram, R. J. (2012). Thin thermoelectric generator system for body energy harvesting. *Journal of Electronic Materials*, *41*(6), 984–988.
16. Hsiao, Y., Chang, W., & Chen, S. (2010). A mathematic model of thermoelectric module with applications on waste heat recovery from automobile engine. *Energy*, *35*(3), 1447–1454.
17. Bergqvist, U., Friedrich, G., Hamnerius, Y., Martens, L., Neubauer, G., Thuroczy, G., Vogel, E., & Wiart, J. (2000). Mobile telecommunication base stations–exposure to electromagnetic fields. Report of a short term mission within COST-244bis, COST-244bis short term mission on base station exposure.
18. Visser, H. J., Reniers, A. C., & Theeuwes, J. A. (2008). Ambient rf energy scavenging: Gsm and wlan power density measurements. In *Microwave Conference, 2008. EuMC 2008. 38th European* (pp. 721–724). IEEE.
19. Fan, Y., Hu, H., & Liu, H. (2007). Sustainable power generation in microbial fuel cells using bicarbonate buffer and proton transfer mechanisms. *Environmental Science & Technology*, *41*(23), 8154–8158.

20. Hou, J., Liu, Z., & Zhang, P. (2013). A new method for fabrication of graphene/polyaniline nanocomplex modified microbial fuel cell anodes. *Journal of Power Sources*, *224*, 139–144.
21. Bu, L., Wu, X., Wang, X., & Liu, L. (2013). Liquid encapsulated electrostatic energy harvester for low-frequency vibrations. *Journal of Intelligent Material Systems and Structures*, *24*(1), 61–69.
22. Despesse, G., Jager, T., Chaillout, J.J., Lger, J.M., Vassilev, A., Basrour, S., & Charlot, B. (2005). Fabrication and characterization of high damping electrostatic micro devices for vibration energy scavenging. In *Proceedings of DTIP MEMS MOEMS Conference* (pp. 386–390).
23. Dudka, A., Galayko, D., & Basset, P. (2009). Smart adaptive power management in electrostatic harvester of vibration energy. In *Proceedings of power-MEMS conference* (pp. 257–260).
24. Galayko, D., & Basset, P. (2011). A general analytical tool for the design of vibration energy harvesters (VEHs) based on the mechanical impedance concept. *IEEE Transactions on Circuits and Systems I: Regular Papers*, *58*(2), 299–311.
25. Naruse, Y., Matsubara, N., Mabuchi, K., Izumi, M., & Suzuki, S. (2009). Electrostatic micro power generation from low-frequency vibration such as human motion. *Journal of Micromechanics and Microengineering*, *19*, 094,002.
26. de Queiroz, A. C. M. (2010). Electrostatic vibrational energy harvesting using a variation of Bennet's doubler. In *Circuits and Systems (MWSCAS), 2010 53rd IEEE International Midwest Symposium on* (pp. 404–407). IEEE.
27. Roundy, S., Wright, P., & Pister, K. (2002). Micro-electrostatic vibration-to-electricity converters. In *Proceedings of 2002 ASME International Mechanical Engineering Congress*.
28. Tvedt, L. G. W., Nguyen, D. S., & Halvorsen, E. (2010). Nonlinear behavior of an electrostatic energy harvester under wide-and narrowband excitation. *Journal of Microelectromechanical Systems*, *19*(2), 305–316.
29. Knight, C., Davidson, J., & Behrens, S. (2008). Energy options for wireless sensor nodes. *Sensors*, *8*(12), 8037–8066.
30. Rabaey, J. M., Ammer, M. J., da Silva, J. L., Patel, D., & Roundy, S. (2000). Picoradio supports ad hoc ultra-low power wireless networking. *Computer*, *33*(7), 42–48.
31. Stordeur, M., & Stark, I. (1997). Low power thermoelectric generator-self-sufficient energy supply for micro systems. In *Proceedings ICT'97. XVI International Conference on Thermoelectrics, 1997* (pp. 575–577). IEEE.
32. Neri, I., Travasso, F., Mincigrucci, R., Vocca, H., Orfei, F., & Gammaitoni, L. (2012). A real vibration database for kinetic energy harvesting application. *Journal of Intelligent Material Systems and Structures*, 1045389X12444488.
33. Roundy, S., Wright, P. K., & Rabaey, J. (2003). A study of low level vibrations as a power source for wireless sensor nodes. *Computer Communications*, *26*(11), 1131–1144.
34. Ching, N. N., Wong, H. Y., Li, W. J., Leong, P. H., & Wen, Z. (2002). A laser-micromachined vibrational to electrical power transducer for wireless sensing systems. *Sensors and Actuators A: Physical*, *97–98*, 685–690.
35. Li, W. J., Wen, Z., Wong, P., Chan, G., & Leong, P. (2000). A micromachined vibration-induced power generator for low power sensors of robotic systems. In *World Automation Congress: 8th International Symposium on Robotics with Applications* (pp. 16–21).
36. Beeby, S., Torah, R., Tudor, M., Glynne-Jones, P., O'Donnell, T., Saha, C., et al. (2007). A micro electromagnetic generator for vibration energy harvesting. *Journal of Micromechanics and Microengineering*, *17*, 1257.
37. Glynne-Jones, P., Tudor, M. J., Beeby, S. P., & White, N. M. (2004). An electromagnetic, vibration-powered generator for intelligent sensor systems. *Sensors and Actuators A: Physical*, *110*(1), 344–349.
38. Beeby, S. P., Tudor, M. J., & White, N. M. (2006). Energy harvesting vibration sources for microsystems applications. *Measurement Science and Technology*, *17*, 175.
39. Anton, S. R., & Sodano, H. A. (2007). A review of power harvesting using piezoelectric materials (2003–2006). *Smart Materials and Structures*, *16*(3), R1.
40. Glynne-Jones, P., Beeby, S., James, E., & White, N. (2001). The modelling of a piezoelectric vibration powered generator for microsystems. In *Proceedings of the 11th International Conference on Solid-State Sensors and Actuators, Transducers* (Vol. 1, pp. 46–49).

41. Glynne-Jones, P., Beeby, S. P., & White, N. M. (2001). Towards a piezoelectric vibration-powered microgenerator. *IEE Proceedings-Science, Measurement and Technology, 148*(2), 68–72.
42. Roundy, S., & Wright, P. (2004). A piezoelectric vibration based generator for wireless electronics. *Smart Materials and Structures, 13*, 1131.
43. Pillatsch, P., Yeatman, E., & Holmes, A. (2013). Real world testing of a piezoelectric rotational energy harvester for human motion. In *Journal of Physics: Conference Series* (Vol. 476, p. 012010). IOP Publishing.
44. Meninger, S., Mur-Miranda, J., Amirtharajah, R., Chandrakasan, A., & Lang, J. (2001). Vibration-to-electric energy conversion. *Very Large Scale Integration (VLSI) Systems, IEEE Transactions on9*(1), 64–76.
45. Basset, P., Galayko, D., Cottone, F., Guillemet, R., Blokhina, E., Marty, F., & Bourouina, T. (2014). Electrostatic vibration energy harvester with combined effect of electrical nonlinearities and mechanical impact. Journal of Micromechanics and Microengineering *24*(3), 035,001.
46. Pons-Nin, J., Gorreta, S., Dominguez, M., Blokhina, E., O'Connell, D., & Feely, O. (2014). Design and test of resonators using Piezo MUMPS technology. In *proceedings of Design, test, integrations and packaging of MEMS/MDEMS (DTIP) Symposium on 2014* (pp. 1–6). Cannes, France, 1–4 April 2014
47. Lyapunov, A. M. (1966). *Stability of motion*. New-York and London: Academic Press.
48. Poincaré H., J.. J. P... T.. M. V... Y...:
49. Andronov, A. A. (1987). Theory of oscillators (Vol. 4). Courier Dover Publications.
50. Nayfeh, A. H., & Mook, D. T. (2008). Nonlinear oscillations. John Wiley & Sons.
51. Nejmark, I., & Landa, P. (1992). Stochastic and chaotic oscillations (Vol. 77). Springer.
52. Rabinovich, M. I. (1989). Oscillations and waves: in linear and nonlinear systems (Vol. 50). Taylor & Francis.
53. Andò, B., Baglio, S., Trigona, C., Dumas, N., Latorre, L., & Nouet, P. (2010). Nonlinear mechanism in MEMS devices for energy harvesting applications. *Journal of Micromechanics and Microengineering20*, 125,020.
54. Barton, D. A. W., Burrow, S. G., & Clare, L. R. (2010). Energy harvesting from vibrations with a nonlinear oscillator. Journal of Vibration and Acoustics *132*, 0210,091–0210,097.
55. Daqaq, M. F. (2010). Response of uni-modal Duffing-type harvesters to random forced excitations. *Journal of Sound and Vibration, 329*, 3621–3631.
56. Erturk, A., Hoffmann, J., & Inman, D. J. (2009). A piezomagnetoelastic structure for broadband vibration energy harvesting. *Applied Physics Letters94*, 254,102.
57. Miki, D., Honzumi, M., Suzuki, Y., & Kasagi, N. (2010). Large-amplitude mems electret generator with nonlinear spring. In *Proceedings of IEEE Conference on Microelectromechanical Systems (MEMS) 2010*, 24–28 January, 2010, Wanchai, Hong Kong, (pp. 176–179).
58. Stanton, S. C., McGehee, C. C., & Mann, B. P. (2010). Nonlinear dynamics for broadband energy harvesting: Investigation of a bistable piezoelectric inertial generator. *Physica D: Nonlinear Phenomena, 239*, 640–653.
59. Zhu, D., Tudor, M. J., & Beeby, S. P. (2010). Strategies for increasing the operating frequency range of vibration energy harvesters: A review. Measurement Science and Technology, *21*, 022,001.
60. Nguyen, D., Halvorsen, E., Jensen, G., & Vogl, A. (2010). Fabrication and characterization of a wideband MEMS energy harvester utilizing nonlinear springs. *Journal of Micromechanics and Microengineering, 20*(12), 125,009.
61. Cottone, F., Frizzell, R., Goyal, S., Kelly, G., & Punch, J. (2014). Enhanced vibrational energy harvester based on velocity amplification. Journal of Intelligent Material Systems and Structures, *25*, 443–451.
62. Fu, B., Blokhina, E., O'Connell, D., Feely, O., & Frizzell, R. (2014). A wideband 2-dof resonator for electromagnetic energy harvesting systems. In *IEEE International Conference on Electronics, Circuits and Systems (ICECS)*, 7–10 December, 2014, Marseille, France (pp. 878–881).

63. Cottone, F., Basset, P., Marty, F., Galayko, D., Gammaitoni, L., Bourouina, & T. (2014). Electrostatic generator with free micro-ball and elastic stoppers for low-frequency vibration harvesting. In *IEEE International Conference on Micro Electro Mechanical Systems (MEMS)*, 26–30 Jan 2014, San Francisco, CA, USA (pp. 385–388).
64. Le, C.P., & Halvorsen, E. (2012). Mems electrostatic energy harvesters with end-stop effects. *Journal of Micromechanics and Microengineering22*, 074,013.
65. Le, C.P., Halvorsen, E., Sorasen, O., & Yeatman, E. M. (2012). Microscale electrostatic energy harvester using internal impacts. *Journal of Intelligent Material Systems and Structures* 1045389X12436739.
66. Galayko, D., Dudka, A., Karami, A., O'Riordan, E., Blokhina, E., Feely, O., & Basset, P. (2015). Capacitive energy conversion with circuits implementing a rectangular charge-voltage cycle—part 1: Analysis of the electrical domain. *IEEE Transactions on Circuits and System I62*(11), 2652–2663.
67. O'Riordan, E., Dudka, A., Galayko, D., P. Basset, O.F., & Blokhina, E. (2015). Capacitive energy conversion with circuits implementing a rectangular charge-voltage cycle—part 2: Electromechanical and nonlinear analysis. *IEEE Trans. on Circuits and Systems I.*

Chapter 2
MEMS Technologies for Energy Harvesting

Manuel Domínguez-Pumar, Joan Pons-Nin and Juan A. Chávez-Domínguez

2.1 Introduction

The technology of Microelectromechanical Systems, or MEMS, is generally defined as the miniaturization of mechanical and electromechanical structures that are fabricated using standard processes from the integrated circuit industry and other compatible processes usually aimed at 'sculpting' 3D structures. MEMS can in particular include moveable parts such as cantilevers, beams, membranes, plates, etc. They may present interactions with biological, chemical and thermal phenomena, including interaction with fluids. At the same time MEMS devices usually interact with fields and forces that are not electromagnetic, such as mechanical forces, piezoelectric and thermoelectric forces, among others. This has promoted MEMS technology to be an excellent tool for the miniaturization of energy harvesters.

Common physical dimensions of MEMS can range from below 1 μm (in this case they are usually called Nanoelectromechanical), up to the mm scale. It is remarkable that generally the MEMS equivalent of a sensor/actuator macrosystem outperforms the latter. Among other factors, the repeatability and high reliability of MEMS batch fabrication processes can clearly contribute to this improvement, while at the same time even reducing the unit cost. Furthermore, the recent merging between MEMS and CMOS technology has opened new platforms on which both the microelectromechanical parts and the sensing/actuation circuits can be found within the same silicon die.

M. Domínguez-Pumar (✉) · J. Pons-Nin
MNT-DEE, Technical University of Catalonia, Barcelona, Spain
e-mail: manuel.dominguez@upc.edu

J. Pons-Nin
e-mail: joan.pons@upc.edu

J.A. Chávez-Domínguez
GSS-DEE, Technical University of Catalonia, Barcelona, Spain
e-mail: juan.antonio.chavez@upc.edu

E. Blokhina et al. (eds.), *Nonlinearity in Energy Harvesting Systems*,
DOI 10.1007/978-3-319-20355-3_2

This chapter is divided into three main parts. Section 2.2 introduces the most common fabrication processes used in MEMS technology. Special attention is paid to the processes that are specific to MEMS such as anisotropic etching. Section 2.3 introduces the main design procedures that have been used in MEMS to generate nonlinear actuation and sensing in energy harvesters. In particular the creation of bistable potentials, impact energy transfer and nonlinear springs is addressed. Recent examples found in the literature are presented and linked with the fabrication of the devices. Finally, due to its growing relevance, an introduction to piezoelectricity is given in Sect. 2.5. Some examples also found in the literature of energy harvesters designed using piezoelectric actuation and sensing are also introduced.

2.2 MEMS Fabrication Processes

This section provides a brief guide to the fabrication processes used in MEMS technology. As MEMS fabrication developed out of integrated circuits (IC) fabrication, we first focus on the main differences between MEMS and IC technologies. A description of the deposition and etching techniques most frequently used for MEMS follows. Special emphasis is made on those techniques that are MEMS-specific, such as anisotropic etching, wafer bonding, and, in particular, deep reactive-ion etching. The next focus is on MEMS fabrication strategies, including surface micromachining, bulk micromachining and silicon on insulator (SOI) based techniques. Finally, an example is provided of a commercial process aimed at the fabrication of piezoelectric MEMS devices that can be suitable for energy harvesting applications, among others.

2.2.1 *IC Versus MEMS Fabrication*

IC and MEMS fabrication possess a common series of process steps (e.g. photolithography, etch, oxidation, diffusion, LPCVD or sputter deposition) and materials (e.g. silicon, polysilicon, silicon nitride, silicon oxide or metals). In fact, IC and MEMS fabrication technologies can be seen as a complex sequence of deposition, material growth, lithography and etching processes. The objective of lithography is to translate a geometrical pattern onto a given material layer. The basic steps of standard contact lithography are shown in Fig. 2.1, where the layer to pattern is on the substrate. First, a thick layer of photoresistive polymer (photoresist) is deposited on top of the layer to be patterned. Next, the photoresist is exposed to light through a mask with opaque areas corresponding to the desired pattern. Due to such illumination, the solubility of the photoresist to developers becomes very different between shadowed and illuminated areas, Fig. 2.1a. The photoresist is then developed to obtain the desired pattern, which may be that of the mask (positive photoresist) or its complement (negative photoresist), Fig. 2.1b. The pattern in the photoresist is then transferred to the target

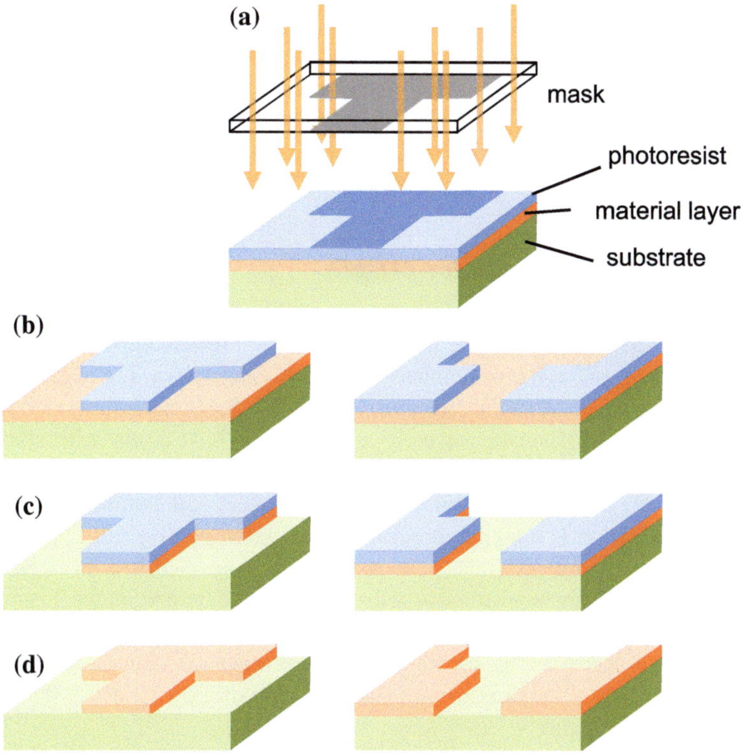

Fig. 2.1 Photolithography-etch process with either positive (*left*) or negative (*right*) photoresist: **a** exposure to light through the mask, **b** photoresist develop, **c** layer etch, **d** photoresist removal

layer using selective etching techniques, Fig. 2.1c. Finally, the remaining photoresist is chemically stripped, Fig. 2.1d.

MEMS technology, on the other hand, has unique requirements and a growing set of applications that makes it clearly diverge from IC fabrication [1, 2]. The first obvious distinction is that MEMS can make use of thicker deposited or grown layers and deeper etchings. It may well be the case that even the whole substrate is to be etched in some parts of the wafer. Second, MEMS fabrication involves a wider variety of materials, including quartz, ceramics, polymers, glass, piezoelectric and magnetic materials, etc. As a consequence, processes such as electroplating, wafer bonding, molding, anisotropic wet etching or deep reactive-ion etching, are more common, or specific, to MEMS fabrication. Third, some devices must be processed on both sides of the wafer, thus adding front-backside alignment as a technological necessary requirement for certain devices. A fourth difference is that MEMS devices include moveable mechanical parts such as beams and membranes that must be released. Additionally, the mechanical properties must be carefully controlled to avoid release distortion due to phenomena such as residual stress or stiction. Obviously controlling the mechanical properties of layers requires considerable effort. Putting all this together, a remarkable characteristic of MEMS technology is the existence of multiple and very diverse MEMS fabrication processes. This is especially true in the area of energy harvesting, where many different processes are designed in order to obtain tailored mechanical properties of the devices. This fact represents in itself a large divergence from what is, generally, the more common situation in IC technology.

2.2.2 Addition Processes for MEMS

Addition of materials onto a wafer is basic in MEMS. Process steps such as doping, thermal oxidation and epitaxy are inherited from IC technology. For instance, diffusion and ion implantation techniques are used in MEMS, but not in the same fashion as in IC technology. In the latter case, the objective is to have a very strict control on the electrical properties of the devices. In the MEMS case, though, the goal can be simply to convert a resistive layer into quite a good electrical conductor.

Thermal oxidation is a simple process used to grow good-quality thin films of silicon oxide on silicon substrates. The substrate is immersed into an oxygen-rich environment, which can be wet/liquid or dry/vapour. High temperatures, from 800 to 1100 °C, speed up the chemical oxidation of the substrate. However, as the thickness of the oxide layer increases, the substrate surface becomes harder to reach for the oxygen molecules. This effect causes nonlinear reduction of growth rate for thicknesses greater than 100–200 nm.

Other addition, or deposition, processes also come from IC technology, but due to the above-mentioned expansion to new applications the variety of specific MEMS processes is growing continuously. The next sections will focus on the physical and chemical deposition processes widely used for MEMS.

Physical Deposition

Physical deposition implies the direct transfer of a material, from a certain source, on the wafer surface. Examples are evaporation, sputtering and casting. Evaporation and sputtering are often used to deposit metals such as copper, gold and aluminium, among others. Evaporation is also used to deposit polymers. Casting is a common method to deposit polymers and glass. The thickness of deposited layers can range from a few hundreds of nanometres to microns. These processes are made at low temperatures, and may therefore be compatible with previous process steps executed on the wafer. On the other hand, the quality of the layers (e.g. density of defects, resistivity, etc.) is not as good as when deposited by chemical methods.

- In **Evaporation** the source material and the wafer are placed inside a vacuum chamber. There, the source material is heated until it boils and evaporates. Being in vacuum, the evaporated molecules freely travel and condensate on all surfaces inside the chamber, including the wafer. A method to heat the source material consists of placing it in a tungsten chest and then applying a high current. A high-energy electron beam or a laser targeting the source material are other common heating methods that serve to speed up the process.

- **Sputtering** is also performed inside a vacuum chamber. The chamber contains the source material, called target, the destination wafer and gas plasma at low pressure. RF power is applied to generate gas ions and accelerate them towards the target. This causes extraction, or sputtering, of target atoms. Sputtered atoms travel and condense on all surfaces inside the chamber. Sputtering is performed at lower temperatures than evaporation.
- In **Casting** the source material is first dissolved in a liquid solvent. Then the solution is dropped on the wafer either by piping or spraying techniques. Next, the wafer is spun to spread the solution uniformly over the entire surface (spin coating). A material layer, with thickness that can range from the nanoscale to tens of microns, is obtained once the solvent has evaporated. Due to their solubility in organic solvents, casting is a common practice for the deposition of polymers, including photoresists used in photolithography.

Chemical Deposition

In this case, the source material is obtained from a chemical reaction taking place close to the wafer. Favourable conditions to excite and control the reaction, such as temperature, pressure, electric field or presence of plasma, are provided by specific equipment, which also provides means to remove the byproducts of the reaction.

- **Chemical Vapour Deposition (CVD)** is achieved by placing the wafer inside a reactor. A chemical reaction between two or more gas species produces the source material, which condenses on all surfaces inside the reactor, wafer included. CVD yields thin layers with almost-uniform thickness and good coverage, even on stepped topographies. A variety of materials, including polysilicon, silicon nitride, silicon oxide, phosphosilicate glass (PSG), ceramics and plastics, can be deposited. Some materials are unpopular because of reactor contamination or due to the hazardous byproducts they generate.
  Low Pressure CVD (LPCVD) is performed at high temperature (from 500 to 800 °C). It yields fair material properties and layers with uniform thickness, but slow deposition rates. As an example, polysilicon deposition is achieved by decomposition of silane gas (SiH_4) into solid silicon and hydrogen gas under temperatures around 600 °C. Plasma Enhanced CVD (PECVD) is performed at lower temperatures because plasma adds extra energy to the gas mixture, but the layers have inferior quality compared with LPCVD layers. PECVD is often used for fast deposition of low-quality silicon oxide.
- **Electroplating** is a generic technique with many industrial applications. In MEMS, it is mainly used to deposit metals such as nickel and gold. The surfaces on which we wish to deposit the corresponding metal are previously coated with a conductive material, if they are not conductive themselves. Then the wafer is immersed into a liquid electrolyte solution. By applying a voltage between the wafer and an electrode also immersed in the solution, a reduction–oxidation reaction is excited. As a result, wafer surfaces are coated with the source material. In this way, nearly

uniform metal layers with thickness ranging from microns to tens of microns can be deposited.
Some appropriate chemical mixtures allow the reaction to take place without applying any external voltage, or without placing contacts or electrodes within a liquid. In this case, known as electroless plating, the reaction is excited by the electrochemical potential between the solution and one or more materials on the wafer. However, the reaction behaves spontaneously and results such as deposition rate and layer thickness are difficult to control.

Wafer Bonding

The objective of wafer bonding is to obtain a permanent contact of two wafers. This process can also be performed at die level or at device level. Wafer bonding allows addition of materials obtained from different processes, but is also highly useful for other processing and post-processing purposes, such as planarization, wafer sealing in vacuum, packaging or MEMS-IC integration.

Wafer bonding can be performed in many ways, chemical or physical, depending on materials to join and compatibility with other fabrication processes. In general, three subsequent steps are applied: (1) wafer preprocessing (cleaning and pretreatment of the surfaces to contact and, if needed, addition of intermediate layers), (2) wafer alignment and contact (it usually yields weak adhesion), (3) annealing and bond consolidation (by temperature, pressure, electric field, intermediate layer hardening, etc.).

Let us distinguish among direct and indirect bonding processes

- **Direct bonding** is performed for wafer to wafer contact. The process is activated and enhanced by a temperature or an electric field. For example, silicon-to-silicon bonding is achieved with annealing temperatures around 800 °C, a value often not compatible with other processes. Plasma enhanced bonding allows to decrease the temperature to 400 °C or less. In anodic bonding, the wafers are stacked and placed between two electrodes, where voltage is applied. This low-temperature process is mostly used for glass-to-silicon and glass-to-metal bonding. In thermocompression, pressure force and temperature are applied simultaneously to the stacked wafers until solid diffusion occurs. This process is used for metal-to-metal bonding.
- In **indirect bonding**, an intermediate material is placed between the wafers. This material provides long-time adhesion properties and reduces the relevance of the properties of the surfaces to contact (e.g. defects or topography). In adhesive bonding, polymers such as BCB (benzocyclobuten) or SU-8 (a negative photoresist) are used. Annealing is performed at relatively low temperature, with no voltage required. Since the wafers are not in direct contact, bonding of many different materials is enabled. In glass soldering, a glass layer with low melting point is used as an intermediate material. In eutectic bonding, the intermediate material is a metal alloy. This alloy goes directly from solid to liquid state at temperatures far below the melting points of the metals involved. This way, good-quality aluminium-to-silicon and gold-to-silicon bonding is achieved.

2.2.3 *Etching Processes for MEMS*

Etching processes are crucial in MEMS fabrication. Selectivity of etchant species against protection (masking) materials, substrates and the target material is a major issue, but etch rate, etch uniformity and temperature are also key characteristics. Let us first distinguish between wet and dry etching processes.

Wet Etching

Wet Etching is relatively simple and cheap. It allows to attack a material by immersing the wafer into a liquid etchant. The etchant chemically reacts with the material in unprotected areas. Wet etching can be isotropic or anisotropic.

- **Isotropic wet etching** is a classical technique performed to remove a wide range of materials, including semiconductors, dielectrics, metals or polymers. When used for patterning purposes, isotropic etching causes undercutting: the chemical reaction attacks the walls under the protection mask and produces a lateral etch effect, Fig. 2.2a. The amount of undercutting is the same distance as etch depth, thus it affects pattern transfer precision.
- **Anisotropic wet etching**. Solubility properties of crystalline materials may vary with crystal orientation. In silicon, potassium hydroxide (KOH) and ethylene diamine pyrocatechol (EDP) have different etch rates along the three crystallographic planes. For example, KOH yields etch rate selectivity of 300:1 or higher between the $\langle 100 \rangle$ and $\langle 111 \rangle$ planes. This means that etching can be virtually stopped on certain planes, and therefore substrates can be "sculpted" to produce a variety of structures. As an example, a trench is built in Fig. 2.2b.
  The alignment of the mask against the crystal planes, the etchant used and the process time are key factors. Additionally, precise temperature control and gas reflux systems are required to keep the etchant concentration constant during the process, and therefore the etch rate. Wet etchant and related byproducts are hazardous substances that may cause handling or environmental safety problems.
  Anisotropic wet etching produces characteristic geometrical forms, since planes are not vertical to the surface when etching holes or trenches, as seen in Fig. 2.2b. This implies large silicon consumption, loss of precision in pattern transfer and limits what structures are possible.

Dry Etching

In dry etching, the wafers are placed inside chambers that contain either gas or plasma reacting with the target material. In general, dry etching yields better etch performance than wet etching, but at noticeably higher costs due to the more sophisticated equipment used.

- **Isotropic gas etching** utilizes a gas with high etch selectivity against the mask protection material. As an example, xenon difluoride (XeF_2) has etch selectivity of silicon versus silicon oxide up to 10,000:1, whereas selectivity versus silicon nitride is around 100:1.

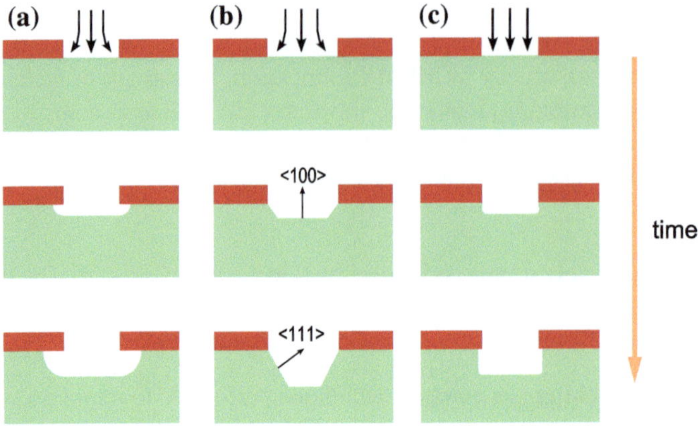

Fig. 2.2 Typical trench profiles obtained in silicon with different etch processes: **a** isotropic etching, **b** wet anisotropic etching—two crystallographic axes shown-, **c** anisotropic dry etching

- **Plasma etching** is performed inside a specific reaction chamber that contains chemically active gas species at low pressure and applies an electric field to the gas. The electric field creates electrically charged gas radicals, which chemically react with the wafer material. In addition, the electric field may cause a physical etch effect: the charged radicals accelerate until they hit and sputter the wafer material. The physical etch component is anisotropic, whereas the chemical component is isotropic and material selective.
- **Reactive Ion Etching** (RIE) is a special kind of plasma etching. In RIE, RF power is applied to increase the physical component of the etching, therefore it can be more anisotropic than traditional plasma etching. Moreover, the chemical-isotropic and physical-anisotropic mechanisms can be balanced to obtain etched features with rounded to nearly vertical sidewalls. However, the process has many parameters to adjust and this implies long-time development until the desired balance is achieved. In practice, RIE is limited in etch depth (e.g. tens of microns for silicon) and etch rate (e.g. typical values around 1 μ/min for silicon). The next section focuses on Deep RIE, a special class of RIE that has gained enormous popularity in modern MEMS fabrication.

Deep Reactive-Ion Etching

Deep Reactive-Ion Etching (DRIE) is a highly anisotropic process aimed at creating deep holes and trenches in silicon, with high aspect ratios and nearly vertical side walls. It was first developed by the German company Robert Bosch GmbH [3, 4]. The process is performed at room temperature. DRIE strongly improves etch performance against traditional anisotropic dry and wet etching techniques: silicon etch depths up to 750 μm at rates up to 25 μm/min are typically achieved. Etching uniformity at wafer scale is also improved.

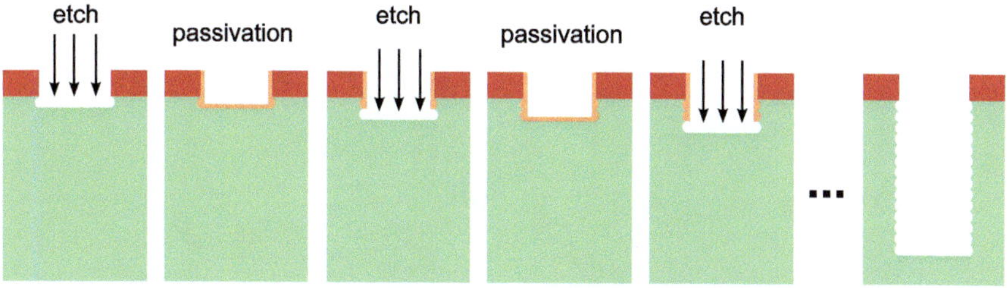

Fig. 2.3 Successive etching and passivation steps of a DRIE process

The DRIE process performs RIE in small depth increments by alternating two different gas mixtures in the reactor

1. The first mixture (e.g. SF_6 for silicon) provides standard RIE etching of the substrate, or another thick layer, through a window opened in a mask material (e.g. photoresist, silicon oxide, silicon nitride and metal, among others). This results in nearly vertical etching of the substrate.
2. The second mixture (e.g. C_4F_8) deposits a passivation layer. This layer dissolves very slowly in the chemical part of the subsequent RIE etch. The RIE directional ions attack the passivation layer at the bottom of the etched area. Then this bottom layer is sputtered off and the substrate becomes exposed to the chemical etch.

An etch-passivation cycle only lasts for seconds. It is repeated many times, resulting in a large number of small etch steps only taking place at the bottom of quasi-vertical sidewalls, Fig. 2.3. However, the sidewalls have slightly undulated shapes, with typical amplitudes from tens to a few hundreds of nanometres. Cycle times can be adjusted for a trade-off between smoother walls and higher etch rates [5].

Specific DRIE recipes have been developed for materials other than silicon, including glass, silica, quartz, InP and polymers. DRIE of glass substrates is performed applying high RF power, but this implies that mask materials must be carefully chosen. Chemically amplified photoresists and polysilicon are typically used as mask materials in DRIE of glass. Metal masks are used in DRIE of polymers, but this is expensive due to the additional deposition and lithography steps they require.

2.2.4 MEMS Fabrication Strategies

Surface Micromachining

MEMS surface micromachining aims at building moveable structures by deposition and patterning of different layers on top of the substrate. The moveable parts of the devices are made of what are called structural materials. These materials are deposited on top of what are called sacrificial layers, previously deposited or grown to hold the structural parts. The sacrificial layers will be removed later during the

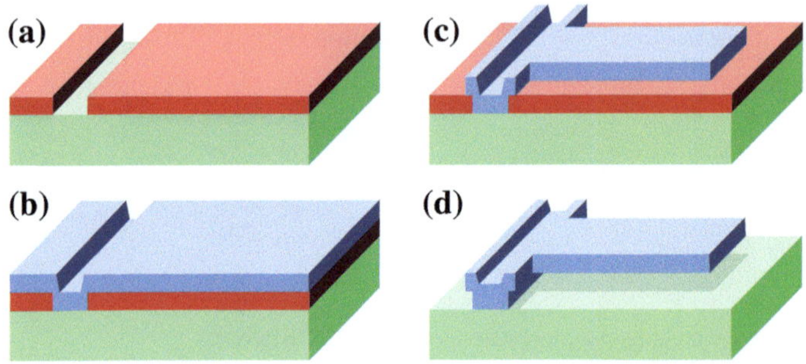

Fig. 2.4 MEMS cantilever fabrication using surface micromachining

fabrication process. The removal of the sacrificial layers is known as the release process.

As an example, Fig. 2.4 summarizes a series of steps performed to build up a cantilever structure. To begin, the sacrificial layer (i.e. silicon dioxide) is deposited and patterned through a photolithography-etch process, Fig. 2.4a. The uncovered area will be used to attach (anchor) the device to the substrate. The structural layer (e.g. polysilicon) is then deposited on top of the wafer, Fig. 2.4b. Next, the structural layer is patterned to define the device, Fig. 2.4c. Finally, the sacrificial layer is etched. As shown in Fig. 2.4d, this creates a void, equal to the thickness of the sacrificial layer, below the beam.

MEMS surface micromachining has evolved directly from IC fabrication, but some strong differences arise. For instance, in surface micromachining only up to 7–8 masks and minimum feature sizes around 0.5–1 μm are typically used; this decreases mask costs. However, in surface micromachining, layers are thicker than the thin films used in IC fabrication. This adds specific challenges to deposition and etch processes, including topography-related issues. As after 7–8 successive layer deposition and etch steps, the vertical dimensions of structures can be anything ranging from zero to tens of microns.

Another difference is that MEMS fabrication must take special care to control the mechanical properties, such as density, stress and Young's modulus, of structures to release. These properties are highly sensitive to the temperatures applied during fabrication. For example, successive heating and cooling can generate residual stress between two stacked layers due to the different thermal expansion coefficients of the materials. If one of such layers is removed, the released structure would not be flat and exhibit either tensile or compressive stress deformation, thus affecting the expected mechanical performance.

Finally, during the release process the surface tension of wet etchant can force contact between fixed and moveable parts of a device and cause stiction failure: the adhesion forces generated in the contact are strong enough to deform and permanently attract these parts. Dry etching, critical point drying (CPD) during the final phase of wet etching, increasing the surface roughness or coating the layer with low surface energy materials are techniques used to prevent stiction. For devices with large areas

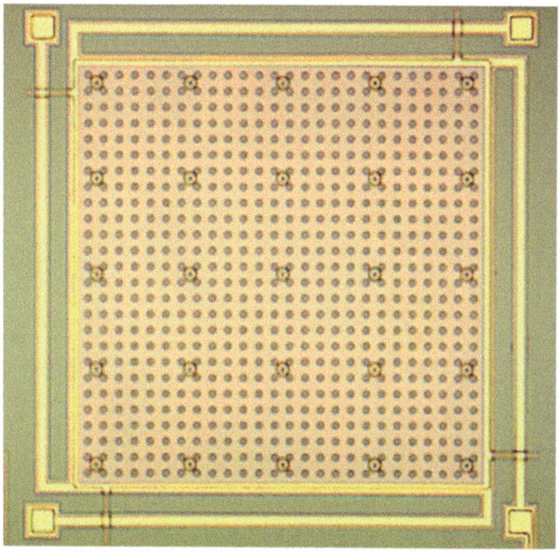

Fig. 2.5 *Top view* of an electrostatic MEMS, taken from [6], fabricated with surface micromachining. A $360 \times 360\,\mu\text{m}^2$ polysilicon plate is held by four arms in "L" anchored to the substrate at the other ends. The plate is suspended over the substrate through a $2\,\mu\text{m}$ air gap and a $0.65\,\mu\text{m}$ silicon nitride layer

to release, as the one shown in Fig. 2.5, it is common practice to etch regular grids of holes in the structural layer. This enables efficient diffusion of wet etchant during the etch of the sacrificial layer below.

Since surface micromachining is relatively close to IC fabrication, it constitutes the most frequent basis to develop monolithic microsystems, where electronic (e.g. CMOS) and mechanical components are fabricated on the same substrate. The properties of the substrate are less important in surface micromachining than in other MEMS fabrication strategies, such as bulk or SOI micromachining. In this case, surface micromachining enables the use of substrate materials other than silicon, such as glass, quartz, plastics or organic polymers.

On the other hand, surface micromachining is not suitable to fabricate structures with large moveable parts, to use layers thicker than 2–5 μm, or to build features with the high aspect ratios that some MEMS applications demand. Additionally, some geometrical structures are not possible to achieve, giving opportunities to other fabrication strategies, such as bulk and SOI micromachining.

Bulk and SOI Micromachining

Unlike surface micromachining, bulk micromachining aims to create MEMS structures within the substrate. It is based on selective deep etching of materials such as silicon or glass. Bulk micromachining fabrication processes that combine anisotropic and isotropic wet etching have been widely used to obtain a variety of mechanical structures in silicon.

Figure 2.6 illustrates the fabrication of a silicon cantilever. A mask material (e.g. silicon dioxide) is first deposited on top of a silicon wafer. Then it is patterned by a lithography-etch process, Fig. 2.6a. A subsequent boron diffusion is performed in

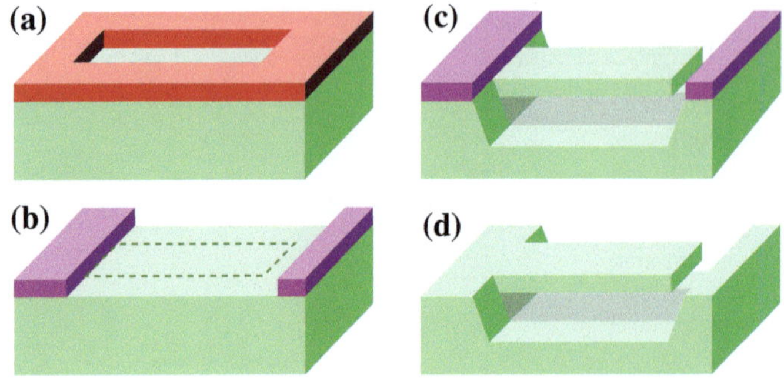

Fig. 2.6 MEMS cantilever fabrication in a silicon substrate using bulk micromachining

the uncovered substrate area. The p^+ doped area becomes resistant against wet etch species.

Once the initial mask removed, a second protection layer (e.g. polymide photoresist) is deposited and patterned, see Fig. 2.6b, where the dashed line encloses the boron diffusion. Next, anisotropic wet etching is applied to chemically attack the uncovered surface. Due to different etch selectivity along crystallographic planes and to the resistance of the p^+ area, the substrate trench and the suspended structure shown in Fig. 2.6c are obtained. The last step is the removal of the protection layer, Fig. 2.6d.

Bulk micromachining techniques based on wet anisotropic etch of silicon are widely used in MEMS. However, these techniques have some issues. Notably, the cost in silicon is high, since large wafer areas are usually etched to obtain a unique device. A second issue are the process limits for which geometrical structures can be done and which are not achievable. Third, wet etching is unreliable to achieve structures with sizes below the micron scale.

In recent years, MEMS processes that use silicon on insulator (SOI) wafers in place of conventional silicon wafers have gained great popularity [1]. A SOI wafer is a stack of silicon, insulator and silicon layers. The MEMS structures are mainly fabricated in the top silicon layer, usually known as the device, or active, layer. This is good-quality crystalline silicon with well-known electrical and mechanical properties. In SOI wafers for MEMS the typical thickness of the device layer is tens of microns, well above the 1 μm or less of SOI wafers used in IC fabrication. The buried insulating layer is typically silicon oxide, with thickness of microns or below. The bottom silicon layer, with typical thickness of hundreds of microns, is known as the handle substrate.

SOI-based MEMS processes allow one combine both bulk and surface micromachining techniques to obtain a rich set of structures. For example, the buried insulator can provide built-in stop for dry and wet etching processes performed in the device layer, but it can also be used to stop deep etch processes performed in the handle substrate. This way, features such as through-wafer holes, front or backside cavities of controlled depth, structure release from both sides of the wafers, full-wafer thickness suspended structures, etc. can easily be achieved. Surface micromachining techniques can be also used, mostly on the device layer. A commercial SOI-based MEMS process is described in Sect. 2.2.4.

Multi-project Wafer Processes

Some MEMS foundries offer multi-project wafer (MPW) processes. These are robust closed fabrication processes that provide well-known and reliable results. Additionally, MPW processes can be inexpensive, since wafers can be shared among several users. Their use facilitate users to focus on the device design, avoiding challenges associated with process development. MPW processes offer CAD tools such as device templates, design rules for layout checking and electrical and mechanical parameters for simulation. In general, MPW processes have not been targeted for commercial production, but used for research and prototyping. In recent years, some foundries have started providing specialized MPW services for low-volume production customers trying to reduce their design cost and time to market.

The first MPW process offered was a surface micromachining process developed by Howe and Muller [7, 8], with three polysilicon structural layers. Today, a number of MPW processes that use surface, bulk and SOI micromachining, electroplating and CMOS integration at wafer level are available. Some examples include

- MEMSCAP's MUMPS (Multi-User MEMS Processes) [9] offers PolyMUMPS, SOIMUMPS, MetalMUMPS and PiezoMUMPS. PolyMUMPS includes surface micromachining with three polysilicon structural layers and two silicon oxide sacrificial layers. SOIMUMPS combines SOI bulk and surface silicon micromachining. MetalMUMPS combines electroplating, surface and bulk micromachining, with nickel and polysilicon as structural materials. Finally, PiezoMUMPS adds a piezoelectric layer to a process similar to SOIMUMPS.
- Sandía SUMMiT (Sandia Ultra-planar, Multi-level MEMS Technology) V [10] is a surface micromachining process similar to PolyMUMPS, but with five structural layers and including planarization.
- InvenSense's NF Shuttle process [11] offers MEMS-CMOS integration. MEMS SOI devices and CMOS circuitry are fabricated on separated wafers. Then the wafers are bonded together using low-temperature eutectic bonding.
- Teledyne-DALSA's High Voltage CMOS/DMOS Technology with MEMS Post-Processing [12]. This process combines high-voltage CMOS circuitry and simple MEMS structures, with metal as the structural material.
- Lionix's TriPleX MPW and Fluidic MPW processes [13] are specifically targeted to devices for either integrated optics or microfluidic applications.

The PiezoMUMPS Process

The PiezoMUMPS process [14] was introduced by MEMSCAP in 2013 as a response to the growing interest in piezoelectric MEMS. This low cost and low temperature process evolved from the previous SOIMUMPS. It uses the same SOI wafers and inherits some well-known bulk and surface micromachining features. The SOI wafer is a stack of, from bottom to top, a silicon substrate layer ($400 \pm 5\,\mu m$), a buried silicon oxide layer ($1 \pm 0.05\,\mu$m) and the SOI device layer ($10 \pm 1\,\mu$m). A thin oxide layer is present on the bottom surface.

The process includes deposition and patterning of an aluminium nitride (AlN) piezoelectric film on top of the SOI device layer. It also includes deposition and patterning of additional metal and silicon oxide layers. Five mask levels are used. The minimum feature size for the SOI layer is $2\,\mu$m, making structures with aspect ratios up to 5:1 possible.

In order to illustrate the process flow, let us apply it to obtain the unimorph piezoelectric cantilever shown in Fig. 2.7, a structure rather similar to resonators used in energy harvesting applications.

The process starts with LPCVD deposition of PSG on top of the SOI layer, Fig. 2.8a, followed by thermal annealing in argon atmosphere, 1050 °C for 1 h. Due to the corresponding phosphorous diffusion, the surface of the SOI device layer becomes highly conductive (15–25 Ω/sq). After that, the PSG layer is completely removed by wet etching. Next, a $0.2\,\mu$m layer of thermal oxide is grown on top surface. Through a photolithography-etch process using positive photoresist and RIE, the oxide is patterned accordingly to the PADOXIDE mask, Fig. 2.8b.

The first material deposited is a $0.5\,\mu$m thick AlN piezoelectric layer. A reactive sputtering process is used for this purpose. Then, using photolithography with positive photoresist and wet etching, the PZFILM mask pattern is transferred to the AlN layer. That is followed by a photoresist strip, Fig. 2.8c.

Next, a metal stack composed of 20 nm of Chrome and $1\,\mu$m of Aluminium is deposited and patterned using the following lift-off process: the wafer is coated with negative photoresist and exposed to light through the PADMETAL mask, then the photoresist is developed and the metal layer is deposited over the entire surface by evaporation; finally, the photoresist is dissolved to leave behind metal only in the uncovered (developed) areas, Fig. 2.8d. This process allows one define $3\,\mu$m minimum metal features and spaces with $3\,\mu$m tolerance alignment.

The $10\,\mu$m SOI device layer is patterned using positive photoresist and the SOI mask. The pattern is transferred to silicon using a specific DRIE process, based on inductively coupled plasma (ICP) technology, that prevents undercutting of the

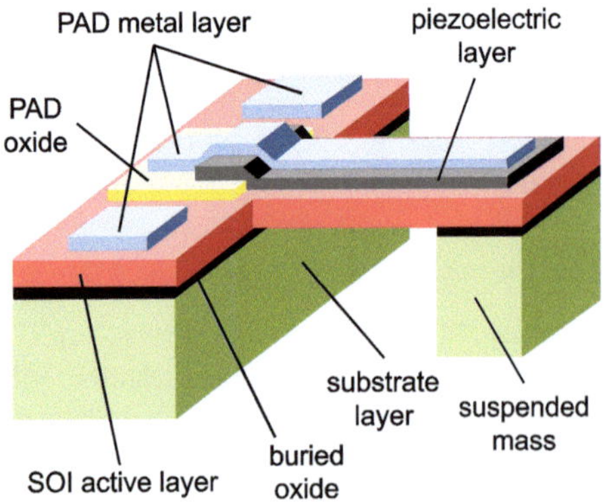

Fig. 2.7 Unimorph piezolectric cantilever structure with a full-wafer thickness mass suspended at the free end. The two outer metal PADs provide electrical contacts to the SOI active layer, while the one in the middle contacts the piezoelectric film

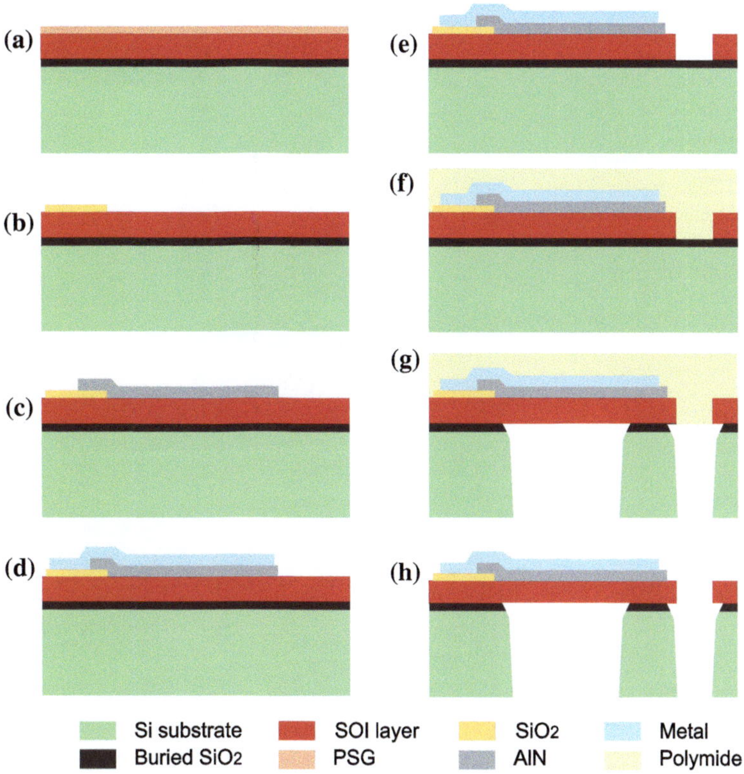

Fig. 2.8 PiezoMUMPS fabrication process steps: **a** silicon doping, **b** thermal oxide deposition and patterning (PADOXIDE), **c** piezoelectric layer deposition and patterning (PZFILM), **d** metal layer lift-off (PADMETAL), **e** silicon device layer patterning (SOI), **f** polymide coating, **g** substrate layer patterning (TRENCH), **h** structure release

silicon when the etch reaches the buried oxide layer. After etching, the photoresist is stripped, Fig. 2.8e. Since the surface of the device layer is heavily doped, the separated metal features deposited on this layer will be electrically contacted unless they are also separated by a trench feature defined by the SOI mask.

Next, a thick polymide coating is applied onto the front surface of the wafer, Fig. 2.8f. This coating holds the wafer together through the next process steps. The wafer is then reversed and negative photoresist is deposited on the bottom surface. Next, the photoresist is patterned accordingly to the TRENCH mask. A RIE process is then used to remove the bottom oxide in the uncovered areas. Subsequent DRIE is performed to etch the TRENCH mask patterns through the substrate, until the buried oxide is reached. When the etch is completed, the remaining photoresist is stripped. Finally, wet etching is used to remove the oxide layer in the uncovered areas. As a result, the substrate layer is patterned accordingly to the TRENCH mask, Fig. 2.8g.

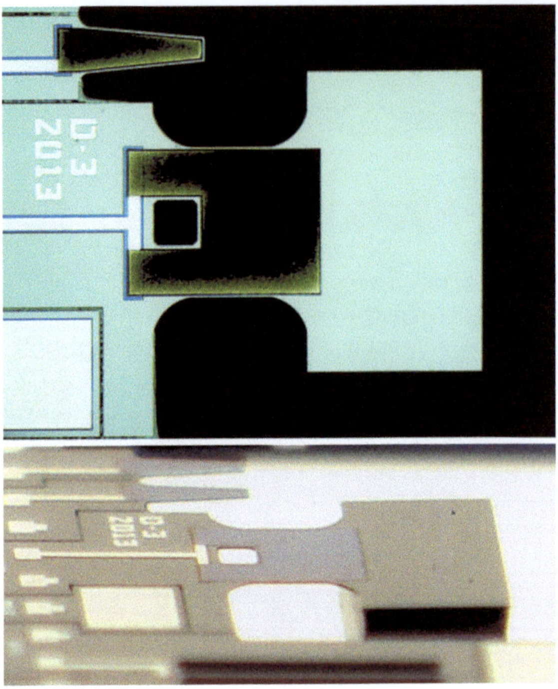

Fig. 2.9 *Top* and *lateral view* of a piezoelectric MEMS resonator fabricated in the PiezoMUMPS batch reported in [15]. Cantilever dimensions are 1100 μm long and 500 μm wide. The mass suspended at the free end has an area of 600 μm x 1000 μm and full-wafer thickness (410 μm). The first mechanical resonance mode is at 1.22 kHz

Finally, the front side protection coating is removed using a dry etch process. This releases the mechanical structures in the SOI active layer located over through-substrate holes, Fig. 2.8h. Figure 2.9 shows two photographs of a low-frequency MEMS resonator fabricated with PiezoMUMPS technology.

2.3 Nonlinear Mechanisms Used in Energy Harvesting with MEMS

The first approach made to energy harvesting was based on the use of linear resonators. The main disadvantage of linear devices is that they present a small bandwidth and therefore are only able to retrieve energy from a part of the mechanical excitations spectrum. To solve this issue, the use of resonators exploiting nonlinear mechanisms has been proposed as a way to, in some cases, dramatically enhance the performance of these devices. In this section, we introduce the main design techniques used in MEMS devices to implement nonlinear mechanisms such as bistable potentials, nonlinear springs and impact energy transfer.

2.3.1 Bistable Potentials

Bistable Potentials Obtained with Permanent Magnets

One of the methods used in the literature to produce bistable potentials is achieved by suitably placing a permanent magnet on top of a cantilever and another one fixed to the structure frame. By placing both magnets so that in the rest position they face each other with the same polarity, a bistable potential is easily generated. In [16] a MEMS device is presented on which the bistable potential is achieved by placing the permanent magnet at the tip of a BESOI (bulk etched SOI) cantilever, opposite another permanent magnet such that the polarity of both magnets is opposed at the nearest point, see Fig. 2.10. This configuration of magnets provides a repulsive force such that the rest position of the cantilever is no longer stable.

Regarding the fabrication, the authors use a BESOI technology with a bulk 450 μm thick, 2 μm of buried oxide, a 15 μm crystalline silicon active layer and several deposited layers with oxide, polysilicon and metal. The cantilevers are 2000 μm long with different widths in the range below 800 μm. The permanent magnet deposited at the tip of the cantilever has a cylindrical shape and is made of NdFeB. Generally, the whole structure is manually assembled or microassembled, [17], since the fabrication process of the magnets is not compatible with the MEMS fabrication. This difficult placement can result in either non-repeatability or high production cost.

The obtained bistable potential is assumed to be that of a nonlinear pendulum governed by a double well nonlinearity

$$U(x) = kx^2 + (ax^2 + b\Delta^2)^{-3/2} + c\Delta^2 \tag{2.1}$$

where Δ is the distance between the south poles of both, see Fig. 2.10, and parameters a, b and c depend on the specific geometry of the cantilever and its mechanical properties. Taking into account this kind of potential, the equation governing the dynamics of the inverted pendulum with the bistable potential is [18]

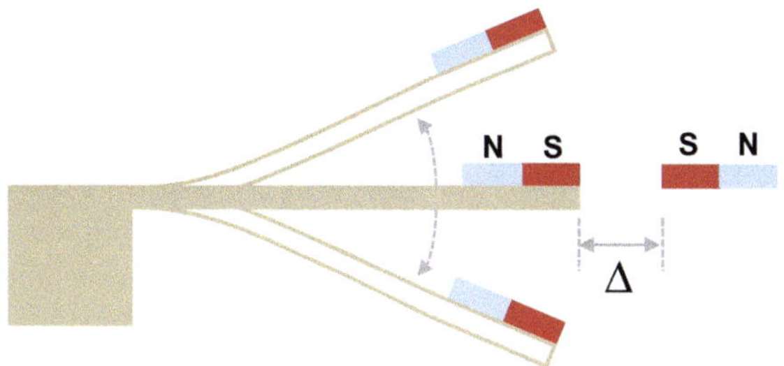

Fig. 2.10 Schematics of the structure proposed in [16]: a cantilever with a fixed magnet at the tip, facing another magnet with south pole polarities facing each other. The rest position of the cantilever is no longer stable and whole structure is bistable

$$m\ddot{x} = \frac{dU(x)}{dx} - \gamma\dot{x} - K_\upsilon V(t) + \sigma\xi(t) \tag{2.2}$$

where the $\gamma\dot{x}$ term represents mechanical damping, $-K_\upsilon V(t)$ represents the energy transfer to the electrical domain and $\sigma\xi(t)$ corresponds to the external vibration force driving the pendulum.

The main disadvantage cited against this type of structure is the presence of moving magnets, since they can generate fluctuations in the magnetic field which can affect other parts of the circuit, [19]. Furthermore, another disadvantage is the placement itself of the permanent magnets on the MEMS structures, which is usually made microassembled by hand. Electroplating [20], though, has been proposed as an alternative means to produce MEMS compatible permanent magnets, without manual assembly [21, 22].

Bistable Potentials Obtained by Buckling or Snap-through Instability

Buckling has also been used in energy harvesters to obtain a bistable potential. As it has been very well explained in [23], a bistable system is created by simply holding a business card between fingers and bowing it. A force normal to the surface of the card can make it snap from one stable position to the other. In order to obtain this kind of behaviour in MEMS structures, the axial load necessary for this buckling behaviour may come from residual stress of the fabrication process, or from the actuation with comb drives conveniently placed in the device to transmit an axial load to the corresponding beam [24]. Other mechanisms have also been established to create bistable mechanical structures such as: a clamp mechanism with an actuator to switch between stable states [25], or two curved parallel beams clamped in the centre [23] not relying on residual stress to obtain the buckling behaviour.

In [24] a tunable bistable mechanism is presented where an actuator generates a tunable axial compressive force on a beam, which is also subjected to a transversal force (see Fig. 2.11). The actuator is used to establish the necessary compressive force on the buckling beam to obtain the bistable potential. The beam is located between points A and B in the schematics of Fig. 2.11a. The actuator force, proportional to the square of the voltage applied on the comb drive (Fig. 2.11b), is shared between

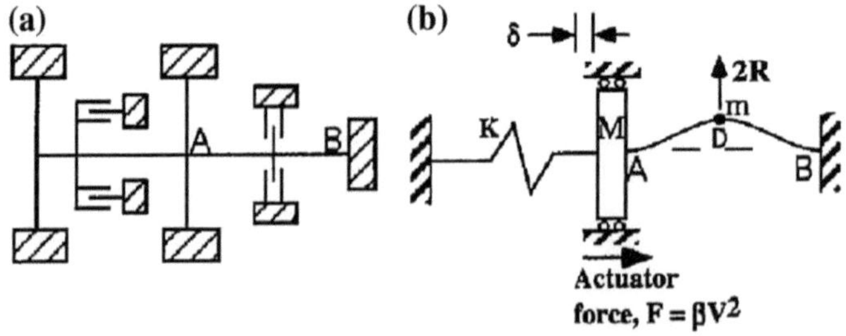

Fig. 2.11 Schematics of the structure proposed in [24]: **a** principle of actuation for a tunable bistable mechanism; **b** model of the structure

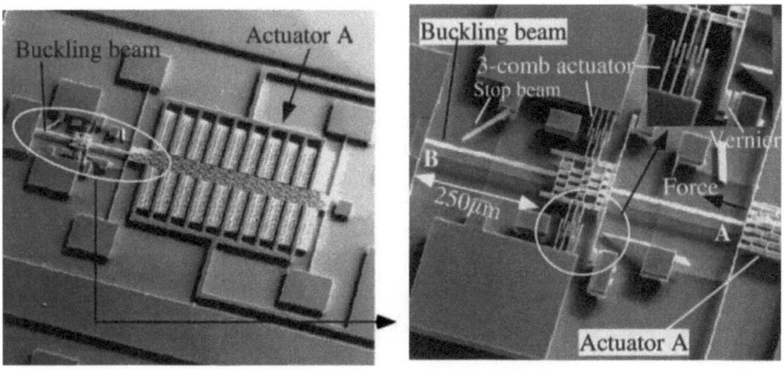

Fig. 2.12 *Left*: overall photograph of the actuator and the bistable buckling beam of [24]. *Right*: detail of the beam and up and bottom comb drives

its own springs (located to the left in Fig. 2.11a) and the buckling beam. The force transverse to the beam is generated by two comb drives vertically arranged between points A and B in Fig. 2.11a). A photograph of the device can be observed in Fig. 2.12. The main actuator spans an area of $1 \times 1\ \text{mm}^2$, has 800 comb drives and is supported by six beams, held by anchors. It has been designed to generate only compressive stress on the buckling beam. The fact that this compressive force is exerted by the comb drives of the main actuator is what makes this device tunable, since the bistable potential can be modified as a function of the voltage applied to the main actuator.

In [23] the authors present another bistable device that does not rely on prestress of the beam to produce buckling. They use two curved centrally clamped parallel beams to generate the bistable potential, see Fig. 2.13. A similar approach is taken in [26] on which preshaped buckled beams are designed [27]. In general, the physics of curved arches requires additional analysis in order to predict the nonlinear effects, due to the curvature of the structures and the snap-through instability [28, 29].

In [19] a bistable potential is obtained by a different structure on which a central mass is connected with two slightly slanted rigid fixed links through flexural pivots. In this way, two symmetrical stable positions can be achieved and therefore the structure presents a bistable potential. The equivalent mechanical model for the potential energy can be observed in Fig. 2.14. The slant angle, namely θ_0 plays an important role in the definition of the elastic potential of the structure, since it is used to generate the bistable potential (for $\theta_0 = 0$ the potential is monostable), while at the same time it will generate an asymmetry in the elastic potential. The reason for this is that the slant angle makes the snap easier from one of the states than from the other. The potential energy of the mechanical structure is difficult to analyze and the authors use FEM simulations to obtain the elastic potential. However, in the final proposed structure, double fixed links are connected in a H configuration with a central mass. The whole thickness of the BESOI wafer (467 μm) is used in the implementation of the arms. This is done in order to improve the mechanical stability of the design, specifically avoiding the appearance of unwanted vibration modes. The slant angle

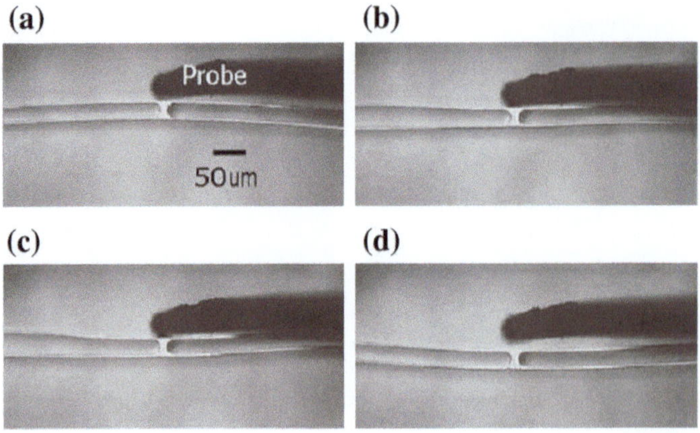

Fig. 2.13 Bistable mechanism of double curved beams, [23], **a** initial stable position, **b** applied force deflects the double beams, **c** more force generates more deflection, **d** beams at the second stable position

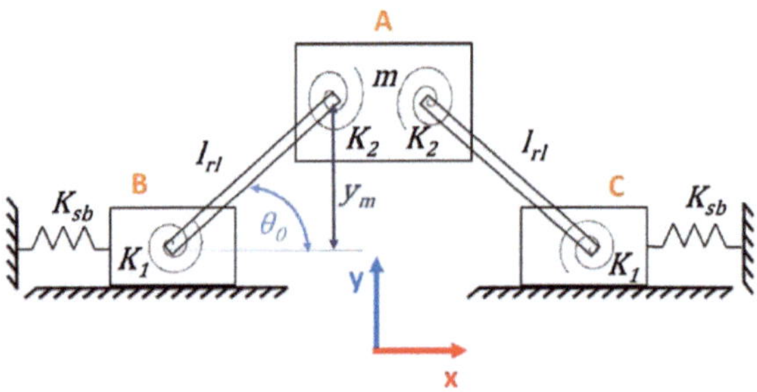

Fig. 2.14 Pseudo-Rigid body model of the first microstructure proposed in [19]

is finally chosen to be 0.5 °. The sensing electrodes, consisting of a comb drive, are placed perpendicular to the direction of the fixed links and extract the mechanical energy harvested by the mechanical structure, see Fig. 2.15.

2.3.2 *Nonlinear Springs*

The use of nonlinear springs has been proposed to increase the effective bandwidth in energy harvesters. Nonlinearity in the spring stress–strain relationship can generate either a hardening, [30, 31], or a softening behaviour, [32, 33]. In the first case, the resonant frequency increases when the oscillation amplitude increases, while in the second case the resonant frequency decreases with increasing oscillation amplitude. In general both phenomena can occur in a device, but depending on some condition (applied voltage, dimensions of the structure, etc.) one of them dominates the other, [34]. Nonlinearity in springs appears for large deflections. For example, in the

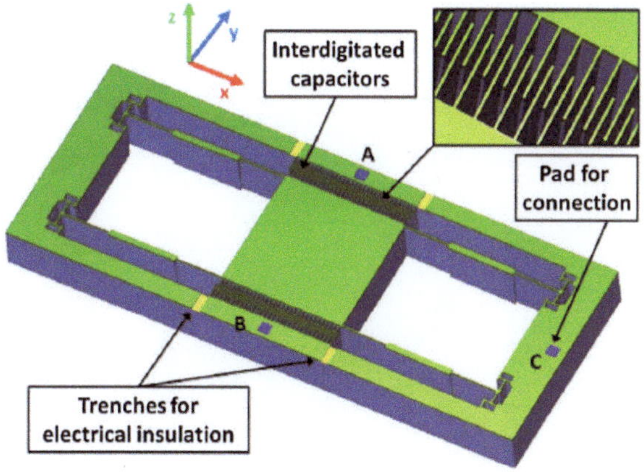

Fig. 2.15 Model of the final structure proposed in [19]. Pads "A", "B" and "C" are for the electrical connection of the comb capacitors

case of thin fixed-fixed beams, or tethers, it appears when the displacement of the proof mass is comparable or larger than the thickness of the springs, resulting in an increase of the strain energy due to stretching of the neutral surface, [34–36]. This is obviously an unwanted phenomenon in linear devices whereas it is a desired feature in some applications, such as energy harvesting, in which nonlinearity is sought as a means to increase bandwidth.

The authors in [32] present a wafer thick resonator that has nonlinear springs for enhancing energy harvesting. The fabrication process consists mainly of a full-wafer-thickness dry etch that serves to delimit the inertial mass and the supporting nonlinear springs. This wafer will be anodically bonded to a Pyrex glass wafer that has been previously wet etched so that the inertial mass is free to move. Special care has been taken by placing dummy protective structures close to springs and capacitor fingers. This is to limit inward sloping sidewalls in places where the etch opening is of the order of the etch depth or larger. In this design the spring restoring force is approximated by a seventh order polynomial. Figure 2.16 shows a model of the proposed structure, a photograph of the nonlinear spring and also the relation between the force versus displacement in the nonlinear spring obtained by FEM simulations. As can be observed in Fig. 2.16ii the softening behaviour is obtained only in one direction (negative displacement).

The authors in [37] propose to use a clamped–clamped configuration for the supporting thin beams of a proof mass. Each beam is clamped between the proof mass and the supporting structure. In this way, the displacement of the proof mass generates stretching. The only existence of bending would require lateral motion and, since the proof mass is in the centre of a clamped–clamped beam, this is not possible. This effect is illustrated in Fig. 2.17a. The fabrication process includes several steps comprised of both surface and bulk micromachining. The structural layer of the beams is composed of LPCVD silicon nitride and low temperature silicon oxide. The active layer consists of Ti/Al interdigitated electrodes, a PZT thin film and a layer of ZrO_2. A PECVD passivation layer is finally added to protect the active layer and compensate the residual stress. After the structure has been fabricated external

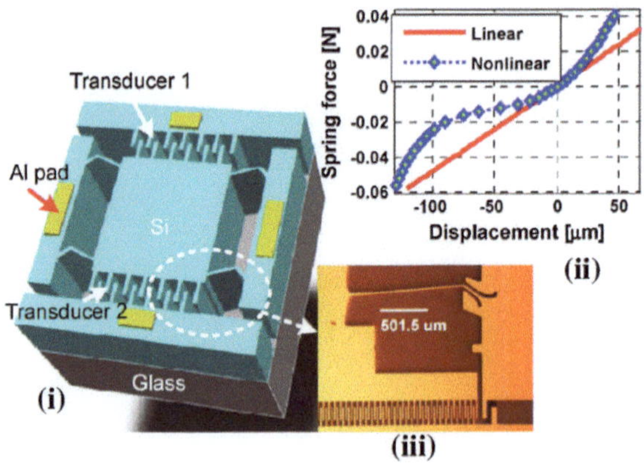

Fig. 2.16 **i** Geometry of the device using nonlinear springs proposed in [33], **ii** spring force as a function of displacement of the nonlinear spring calculated by FEM, **iii** photograph of one the nonlinear springs in the device

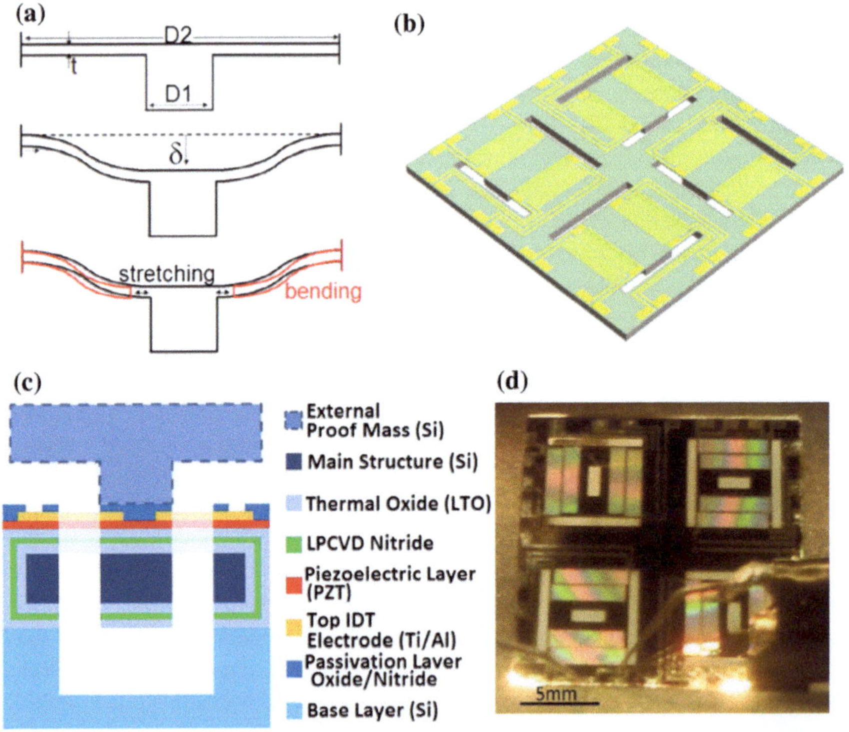

Fig. 2.17 Device proposed in [37]: **a** displacement of the proof mass generates nonlinear bending and stretching strain in the supporting beams, **b** general overview of the device with four beams supporting the main structure of the bulk silicon wafer. An external mass will be attached to the upper surface of the beams afterwards, **c** cross section of the structure, **d** photograph of the final device

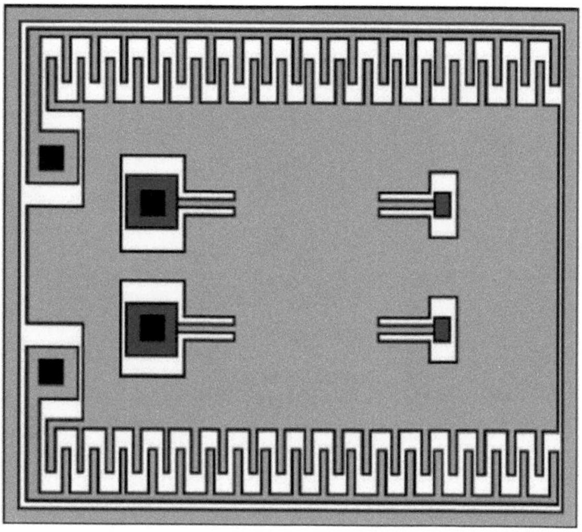

Fig. 2.18 Schematic of the device proposed in [38]. *Dark grey rectangles* are mass-spring anchorages and *black squares* represent electrical pads. The displacement of the proof mass is in plane between comb drives

masses are attached to the upper surface of the beams to achieve a higher nonlinearity of the structure.

The authors in [38] propose an electrostatic harvester using a large mass subjected by nonlinear springs to increase the effective bandwidth. A schematic of the device can be seen in Fig. 2.18 on which two comb drives are at the top and bottom, while the inertial mass is anchored in four points (dark grey areas in the middle of the inertial mass). The device has been fabricated with a commercial SOI process from Tronics Inc. The nonlinear regime is obtained at large displacements of the inertial mass. The authors have fitted with a seventh order polynomial the nonlinear terms of the spring.

2.3.3 Nonlinear Energy Transfer by Impact

The first proposed use of impact in energy harvesting application can be traced back to [39]. Impact actuation can be seen as an example of pulsed actuations, i.e. an actuation on which the velocity of the device is almost instantly changed. This type of excitation has been used to obtain self-sustained oscillations in MEMS resonators, [40–42].

An example of energy transfer by impact applied to energy harvesting can be found in [43, 44] where the authors propose the use of two cantilevers, one with a low resonant frequency (LRF) and another with a higher resonant frequency (HRF) that may impact each other. The structure of the proposed device can be observed in Fig. 2.19. The LRF cantilever has a silicon inertial mass and has to be stopped by the HRF cantilever. For this purpose, a specific packaging has been designed. The LRF cantilever excites the second one, the HRF cantilever. This results in a

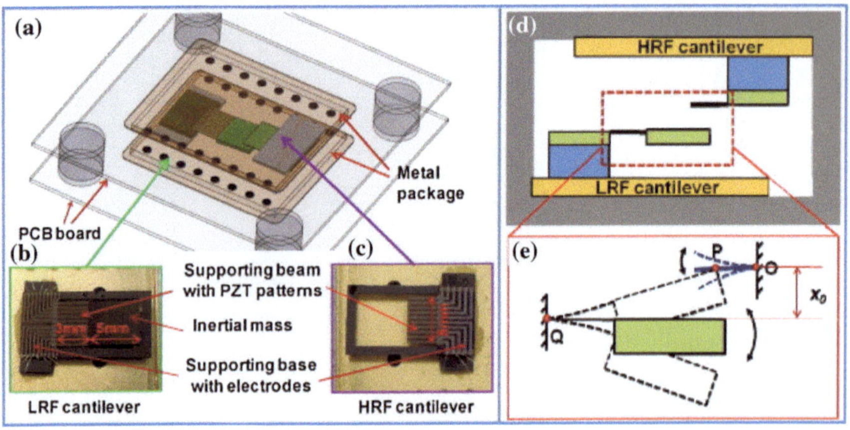

Fig. 2.19 Device proposed in [43]: **a** Schematic of the overall system, **b** LRF cantilever, **c** HRF cantilever, **d** schematic of the arrangement of both cantilevers inside the packaging, **e** illustration of the impact between cantilevers

FUC, Frequency Up-Conversion, on which the energy is transferred from the LRF cantilever to the HRF.

The simplified equations of motion are

$$\begin{aligned}(m_0 + m_1)\ddot{z} + (c_0 + c1)\dot{z} + (k_0 + k_1)z - k_1 x_0 &= -(m_0 + m_1)\ddot{y}, \qquad (z \geq x_0)\\ m_0\ddot{z} + c_0\dot{z} + k_0 z &= -m_0\ddot{y}, \qquad (z < x_0)\end{aligned} \tag{2.3}$$

The excitation cantilever, LRF, has as a proof mass m_0, a damping coefficient c_0 and a spring constant k_0. At a distance x_0 the second cantilever is placed with a mass m_1, a damping coefficient c_1 and a spring constant k_1. Both cantilevers interact when the displacement of the first one is greater than the distance between both resonators, $z \geq x_0$.

The fabrication process of the piezoelectric cantilever, [45], is started from a SOI wafer with a 5 μm thick silicon active layer, a buried oxide of 1 μm and the bulk of the wafer, 400 μm. On top of the active Si layer, multiple layer depositions are made (the first one of Pt/Ti for the bottom electrode, second one of 3 μm of (100)-oriented PZT, and a third metal deposition composed of Ti/Pt/Ti for the configuration of the top electrode). The metal depositions are made using sputtering and the PZT is deposited with a sol-gel process. The etching of the multilayer is made with an Ar-ion beam for the metals and the PZT layer is wet etched. The whole structure is passivated with a SiO_2 layer deposited by sputtering. Holes are made on the oxide layer to place contacts. The bulk of the SOI wafer is later etched from the backside to release the moveable parts while preserving the inertial mass.

For this application, special care must be taken with the packaging because it must provide the correct separation between the cantilevers while at the same time one of them is placed upside-down, see Fig. 2.19.

2.4 Transduction Principles

The objective of this section is to present the specific requirements of electrostatic and electromagnetic transduction when used in energy harvesters. In particular the electrostatic transduction, in some devices, requires the presence of electrets. Electrets are dielectric materials on which permanent charge has been introduced. The movement of the electrets due to external vibrations can trigger currents in some parts of the device from which energy can be retrieved. Piezoelectric transduction will be treated later in a separate section.

2.4.1 Electrostatic Transduction

In electrostatic harvesters, the vibration energy is recovered by means of electrostatic force, which acts as a damping mechanical force. These kinds of devices require an initially precharged surface, so that the movement of a seismic mass generates a current flow in some part of a circuit. In order to provide this initial charge, several alternatives have been proposed such as insulated floating electrodes, [46], electrets [47, 48] and also electret-less circuits that require an external starting voltage or current, [49]. Electrets have been extensively used although they may require unconventional steps in the fabrication process.

On the other hand, some papers in the literature present electrostatic energy harvesters avoiding the presence of electrets or even external circuits necessary for precharging the structure. This kind of papers focus on the exploited mechanical nonlinearity or other features and use standard power sources to harvest. In this subsection, we will focus on electrets and we will present some examples of energy harvesters that make use of them.

Electrets are dielectrics on which a quasi-permanent charge has been injected, [50]. The lifetime of these charges can be very large. As an example, experiments on which a charge density has remained unchanged for 35 years are reported in [51]. Electrets present this charge, and therefore a stable built-in voltage, due to some previous ion injection, poling or application of a voltage. Among the materials most commonly used for electrets we find inorganic layers made of SiO_2, a combination of SiO_2 and Si_3N_4 layers, [52], and also organic compounds based on polymers. Among the polymers compatible with MEMS processes we may find the amorphous perfluorinated polymer CYTOP (Asahi Glass Col., Ltd), [53–56]. This polymer can provide up to four times more charge density than Teflon.

There are numerous methods used for poling electrets such as: corona charging, ion implantation, contact charging, thermal poling, UV irradiation, soft X-ray irradiation and electron-beam irradiation. Charges have also been successfully injected into electrets with very inexpensive equipment such as ionic hair-dryers, [56]. Corona charging generates principally surface charge whereas electron beam and thermal poling produce volume, surface charges and polarization. Thermal annealing may follow the corona charging to increase charge stability by driving charges to the bulk of the material. Long term, both SiO_2 and CYTOP present good charge stability, [55], which is an essential property.

The more conventional approach for using electrets includes planar electrets, [57], although they can also be deposited on vertical walls, [56, 58]. Vertical electrets are more difficult to produce as, for example, the corona method or ion implantation cannot be used. On the other hand, vertical electrets are compatible with standard electrostatic comb drive actuation.

Examples of recent energy harvesters using electrets can be found in [56, 57, 59]. The authors in [57] propose an electrostatic energy harvester using an electret inside the active gap. Figure 2.20 shows a schematic of the device. The electret has been patterned in stripes on the bottom wafer while the proof mass, which is conductive, moves inside the potential generated by the electret. If a load is connected between electrodes 1 and 2, a current will flow and power will be transferred. Since the displacement of the mass can surpass a single electret stripe, frequency up-conversion may occur. This is because a single period in the proof mass movement can generate several electret stripe crossings and therefore several periods in the generated current.

The authors in [56] propose a 3D electrostatic energy harvester using sidewall electrets. The device allows one retrieve vibration energy in any direction due to the specific design that fixes an angle of 45° between the suspending beams and the comb drive electrodes with electrets, see Fig. 2.21. The pull-in contact between electrodes is avoided by the presence of the mechanical stoppers shown in Fig. 2.21a. The electrets have been vertically placed on the surface of the fixed electrode fingers and are made of SiO_2. The authors have charged the electrets using an hair-dryer that is commercially called "ionic hair-dryer" since it generates ions to compensate for the charge generated in hairs by combing. Electret charging with this method

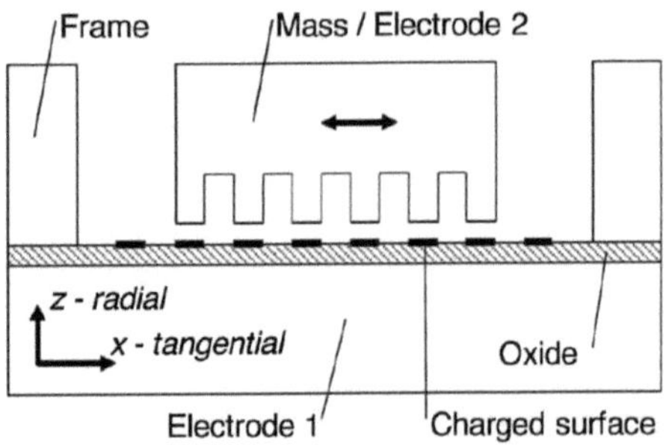

Fig. 2.20 Cross section of the device proposed in [57]. The motion of the proof mass is indicated by the *arrow*. The proof mass is conductive

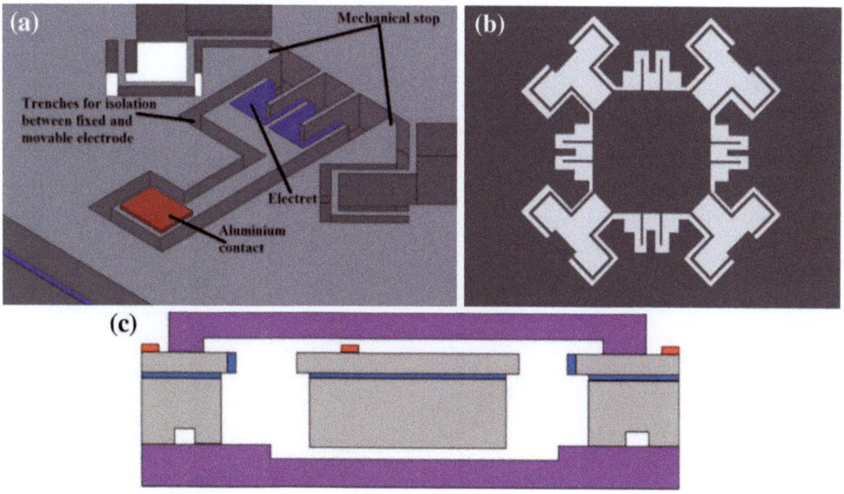

Fig. 2.21 **a** Detail of the device proposed in [56], **b** schematic of the overall device, **c** cross section of the device (suspending beams not shown)

strongly depends on the exposure angle. Nevertheless, the authors are able to charge these vertically placed electrets by exposing them to the air flow for more time.

2.4.2 Electromagnetic Transduction

In this kind of transduction, the energy is recovered in coils present in the device as a function of the variation of a magnetic field. The variation of the magnetic field will be due to some vibration mode activated in the structure by the external acceleration from which energy must be recovered.

The authors in [60] propose a device for electromagnetic energy harvesting. This device uses a SOI wafer for which the thickness of the active layer and the bulk of the wafer are 5 μm and 400 μm, respectively. The design has two thin clamped–clamped beams connected to a large mass by beam joints, see Fig. 2.22. A cross section of the device can be observed in the same figure. The device presents a two fold objective to increase performance of the device: nonlinear force-deflection characteristic, an amplitude-stiffened Duffing-spring and multi-frequency harvesting mechanisms using different resonant modes of the structure (out of plane 70.7 Hz, torsion 85.78 Hz and twist modes 147.92 Hz). The electromagnetic harvesting is accomplished by placing a cylindrical magnet on top of the device that generates a magnetic field perpendicular to the surface of the device. The large mass integrates the EM coils that will be used to retrieve the electromagnetic energy. All three modes can generate large beam deflection and stretching strain, generating therefore a nonlinear response of the device.

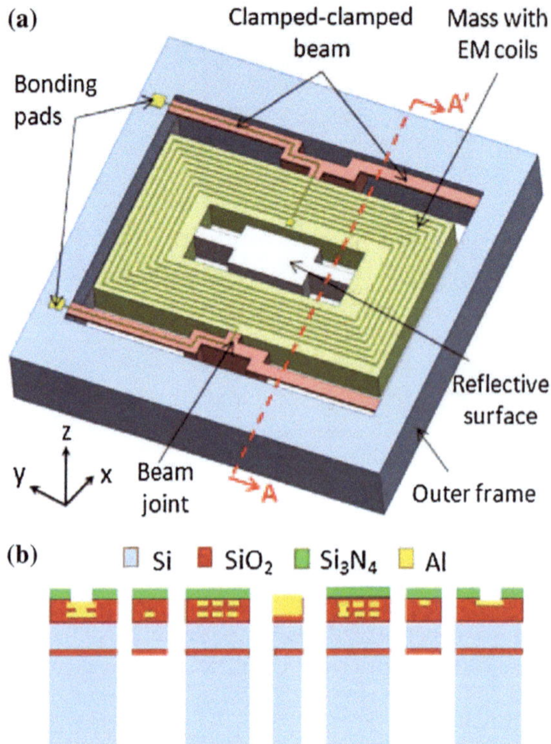

Fig. 2.22 Schematic of the electromagnetic harvester presented in [60]. A cross section of the device is shown at the *bottom* of the figure. The thickness of the silicon active layer, used for the clamped–clamped beams is 5 μm and the thickness of the bulk of the wafer is 400 μm

The expression showing the equivalent force applied in the centre of the clamped–clamped beam of the device is

$$F = \left(\frac{\pi^4}{6}\right)\left[\frac{Ewh^3}{l^3}\right]x + \left(\frac{\pi^4}{8}\right)\left[\frac{Ewh}{l^3}\right]x^3 \tag{2.4}$$

where E is the Young's modulus, w, h and l are the width, thickness and length of the beam. The first term is due to the bending strain and produces a linear spring constant, whereas the second term, related to the stretching strain, produces the cubic term. Observing this expression the nonlinear term will be significant for displacements larger that the thickness of the beam [60].

2.5 MEMS Devices Based on the Piezoelectric Transduction Mechanism

This section describes the structures, materials and main properties of MEMS based on the piezoelectric transduction mechanism. The piezoelectric properties of materials and a description of the most common piezoelectric materials are included. Subsequently, cantilever beam structure with a tip mass is described due to its wide use in MEMS energy harvesters. The unimorph and bimorph cantilever structures are presented and discussed. Bimorph piezoelectric cantilevers allow higher output power while unimorph piezoelectric cantilevers are more manufacturable. Furthermore, the top-bottom electrodes, TPE, and interdigitated, IDE, and electrodes configurations for unimorph cantilevers are also reviewed. These electrode configurations are used when 31 or 33 modes of piezoelectric materials are required. Finally, an application example of a MEMS-based bimorph energy harvester is illustrated.

Piezoelectric transduction is an advantageous mechanism to convert vibration-to-electric energy for small scales as stated by Marin et al. [61] and Cook et al. [62]. Marin studied the relationship between the effective material volume (v) and the output power for different mechanisms. The output power for the piezoelectric mechanism is proportional to $v^{3/4}$, while that for the electromagnetic mechanism is proportional to v^2. At volumes smaller than 0.5 cm^3 the electromagnetic transformation factor reduces abruptly [61]. The piezoelectric mechanism has the highest energy conversion efficiency [63, 64] and is the most appropriate technology for the scale of MEMS devices [65, 66].

2.5.1 Piezoelectricity

The piezoelectric effect was reported by Jacques and Pierre Curie in 1880 [67, 68]. They found that some kind of crystals subjected to mechanical strain became electrically polarized. Furthermore, they found that the relationship between the degree of polarization and the applied strain is proportional. The inverse piezoelectric effect, these materials deform when a voltage is applied, was predicted by Lippmann [69] and confirmed experimentally by the brothers Curie.

Piezoelectric materials present anisotropic behaviour. Consequently, the properties of the material depend upon the direction of the strain and the direction of the polarization and in particular, on the position of the electrodes. It is important to note that the piezoelectric engineered materials, unlike natural materials, are subjected to a process called poling to impart the piezoelectric behaviour. The dipoles of the cells of a macroscopic crystalline structure are originally randomly oriented and the piezoelectric effect is negligible when the material is subjected to a mechanical stress. The poling process consists of applying a very high electric field that permanently orients the dipoles in the direction of the field, Fig. 2.23.

It is important to avoid the depolarization of the piezoelectric material. There are two ways to de-pole a piezoelectric material: (i) applying a very high electric field opposite to the field applied during the poling process, and (ii) heating the material above its Curie temperature.

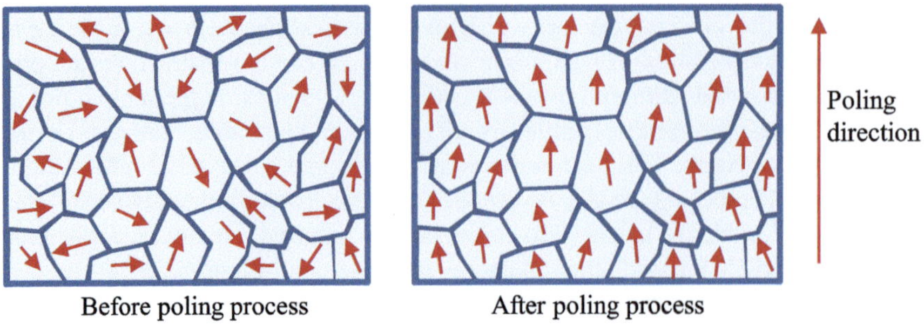

Fig. 2.23 The poling process induces piezoelectric properties in the material. Before the poling process the dipoles are randomly aligned (*left*). After poling process the dipoles tend to be pointed to poling direction (*right*)

2.5.2 Properties of Piezoelectric Materials

The compliance, coupling and permittivity properties for a piezoelectric material are second-order tensors due to the anisotropic behaviour of these materials. As a result, constants defined to characterize the piezoelectric materials have two subscripts, one related to the applied force, stress, and the other related to the change in length, the strain. The specific coordinate system used to indicate the directions of the stress and the strain are shown in Fig. 2.24. Conventionally, the poling direction is the z-axis (or 3), except in quartz where it is x-axis (or 1).

As can be seen, the x, y and z axes are represented by the subscripts 1, 2 and 3, respectively. The subscripts 4, 5 and 6 are used to represent the rotation about these axes.

The constants commonly used to characterize piezoelectric materials are: (i) the piezoelectric strain/charge constant, d, (ii) the piezoelectric stress/voltage constant, g, (iii) the electromechanical coupling coefficient, k, and (iii) the permittivity, ϵ.

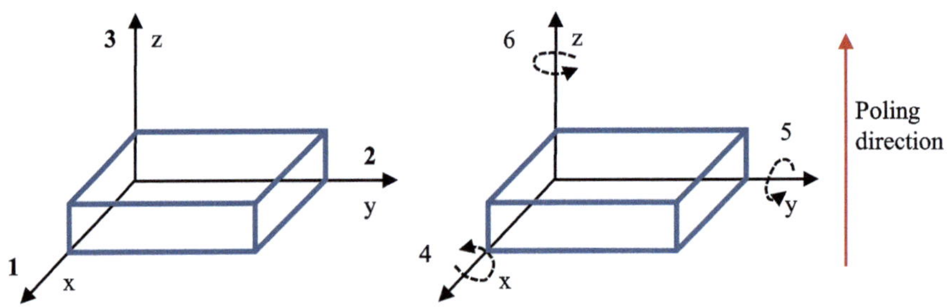

Fig. 2.24 Specific notation for coordinates used in piezoelectric constants and quantities. The z-axis coincides with the poling direction. The numbers 1, 2 and 3 represent the Cartesian coordinates (*left*) and the numbers 4, 5 and 6 represents the shear about Cartesian coordinates (*right*)

The anisotropic piezoelectric mechanical and electrical quantities are linked using double subscripts (i.e. k_{ij}). The subscripts designate the axis parallel to the direction of the excitation and the axis parallel to the direction of the response [70].

The strain/charge constants, d_{ij}, express the relationship between the strain developed along or around an axis and the applied electric field parallel to an axis when all external stresses are constant [71]

$$d_{ij} = \frac{strain\ developed\ along\ j-axis}{applied\ electric\ field\ parallel\ to\ i-axis} \quad \left(\frac{m}{V}\right) \tag{2.5}$$

The strain/charge constant can also be defined as the relationship between the short-circuit charge per unit area flowing between connected electrodes that are perpendicular to an axis and the stress applied along or around an axis when all other external stresses are constant [71]

$$d_{ij} = \frac{short-circuit\ charge\ flowperpendicular\ to\ i-axis}{applied\ stress\ along\ \ j-axis} \quad \left(\frac{C}{N}\right) \tag{2.6}$$

The stress/voltage constant, g_{ij}, is the relationship between the open circuit electric field developed along an axis and the stress applied along or around an axis when all other external stresses are constant [71]

$$g_{ij} = \frac{open\ circuit\ field\ along\ i-axis}{stress\ applied\ along\ j-axis} \quad \left(\frac{Vm}{N}\right) \tag{2.7}$$

The stress/voltage constant can also be defined as the relationship between the strain developed along an axis and the electric charge density applied to electrodes that are normal to an axis [71]

$$g_{ij} = \frac{strain\ developed\ along\ j-axis}{applied\ charge\ density\ normal\ to\ j-axis} \quad \left(\frac{m^2}{C}\right) \tag{2.8}$$

The electromechanical coupling coefficients, k_{ij}, are the energy ratios describing the conversion from mechanical energy to electrical energy or vice versa [64]. The ratio of energy stored, mechanical or electrical, to energy, mechanical or electrical, applied can be calculated by squaring the electromechanical coupling factor, k^2. The first subscript, i, indicates the direction of the electrical field and the second designates the direction of the mechanical strain. These coefficients have no dimensions. Equation 2.9 shows the relationship between the mechanical input energy in the j-axis, W_j^m, and the electrical energy stored in the i-axis, W_i^e.

$$k_{ij}^2 = \frac{W_i^e}{W_j^m} \quad (dimensionless) \tag{2.9}$$

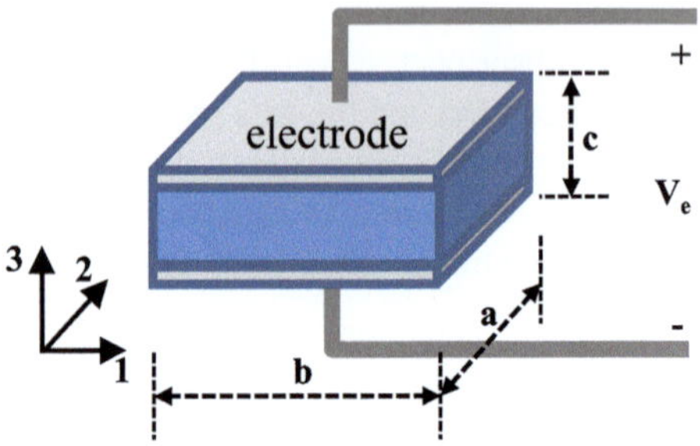

Fig. 2.25 Piezoelectric material with two electrodes on its opposite sides. V_e is the voltage across the electrodes and a, b an c are the dimensions of the piezoelectric element

The most common piezoelectric constants and coefficients used in energy harvesting systems are listed below using the dimensions and the reference system shown in Fig. 2.25.

- d_{31}: This constant represents the relationship between the charge developed, Q, flowing between the shorted electrodes, $V_e = 0$ V, that are orthogonal to the poling axis (3 – z), and a force, F, applied along the axi 1. Therefore, the charge developed Q can be calculated as [72]:

$$Q(V_e = 0\ V) = d_{31} F \frac{b}{c} \tag{2.10}$$

- d_{33}: This constant represents the relationship between the charge developed, Q, flowing between shorted electrodes that are orthogonal to the poling axis (3 – z), and a force, F, applied in the direction of the polarization axis (3 – z).Therefore, Q can be calculated as [72]:

$$Q(V_e = 0\ V) = d_{33} F \tag{2.11}$$

- g_{31}: This constant expresses the relationship between the induced electric field is along the poling axis(3 – z) and the force applied in the direction orthogonal to the polarization axis (1 – x), Fig. 2.26. Therefore, the induced voltage can be calculated as [72]

$$V_e(Q = 0\ C) = g_{31} \frac{F}{a} \tag{2.12}$$

- g_{33}: This constant expresses the relationship between the induced electric field along the poling axis(3 – z) and the force applied in the direction of the polarization axis (3 – z), Fig. 2.27. Therefore, voltage induced can be calculated as [72]

$$V_e(Q = 0\ C) = g_{33} F \frac{c}{ab} \tag{2.13}$$

Therefore, the spatial orientation of the materials is very important. It must be taken into account when a system is modelled, analyzed or simulated.

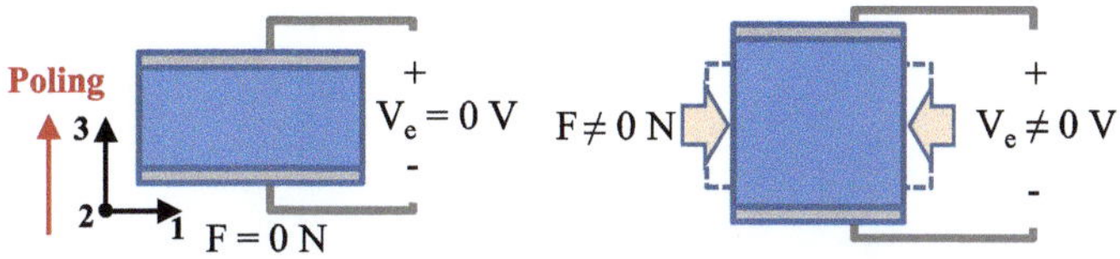

Fig. 2.26 The 31 mode operation for piezoelectric materials. No voltage is induced when no force is applied (*left*) and voltage is induced when force is applied (*right*)

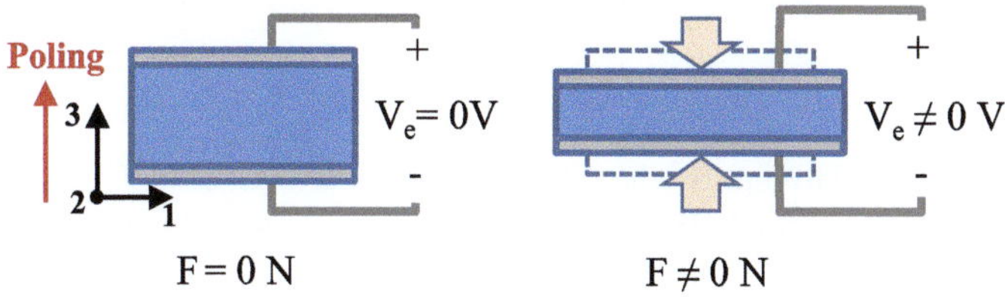

Fig. 2.27 The 33 mode operation for piezoelectric materials. Effect when no force is applied (*left*) and voltage is induced when force is applied along the poling axis (*right*)

2.5.3 Piezoelectric Materials

Piezoelectric materials are used in a range of applications such as pressure sensing, data storage, mechanical actuation, ultrasonic wave generation and energy harvesting. Owing to this wide range of applications, a large number of materials have been developed. The following classification includes the most available forms of piezoelectric materials [73, 74]:

- Crystals: Quartz (SiO_2), Berlinite ($AlPO_4$), Gallium Ortophosphate $GaPO_4$, Turmaline, etc.
- Polycrystalline ceramics: Barium titanate ($BaTiO_3$), Lead Zirconate Titanate (PZT) [75]. These materials are characterized by its perovskite tungsten-bronze structure.
- Thin film non-ferroelectric materials: Sputtered zinc oxide (ZnO), Aluminium Nitride (AlN) [76].
- Polymeric materials: Polyvinylidine fluoride (PVDF) [77].
- Screen printable thick-films based upon piezoceramic powders [78] and composites such as polyvinylidene-trifluoroethylene-PZT (PVDF-TrFE) [79]
- Organic crystals: single-crystal diisopropylammonium chloride (DIPAC) and diisopropylammonium bromide (DIPAB) [80, 81]

MEMS devices require thin and thick film technologies since the layers of the materials must be below 100 μm. Thin-film technologies, physical or chemical deposition, are used to fabricate films with thickness lower than 5 μm. Thick layers with thickness up to 100 μm can be fabricated using the screen printing method. Maas et al., [82], describe PZT printing onto silicon, the powdered PZT is mixed with borosilicate glass powder and an organic vehicle to make a paste, it is the ink. The

Table 2.1 Coefficients of common piezoelectric materials [64, 84–86]

Property	PZT-5H	PZT-5A	$BaTiO_3$	PVDmF	AlN	GaN	ZnO
d_{33} (10^{-12} C N^{-1})	593	374	149	−33	5	3.7	12.4
d_{31} (10^{-12} C N^{-1})	−274	−171	78	23	−2	−1.9	−5
g_{33} (10^{-3} V m N^{-1})	19.7	24.8	14.1	330	–	–	–
g_{31} (10^{-3} V m N^{-1})	−9.1	−11.4	5	216	–	–	–
k_{33} (dimensionless)	0.75	0.71	0.48	0.15	0.23	–	0.48
k_{31} (dimensionless)	0.39	0.31	0.21	0.12	–	–	0.182

layer is printed, dried to remove the solvent and finally fired. The aluminium nitride, AlN, has recently been included on the list of materials used in energy harvesting. Schaijk et al. justify the selection of AlN as piezoelectric material as it is easier to process than PZT [83]. The piezoelectric constants of PZT are better than AlN but the main advantage of AlN is its compatibility with CMOS processing. This is because it is foreseeable that in the near future, new investigations will result in better performance devices based on AlN. Finally, the piezoelectric constants for common materials used in MEMS devices are shown in Table 2.1.

2.5.4 *Piezoelectric Energy Harversters*

Energy harvesting MEMS devices based on piezoelectric materials generally have a cantilever beam structure as shown in Fig. 2.28. The beam is clamped at one end and the mechanical vibrations are generated principally along the third axis through bending. As can be seen, the vertical movement of the mass induces a strain along the x-axis (or 1). Accordingly, if a 31 mode piezoelectric material is deposited on the beam, then a voltage would be induced between the two sides of this material when it is exposed to vibration sources.

The structure shown at Fig. 2.28 creates equal and opposite strains on the beam. Therefore, it is not possible to use a beam based only on a piezoelectric material as the total voltage or current generated would be null. To be effective as a generator a piezoelectric layer is fixed to a non-piezoelectric elastic layer, Fig. 2.29. Thereby the

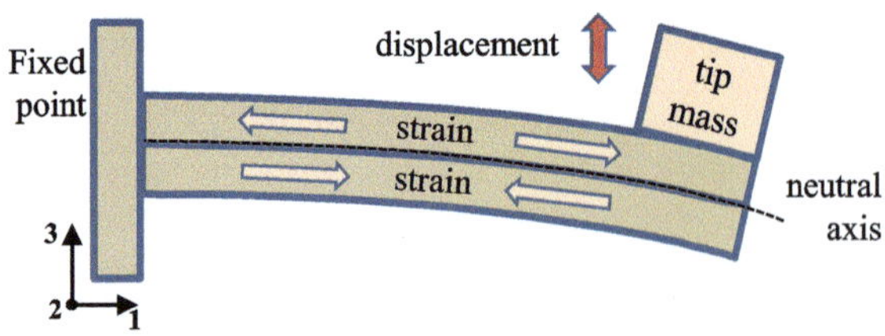

Fig. 2.28 Cantilever beam structure with a tip mass subjected to bending. The strain developed above the neutral axis is opposite to the strain developed below the neutral axis

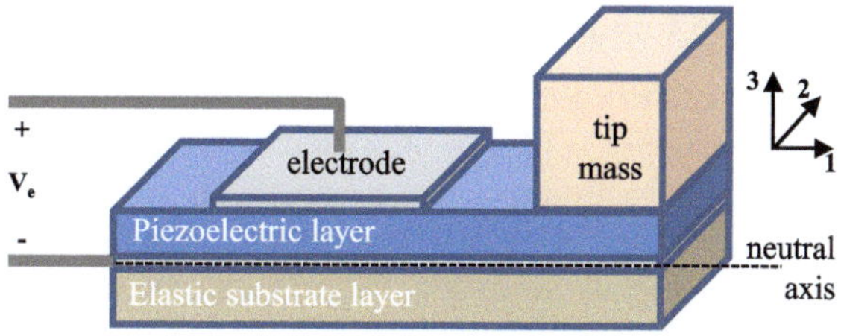

Fig. 2.29 The unimorph structure based on a piezoelectric layer and a inactive substrate. The generated voltage, Ve, depends only on the strain on the piezoelectric layer

neutral axis, the line in which the beam does not change in length, is not in the centre of the piezoelectric layer. This structure is named unimorph because it is composed of a unique piezoelectric active layer.

The 33 mode of piezoelectric materials is often more favourable than 31 mode, see Table 2.1. This would be a reason to use this mode but the structure shown at Fig. 2.29 is not feasible as the direction of the induced electric field in 33 mode is parallel to the stress direction. The interdigitated electrode, IDE, configuration presented by Jeon et al. at [66] and Park et al. at [87] enables one employ the 33 mode and discards the bottom electrode layer, Fig. 2.30.

IDE configuration has the additional advantage of allowing a designer adjust the electrode spacing and, consequently, output voltage since the open circuit voltage is proportional to the distance between electrodes, see Eq. 2.13. A comparison between the structures proposed using the 31 and 33 modes is shown in Fig. 2.31.

Therefore, the electrode lengths and shapes are important parameters that affect the output voltage due to the nonuniformity of the strain along the beam [88]. Kim et al. have fabricated and compared piezoelectric energy harvesters based on d_{31} and d_{33} modes [89]. These authors have demonstrated that the output power of the d_{33} mode depends strongly on the dimensions of IDE.

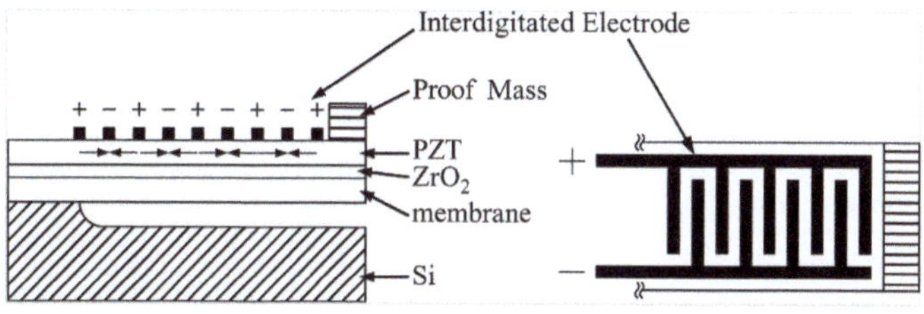

Fig. 2.30 Structure of the unimorph cantilevered beam with interdigitated electrode proposed by Jeon et al. [66]

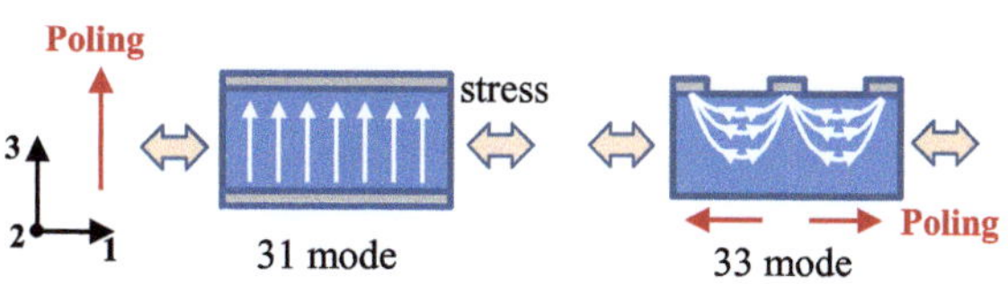

Fig. 2.31 Relationship between applied stress and induced field, white arrows, for 31 and 33 modes

Bimorph Cantilever Structure

The structure of a bimorph piezoelectric energy harvester based on a cantilever beam is shown in Fig. 2.32. Two piezoelectric layers are fixed to a non-piezoelectric elastic layer. The electrodes can be connected in series or in a parallel configuration. The main advantage of a bimorph structure is that there is a lower loss in the mechanical support materials since only the piezoelectric layer is strained.

Currently, the optimization of energy harvesters is one of the most important challenges due the low energy extracted. In order to analyze and optimize designs, several models of the piezoelectric energy harvesters have been proposed. Lumped-parameter models allow simple expressions but with limited accuracy. Models based on distributed-parameters offer higher accuracy, increasing the complexity of the expressions. Erturk et al. at [90] present a detailed electromechanical modelling of piezoelectric energy harvesters.

Application Example

A MEMS vibration energy harvester based on a PZT/PZT thick film bimorph with an integrated silicon tip mass is described in [91], Fig. 2.33.

The authors propose a new fabrication process that improves the quality of the thick film. The PZT layers are screen printed, treated with a high pressure process and finally sintered (high-temperature heating without melting the material). The

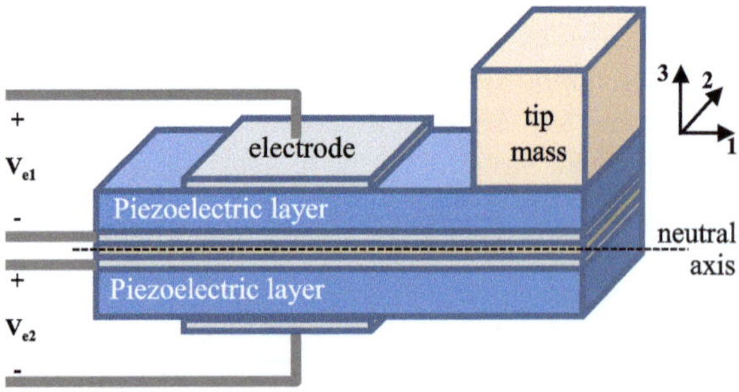

Fig. 2.32 Structure of a bimorph piezoelectric energy harvester

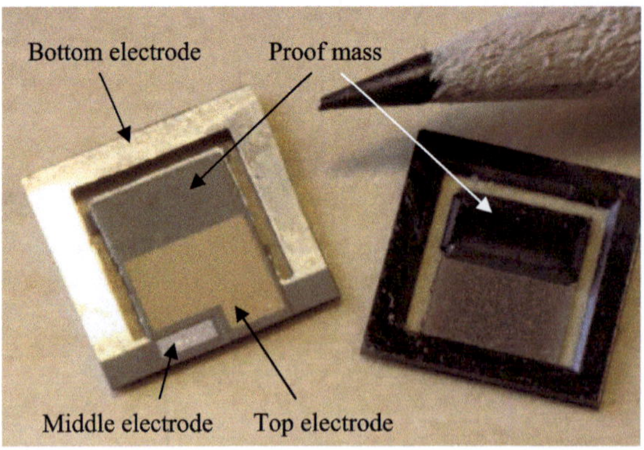

Fig. 2.33 *Front* and *back* of a MEMS-based PZT/PZT thick film bimorph vibration energy harvester fabricated by R. Xu et al [91]

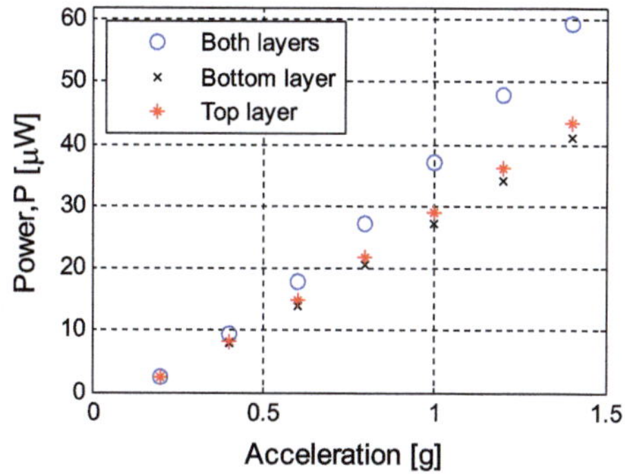

Fig. 2.34 RMS power output as a function of the input acceleration [91]

characterization of the power output as a function of the acceleration is shown in Fig. 2.34. The output power reaches 37.1, μW when the energy harvester is subjected to an acceleration of 1 g. It is important to denote that the power output due to the bottom layer and the power due to the top layer are very similar.

References

1. Liu, C. (2012). *Foundations of MEMS* (2nd ed.). Pearson Education.
2. Ghodssi, R., & Lin, P. (2012). *MEMS materials and processes handbook*. Springer.
3. Laermer, F., & Schilp, A. (1996). Method for anisotropic plasma etching of substrates. US Patent 5,498,312.
4. Laermer, F., & Schilp, A. (2003). Method of anisotropic etching of silicon. US Patent 6,531,068.
5. Chen, K. S., Ayon, A. A., Zhang, X., & Spearing, S. M. (2002). Effect of process parameters on the surface morphology and mechanical performance of silicon structures after deep reactive ion etching (DRIE). *Journal of Microelectromech Systems*, *11*(3), 264–275.
6. Gorreta, S., Fernandez, D., Blokhina, E., Pons-Nin, J., Jimenez, V., O'Connell, D., et al. (2012). Pulsed digital oscillators for electrostatic MEMS. *IEEE Transactions on Circuits and Systems I: Regular Papers*, *59*(12), 2835–2845.
7. Howe, R. T., & Muller, R. S. (1983). Polycrystalline silicon micromechanical beams. *Journal of the Electrochemical Society*, *130*(6), 1420–1423.
8. Bustillo, J. M., Howe, R. T., & Muller, R. S. (1998). Surface micromachining for microelectromechanical systems. *Proceedings of the IEEE*, *86*(8), 1552–1574.
9. http://www.memscap.com/products/mumps.
10. http://www.mems.sandia.gov/tech-info/summitv.html.
11. http://www.invensense.com/invensense-shuttle.
12. https://www.teledynedalsa.com/semi/fab/services/shuttleruns.
13. http://www.lionixbv.nl/mpw.
14. Cowen, A., Hames, G., Glukh, K., & Hardy, B. (2013). *PiezoMUMPS Design Handbook. Revision 1.2*, MEMSCAP Inc.
15. Pons, J., Gorreta, S., Blokhina, E., O'Connell, D., Feely, O., & Domínguez, M. (2014). Design and test of resonators using piezoMUMPS technology. In Symposium on Design, Test, Integration and Packaging of MEMS/MOEMS, Cannes, France, April (pp. 227–230).
16. Andò, B., Baglio, S., Trigona, C., Dumas, N., Latorre, L., & Nouet, P. (2010). Nonlinear mechanism in MEMS devices for energy harvesting applications. *Journal of Micromechanics and Microengineering*, *20*(12), 125020.

17. Guan, S., & Nelson, B. (2006). Magnetic composite electroplating for depositing micromagnets. *Journal of Microelectromechanical Systems*, *15*(2), 330–337.
18. Cottone, F., Vocca, H., & Gammaitoni, L. (2009). Nonlinear energy harvesting. *Physical Review Letters*, *102*, 080601.
19. Ando, B., Baglio, S., L'Episcopo, G., & Trigona, C. (2012). Investigation on mechanically bistable MEMS devices for energy harvesting from vibrations. *Journal of Microelectromechanical Systems*, *21*(4), 779–790.
20. Myung, N. V., Park, D. Y., Yoo, B. Y., & Sumodjo, P. T. A. (2003). Development of electroplated magnetic materials for MEMS. *Journal of Magnetism and Magnetic Materials*, *265*(2), 189–198.
21. Han, M., Yuan, Q., Sun, X., & Zhang, H. (2014). Design and fabrication of integrated magnetic MEMS energy harvester for low frequency applications. *Journal of Microelectromechanical Systems*, *23*(1), 204–212.
22. Sun, X., Yuan, Q., Fang, D., & Zhang, H. (2012). Electrodeposition and characterization of CoNiMnP permanent magnet arrays for MEMS sensors and actuators. *Sensors and Actuators A: Physical*, *188*, 190–197. Selected papers from The 16th International Conference on Solid-State Sensors, Actuators and Microsystems.
23. Qiu, J., Lang, J. H., & Slocum, A. H. (2004). A curved-beam bistable mechanism. *Journal of Microelectromechanical Systems*, *13*(2), 137–146.
24. Saif, M. T. A. (2000). On a tunable bistable MEMS-theory and experiment. *Journal of Microelectromechanical Systems*, *9*(2), 157–170.
25. Hoffmann, M., Kopka, P., & Voges, E. (1999). All-silicon bistable micromechanical fiber switch based on advanced bulk micromachining. *IEEE Journal of Selected Topics in Quantum Electronics*, *5*(1), 46–51.
26. Casals-Terre, J., Fargas-Marques, A., & Shkel, A. M. (2008). Snap-action bistable micromechanisms actuated by nonlinear resonance. *Journal of Microelectromechanical Systems*, *17*(5), 1082–1093.
27. Park, S., & Hah, D. (2008). Pre-shaped buckled-beam actuators: Theory and experiments. *Sensors and Actuators A: Physical*, *148*(1), 186–192.
28. Krylov, S., Ilic, B. R., Schreiber, D., Seretensky, S., & Craighead, H. (2008). The pull-in behavior of electrostatically actuated bistable microstructures. *Journal of Micromechanics and Microengineering*, *18*(5), 055026.
29. Das, K., & Batra, R. C. (2009). Pull-in and snap-through instabilities in transient deformations of microelectromechanical systems. *Journal of Micromechanics and Microengineering*, *19*(3), 035008.
30. Marinkovic, B., & Koser, H. (2009). Smart sand—a wide bandwidth vibration energy harvesting platform. *Applied Physics Letters*, *94*(10).
31. Marzencki, M., Defosseux, M., & Basrour, S. (2009). MEMS vibration energy harvesting devices with passive resonance frequency adaptation capability. *Journal of Microelectromechanical Systems*, *18*(6), 1444–1453.
32. Nguyen, D. S., Halvorsen, E., Jensen, G. U., & Vogl, A. (2010). Fabrication and characterization of a wideband MEMS energy harvester utilizing nonlinear springs. *Journal of Micromechanics and Microengineering*, *20*(12), 125009.
33. Nguyen, D. S., & Halvorsen, E. (2011). Nonlinear springs for bandwidth-tolerant vibration energy harvesting. *Journal of Microelectromechanical Systems*, *20*(6), 1225–1227.
34. Elshurafa, A. M., Khirallah, K., Tawfik, H. H., Emira, A., Abdel Aziz, A. K. S., & Sedky, S. M. (2011). Nonlinear dynamics of spring softening and hardening in folded-MEMS comb drive resonators. *Journal of Microelectromechanical Systems*, *20*(4), 943–958.
35. Gabbay, L. D., & Senturia, S. D. (2000). Computer-aided generation of nonlinear reduced-order dynamic macromodels. i. non-stress-stiffened case. *Journal of Microelectromechanical Systems*, *9*(2), 262–269.
36. Mehner, J. E., Gabbay, L. D., & Senturia, S. D. (2000). Computer-aided generation of nonlinear reduced-order dynamic macromodels. ii. stress-stiffened case. *Journal of Microelectromechanical Systems*, *9*(2), 270–278.

37. Hajati, A., & Kim, S. G., (2011). Ultra-wide bandwidth piezoelectric energy harvesting. *Applied Physics Letters*, *99*(8).
38. Tvedt, L. G. W., Nguyen, D. S., & Halvorsen, E. (2010). Nonlinear behavior of an electrostatic energy harvester under wide and narrowband excitation. *Journal of Microelectromechanical Systems*, *19*(2), 305–316.
39. Umeda, M., Nakamura, K., & Ueha, S. (1996). Analysis of the transformation of mechanical impact energy to electric energy using piezoelectric vibrator. *Japanese Journal of Applied Physics*, *35*(1), Part 1, No. 5B, 3267–3273.
40. Dominguez-Pumar, M., Pons-Nin, J., & Ricart, J. (2008). General dynamics of pulsed digital oscillators. *IEEE Transactions on Circuits and Systems I: Regular Papers*, *55*(7), 2038–2050.
41. Blokhina, E., Pons-Nin, J., Ricart, J., Feely, O., & Dominguez-Pumar, M. (2010). Control of MEMS vibration modes with pulsed digital oscillators part i: Theory. *IEEE Transactions on Circuits and Systems I: Regular Papers*, *57*(8), 1865–1878.
42. Ricart, J., Pons-Nin, J., Blokhina, E., Gorreta, S., Hernando, J., Manzaneque, T., et al. (2010). Control of MEMS vibration modes with pulsed digital oscillators part ii: Simulation and experimental results. *IEEE Transactions on Circuits and Systems I: Regular Papers*, *57*(8), 1879–1890.
43. Liu, H., Lee, C., Kobayashi, T., Tay, C. J., & Quan, C. (2012). Piezoelectric MEMS-based wideband energy harvesting systems using a frequency-up-conversion cantilever stopper. *Sensors and Actuators A: Physical*, *186*, 242–248. Selected Papers presented at Eurosensors XXV Athens, Greece, September 2011.
44. Liu, H., Lee, C., Kobayashi, T., Tay, C. J., & Quan, C. (2012). Investigation of a MEMS piezoelectric energy harvester system with a frequency-widened-bandwidth mechanism introduced by mechanical stoppers. *Smart Materials and Structures*, *21*(3), 035005.
45. Liu, H., Tay, C. J., Quan, C., Kobayashi, T., & Lee, C. (2011). Piezoelectric MEMS energy harvester for low-frequency vibrations with wideband operation range and steadily increased output power. *Journal of Microelectromechanical Systems*, *20*(5), 1131–1142.
46. Ma, W., Zhu, R., Rufer, L., Zohar, Y., & Wong, M. (2007). An integrated floating-electrode electric microgenerator. *Journal of Microelectromechanical Systems*, *16*(1), 29–37.
47. Ahmad, M. R., Khir, M. H., & Dennis, J. O. (2013). Design and modeling of the trapezoidal electrodes array for electrets energy harvester. In *SPIE Defense, Security, and Sensing* (p. 87280Z).
48. Suzuki, Y., Edamoto, M., Kasagi, N., Kashiwagi, K., Morizawa, Y., Yokoyama, T., et al. (2008). Micro electret energy harvesting device with analogue impedance conversion circuit. *PowerMEMS*, *2008*, 7–10.
49. Basset, P., Galayko, D., Paracha, A. M., Marty, F., Dudka, A., & Bourouina, T. (2009). A batch-fabricated and electret-free silicon electrostatic vibration energy harvester. *Journal of Micromechanics and Microengineering*, *19*(11), 115025.
50. Suzuki, Y. (2011). Recent progress in MEMS electret generator for energy harvesting. *IEEE Transactions on Electrical and Electronic Engineering*, *6*(2), 101–111.
51. Takamatsu, T. (1991). Life time of thermal electrets of carnauba wax, esters, fatty acids and alcohols. In *7th International Symposium on Electrets*, (pp. 106–110).
52. Minami, T., Utsubo, T., Yamatani, T., Miyata, T., & Ohbayashi, Y. (2003). SiO2 electret thin films prepared by various deposition methods. *Thin Solid Films*, *426*(1–2), 47–52.
53. Sakane, Y., Suzuki, Y., & Kasagi, N. (2008). The development of a high-performance perfluorinated polymer electret and its application to micro power generation. *Journal of Micromechanics and Microengineering*, *18*(10), 104011.
54. Nimo, A., Mescheder, U., Müller, B., & Elkeir, A. S. A. (2011). 3D capacitive vibrational micro harvester using isotropic charging of electrets deposited on vertical sidewalls. In *SPIE Microtechnologies*. International Society for Optics and Photonics (p. 80661Q).
55. Mescheder, U., Müller, B., Baborie, S., & Urbanovic, P. (2009). Properties of SiO2 electret films charged by ion implantation for MEMS-based energy harvesting systems. *Journal of Micromechanics and Microengineering*, *19*(9), 094003.

56. Mescheder, U., Nimo, A., Müller, B., & Elkeir, A. (2012). Micro harvester using isotropic charging of electrets deposited on vertical sidewalls for conversion of 3D vibrational energy. *Microsystem Technologies*, *18*(7–8), 931–943.
57. Westby, E. R., & Halvorsen, E. (2012). Design and modelling of a patterned-electret-based energy harvester for tire pressure monitoring systems. *IEEE/ASME Transactions on Mechatronics*, *17*(5), 995–1005.
58. Yamashita, K., Honzumi, M., Hagiwara, K., Iguchi, Y., & Suzuki, Y. (2010) Vibration-driven MEMS energy harvester with vertical electrets. In textitProceedings of PowerMEMS, (pp. 165–168).
59. Wang, F., & Hansen, O. (2014). Electrostatic energy harvesting device with out-of-the-plane gap closing scheme. *Sensors and Actuators A: Physical*.
60. Liu, H., How Koh, K., & Lee, C. (2014). Ultra-wide frequency broadening mechanism for micro-scale electromagnetic energy harvester. *Applied Physics Letters*, *104*(5).
61. Marin, A., Bressers, S., & Priya, S. (2011). Multiple cell configuration electromagnetic vibration energy harvester. *Journal of Physics D: Applied Physics*, *44*(29), 295501.
62. Cook-Chennault, K. A., Thambi, N., & Sastry, A. M. (2008). Powering MEMS portable devices –a review of non-regenerative and regenerative power supply systems with special emphasis on piezoelectric energy harvesting systems. *Smart Materials and Structures*, *17*(4), 043001.
63. Morimoto, K., Kanno, I., Wasa, K., & Kotera, H. (2010). High-efficiency piezoelectric energy harvesters of c-axis-oriented epitaxial PZT films transferred onto stainless steel cantilevers. *Sensors and Actuators A: Physical*, *163*(1), 428–432.
64. Beeby, S. P., Tudor, M., & White, N. (2006). Energy harvesting vibration sources for microsystems applications. *Measurement Science and Technology*, *17*(12), R175–R195.
65. Kim, S. G., Priya, S., & Kanno, I. (2012). Piezoelectric MEMS for energy harvesting. *MRS Bulletin*, *37*, 1039–1050.
66. Jeon, Y. B., Sood, R., Jeong, J. H., & Kim, S. G. (2005). MEMS power generator with transverse mode thin film PZT. *Sensors and Actuators A: Physical*, *122*(1), 16–22.
67. Curie, J., & Curie, P. (1880). Développement, par pression, de l'électricité polaire dans les cristaux hémiédres à faces inclinées. *Comptes Rendus des Séances de l'Academie des Sciencies*, *91*, 294–295.
68. Curie, J., Curie, J. P., & Curie, P. (1880). Sur l'électricité polaire dans les cristaux hémiedres à faces inclinées. *Comptes Rendus des Séances de l'Academie des Sciencies*, *91*, 383–386.
69. Lippmann, G. (1881). Principe de la conservation de l'électricité, ou second principe de la théorie des phénoménes électriques. *Journal of Theoretical and Applied Physics*, *10*(1), 381–394.
70. IEEE standard on piezoelectricity. (1988). *ANSI/IEEE Std 176-1987*, pp. 1–66.
71. Ensminger, D., & Stulen, F. B. (2008). *Ultrasonics: data, equations and their practical uses*. Taylor & Francis.
72. Bowen, C. R., Kim, H. A., Weaver, P. M., & Dunn, S. (2014). Piezoelectric and ferroelectric materials and structures for energy harvesting applications. *Energy & Environmental Science*, *7*, 25–44.
73. Haertling, G. H. (1999). Ferroelectric ceramics: History and technology. *Journal of the American Ceramic Society*, *82*, 797–818.
74. Beeby, S., & White, N. (2010). *Energy Harvesting for Autonomous Systems*, Artech House series smart materials, structures, and systems. Artech House, Incorporated.
75. Jaffe, B., Roth, R. S., & Marzullo, S. (1954). Piezoelectric properties of lead zirconate-lead titanate solid-solution ceramics. *Journal of Applied Physics*, *25*(6).
76. Trolier-McKinstry, S., & Muralt, P. (2004). Thin film piezoelectrics for MEMS. *Journal of Electroceramics*, *12*(1–2), 7–17.
77. Lovinger, A. J. (1983). Ferroelectric polymers. *Science*, *220*(4602), 1115–1121.
78. Baudry, H. (2013). Screen printing piezoelectric devices. *Microelectronics International*, *4*(3), 71–74.
79. Dietze, M., & Es-Souni, M. (2008). Structural and functional properties of screen-printed PZTPVDF-TrFE composites. *Sensors and Actuators A: Physical*, *143*(2), 329–334.

80. Fu, D. W., Zhang, W., Cai, H., Ge, J., Zhang, Y., & Xiong, R. (2011). Diisopropylammonium chloride: A ferroelectric organic salt with a high phase transition temperature and practical utilization level of spontaneous polarization. *Advanced Materials, 23*(47), 5658–5662.
81. Fu, D. W., Cai, H. L., Liu, Y., Ye, Q., Zhang, W., Zhang, Y., et al. (2013). Diisopropylammonium bromide is a high-temperature molecular ferroelectric crystal. *Science, 339*(6118), 425–428.
82. Maas, R., Koch, M., Harris, N. R., White, N. M., & Evans, A. G. R. (1997). Thick-film printing of PZT onto silicon. *Materials Letters, 31*(1–2), 109–112.
83. Van Schaijk, R., Elfrink, R., Kamel, T. M., & Goedbloed, M. (2008). Piezoelectric aln energy harvesters for wireless autonomous transducer solutions. In *Sensors, 2008 IEEE* (pp. 45–48).
84. Guy, I. L., Muensit, S., & Goldys, E. M. (1999). Extensional piezoelectric coefficients of gallium nitride and aluminium nitride. *Applied Physics Letters, 75*(26), 4133–4135.
85. Tadigadapa, S., & Mateti, K. (2009). Piezoelectric MEMS sensors: state-of-the-art and perspectives. *Measurement Science and Technology, 20*(9), 092001.
86. Crisler, D. F., Cupal, J. J., & Moore, A. R. (1968). Dielectric, piezoelectric, and electromechanical coupling constants of zinc oxide crystals. *Proceedings of the IEEE, 56*(2), 225–226.
87. Park, J. C., Park, J. Y., & Lee, Y. (2010). Modeling and characterization of piezoelectric d33-mode MEMS energy harvester. *Journal of Microelectromechanical Systems, 19*(5), 1215–1222.
88. Cho, J., Anderson, M., Richards, R., Bahr, D., & Richards, C. (2005). Optimization of electromechanical coupling for a thin-film PZT membrane: I. modeling. *Journal of Micromechanics and Microengineering, 15*(10), 1797.
89. Kim, S. B., Park, H., Kim, S. H., Wikle, H. C., Park, J. H., & Kim, D. J. (2013). Comparison of MEMS PZT cantilevers based on d31 and d33 modes for vibration energy harvesting. *Journal of Microelectromechanical Systems, 22*(1), 26–33.
90. Erturk, A., & Inman, D. J. (2011). *Piezoelectric energy harvesting*. John Wiley & Sons.
91. Xu, R., Lei, A., Dahl-Petersen, C., Hansen, K., Guizzetti, M., Birkelund, M., Thomsen, E., et al. (2012). Fabrication and characterization of MEMS-based PZT/PZT bimorph thick film vibration energy harvesters. *Journal of Micromechanics and Microengineering, 22*(9).

Chapter 3
Oscillators for Energy Harvesting

E. Blokhina and D. Galayko

3.1 Linear Oscillator

3.1.1 Free Linear Oscillator

The linear oscillator is a fundamental model in physics. It is broadly used to model various systems of mechanical, electronic, biological and chemical nature [3, 19–21, 28, 35]. The linear oscillator is the most common model that describes deflections and oscillations of micro-mechanical structures including MEMS oscillators [14, 31] and therefore it has a direct relation to energy harvesting applications. Although in practice, continuous micro-structures such as cantilevers or beams are described using partial differential equations [27], it is possible to reduce them to the model of a linear or nonlinear oscillators [22]. In this way, an oscillator is the most common model of mechanical microsystems.

The simplest example of a linear oscillator is shown in Fig. 3.1. The mechanical oscillator[1] (Fig. 3.1a) is composed of a mass m attached to a spring k, one end of which is fixed, and placed in a viscous medium with the friction (also called damping

[1] In engineering sciences, the systems from Fig. 3.1 are called resonators. In a certain sense, the term resonator is more accurate since it highlights that these systems are passive. Indeed, as we shall see later, sustained oscillations can exist in them only if an external force/driving is applied. In physics and mathematics, such systems and equations describing them are generally called oscillators. To specify, one may say that Fig. 3.1 shows a passive oscillator in contrast to a self-oscillator such as the Van der Pol oscillator [3]. The latter is a system with negative nonlinear damping that can display an oscillatory process without any external forcing. However, this is only a matter of naming, and in this book we will often refer to all oscillator-like systems as simply `oscillators`.

E. Blokhina (✉)
University College Dublin, Dublin, Ireland
e-mail: elena.blokhina@ucd.ie

D. Galayko
UPMC—Sorbonne Universities, Paris, France
e-mail: dimitri.galayko@lip6.fr

E. Blokhina et al. (eds.), *Nonlinearity in Energy Harvesting Systems*,
DOI 10.1007/978-3-319-20355-3_3

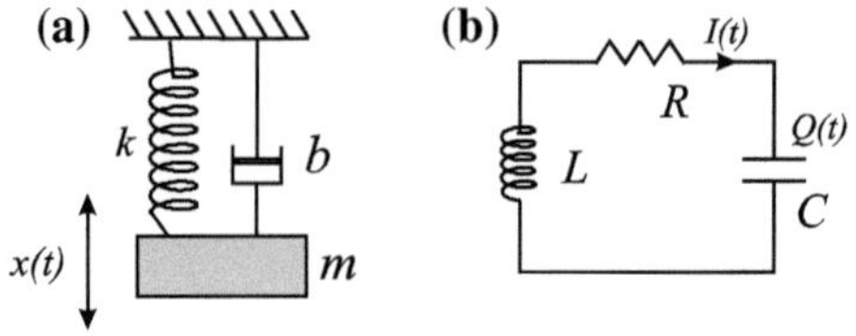

Fig. 3.1 Model of a linear oscillator. **a** Mechanical system composed of a body of mass m suspended on a spring with the coefficient k. The system is subjected to viscous air damping described by the coefficient b. **b** Oscillating circuit composed of a capacitor C, inductor L and resistor R

or dissipation) coefficient b. The electric oscillating circuit (Fig. 3.1b) consists of a capacitor C, inductor L and resistor R.

The displacement $x(t)$ of the mechanical oscillator from Fig. 3.1a, in the absence of external forces, can be obtained from Newton's second law $m\ddot{x} = \sum F_i$ by taking into account all forces acting on the mass m: the spring force $F_{spring} = -kx$ and the viscous friction (damping) force $F_{damping} = -bv = -b\dot{x}$:

$$m\ddot{x}(t) = -kx(t) - b\dot{x}(t) \tag{3.1}$$

By rearranging the above equation, one can obtain the equation describing the displacement $x(t)$ of the mass m

$$\ddot{x}(t) + 2\gamma\dot{x}(t) + \omega_0^2 x(t) = 0 \tag{3.2}$$

where $\gamma = b/(2m)$ is the dissipation (damping, friction) parameter and $\omega_0 = \sqrt{k/m}$ is the natural frequency of the oscillator. The overdot denotes the derivative with respect to time. For instance, the velocity of the mass is the derivative of its displacement $v(t) = \dot{x}(t) = dx/dt$.

A similar equation can be obtained for the oscillating circuit in Fig. 3.1b, from Kirchhoff's voltage law for the charge on the capacitor $Q(t)$:

$$\ddot{Q}(t) + (R/L)\dot{Q}(t) + (1/LC)Q(t) = 0 \tag{3.3}$$

and in this case we may denote $\gamma = R/(2L)$ and $\omega_0 = \sqrt{1/(LC)}$.

Ordinary differential equation (3.2) is known as a linear mass–spring–damper system or a linear oscillator or sometimes also a linear resonator (see the footnote). In particular, in the case $b > 0$ we say that it is a damped linear oscillator or the linear oscillator with dissipation, while in the case $b = 0$ we say it is a conservative or harmonic oscillator.

We will start with the case of a harmonic oscillator when friction is absent (no damper and no resistor in Fig. 3.1, i.e. $\gamma = 0$). Equation (3.2) takes the following form

$$\ddot{x} + \omega_0^2 x = 0 \tag{3.4}$$

The general solution of differential equation (3.4) with the initial conditions $x(0) = x_0$ and $\dot{x}(0) = v_0$ is written in terms of harmonic functions of sine and cosine

$$x(t) = x_0 \cos \omega_0 t + \frac{v_0}{\omega_0} \sin \omega_0 t = A_0 \cos(\omega_0 t + \varphi_0)$$
$$v(t) = v_0 \cos \omega_0 t - x_0 \omega_0 \sin \omega_0 t = -A_0 \omega_0 \sin(\omega_0 t + \varphi_0) \tag{3.5}$$

The quantity $A_0 = \sqrt{x_0^2 + \frac{v_0^2}{\omega_0^2}}$ defines the amplitude of oscillations, while the quantity $\varphi_0 = \arctan(-v_0/\omega_0 x_0)$ defines their phase. This solution is shown in Fig. 3.2a, and it corresponds to the steady-state periodic oscillations of the displacement $x(t)$ and the velocity $v(t)$.

Note that the harmonic oscillator is a conservative system whose total energy remains constant. In order to prove this, let us multiply both sides of Eq. (3.1), recalling that $\gamma = 0$ for the conservative case, by $\dot{x}$. Noting that $\dot{x}\ddot{x} = d(\dot{x}^2)/dt$ and $\dot{x}x = d(x^2)/dt$, we obtain

$$\frac{d}{dt}\left[\frac{m\dot{x}^2}{2} + \frac{kx^2}{2}\right] = 0 \tag{3.6}$$

The expression in the brackets is the total energy W of the oscillator that is the superposition of the kinetic energy $W_k = m\dot{x}^2/2$ and the potential energy $W_p = kx^2/2$. It is clear that the energy of a harmonic oscillator does not change

$$\frac{dW}{dt} = 0 \implies W = \frac{m\dot{x}^2}{2} + \frac{kx^2}{2} = \text{const} \tag{3.7}$$

Now we consider a damped oscillator, $\gamma > 0$. There are two particular cases: 'small' dissipation $\gamma < \omega_0$ and 'large' dissipation $\gamma > \omega_0$. In both cases, one can solve the equation of the linear damped oscillator (3.2) using the standard method, i.e. by substituting $x = \exp(\lambda t)$. Here we give the final result yielded by this method.

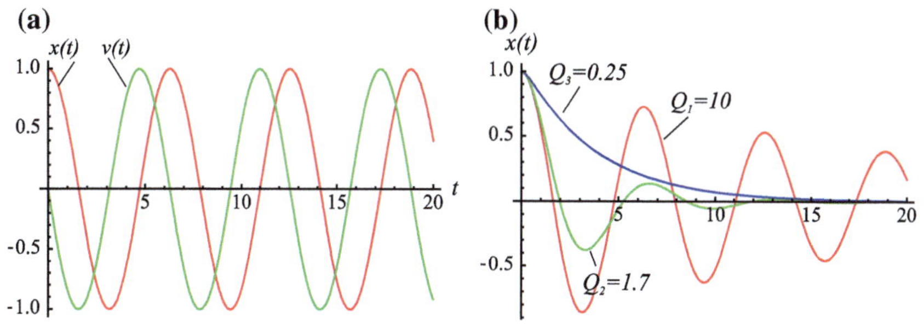

Fig. 3.2 **a** Displacement $x(t)$ and the velocity $v(t)$ of a linear oscillator without damping given by Eq. (3.4) as functions of time, $\omega_0 = 1$. **b** Displacement $x(t)$ of a linear damped oscillator given by Eq. (3.2) for three different dissipation parameters (or three different Q-factors), $\omega_0 = 1$

For sufficiently small dissipation $\gamma < \omega_0$ the solution is

$$
\begin{aligned}
x(t) &= e^{-\gamma t}\left[x_0 \cos \omega t + \frac{v_0 + \gamma x_0}{\omega} \sin \omega t\right] \\
v(t) &= e^{-\gamma t}\left[v_0 \cos \omega t - \frac{\gamma v_0 + \omega_0^2 x_0}{\omega} \sin \omega t\right]
\end{aligned}
\tag{3.8}
$$

where we have introduced a new quantity $\omega = \sqrt{\omega_0^2 - \gamma^2}$ (note that $\omega_0^2 - \gamma^2 > 0$, and therefore ω is a real value). It is sometimes referred to as the frequency of the damped oscillatory process.

For sufficiently large dissipation $\gamma > \omega_0$ on the other hand, the solution is

$$
\begin{aligned}
x(t) &= e^{-\gamma t}\left[x_0 \cosh \varpi t + \frac{v_0 + \gamma x_0}{\varpi} \sinh \varpi t\right] \\
v(t) &= e^{-\gamma t}\left[v_0 \cosh \varpi t - \frac{\gamma v_0 + \omega_0^2 x_0}{\varpi} \sinh \varpi t\right]
\end{aligned}
\tag{3.9}
$$

where we have introduced another quantity $\varpi = \sqrt{\gamma^2 - \omega_0^2}$ (in this case, $\gamma^2 - \omega_0^2 > 0$ and therefore ϖ is a real value). In contrast to Eq. (3.8) that describes decaying oscillating process, Eq. (3.9) describes a straightforward non-oscillatory decay of the displacement and velocity with time since these equations contain hyperbolic sine and cosine functions ($\sinh(\varpi t)$ and $\cosh(\varpi t)$). The displacement $x(t)$ of a damped linear oscillator is shown in Fig. 3.2b, where the red and green lines correspond to a linear oscillator with small dissipation and the blue line corresponds to a linear oscillator with large dissipation. Oscillators with $\omega_0 > \gamma$ are called *underdampled* while oscillators with $\omega_0 < \gamma$ are called *overdampled.*

It should be mentioned that there are several examples of oscillators with negative damping, i.e. $\gamma < 0$. In this case, solutions (3.8) and (3.9) are still valid. According to them, the displacement and velocity of the system will increase infinitely due to the coefficient $\exp(|\gamma|t)$ in front of the brackets. In realistic systems however, the infinite increase of displacement is not possible and it will be later limited by other factors, such as nonlinearities. Oscillators with negative (linear or nonlinear) dissipation belong to the class of self-oscillators [3]. We do not consider self-oscillating systems since they are beyond the scope of this book.

In Fig. 3.2b, we use the quantity Q to distinguish between 'small' and 'large' dissipation. This quantity is a characteristic of the system introduced in the theory of vibrations. It is called the *quality factor* or the Q-factor. Fundamentally, it is given by the relation

$$
Q = 2\pi \frac{\text{Maximal energy stored in the system over a cycle}}{\text{Energy dissipated per one cycle}} = \omega \frac{\text{Energy stored}}{\text{Power loss}}
\tag{3.10}
$$

In general, the quality factor is a frequency-dependent parameter. In the case of a second-order resonator, it is given by:

$$Q = \frac{\omega}{2\gamma} \tag{3.11}$$

Another very common definition of the quality factor that can be found in the literature is

$$\tilde{Q} = \frac{\omega_0}{2\gamma} \tag{3.12}$$

In the context of formula (3.12), it is often said that the case $\tilde{Q} > 1/2$ corresponds to an underdamped oscillator (described by expressions (3.8)) while the case $\tilde{Q} < 1/2$ corresponds to an overdamped oscillator (described by expressions (3.9)). We used this notation in Fig. 3.2b. The solution given by (3.8) is shown in Fig. 3.2 for three different $\tilde{Q}$-factors. It is seen that the higher $\tilde{Q}$-factor is, the longer time is required for oscillations to decay. Eventually, all oscillations of a free linear damped oscillator (3.2) will decay.

In the case of a mechanical resonator, the Q-factor characterises the losses in the resonator, which may be unwanted (for instance, due to friction) or intentionally introduced (for instance, in the case of the energy conversion systems as we will show later in Sect. 3.3). An oscillator with a higher Q-factor will loose less energy during one period of oscillations $T = 2\pi/\omega$ than another oscillator with a lower Q-factor during the same time T.

A linear damped oscillator loses its energy due to friction that is present in the equation in the form of the term $b\dot{x}$ (or $\gamma\dot{x}$, recall the expression linking these two). Let us again multiply Eq. (3.1) by $\dot{x}$. We obtain

$$\frac{d}{dt}\left[\frac{m\dot{x}^2}{2} + \frac{kx^2}{2}\right] = -b\dot{x}^2 \tag{3.13}$$

It is obvious from this equation that the energy does not remain constant and decreases due to the action of dissipation. We can rewrite the above expression in another form. By definition, the power is $P = dW/dt = vF$, where v is the velocity of the system and F is some force. We can write that

$$\frac{dW}{dt} = \frac{d}{dt}\left[W_k + W_p\right] = P_{damping} \tag{3.14}$$

where $P_{damping} = -b\dot{x}^2$ is the power of the friction force.

Note that the same discussion as presented above can be carried out in terms of the oscillating circuit from Fig. 3.1b. In this case we obtain

$$\frac{dW}{dt} = \frac{d}{dt}\left[W_m + W_e\right] = P_{res} \tag{3.15}$$

Table 3.1 The relationship between the variables and parameters of mechanical and electrical oscillators

Mechanical domain	Electrical domain
Displacement x	Charge Q
Velocity v	Current I
Mass m	Inductance L
Spring coefficient k	Inverse capacitance $1/C$
Damping coefficient b	Resistance R
Force $F(t)$	Voltage $V(t)$ (as shown in Fig. 3.3)
Kinetic energy of mass $W_k = \frac{mv^2}{2}$	Energy of the magnetic field in the inductor $W_m = \frac{LI^2}{2}$
Potential energy of spring $W_p = \frac{kx^2}{2}$	Energy of the electric field in the capacitor $W_c = \frac{Q^2}{2C}$
Energy relation $\frac{d(W_k+W_p)}{dt} = -bv^2$	Energy relation $\frac{d(W_m+W_c)}{dt} = -RI^2$

where $W_m = LI^2/2$, $W_e = Q^2/(2C)$ and $P_{res} = -RI^2$. The relationships between the variables and parameters of the mechanical and electrical oscillators shown in Fig. 3.1 are presented in Table 3.1.

3.1.2 Forced Oscillator and Linear Resonance

The case that is of particular interest for energy harvesting is a driven linear oscillator, or an oscillator actuated by an external force. Figure 3.3a shows a mechanical oscillator. We will consider the fundamental case of a harmonic external force

$$\ddot{x} + 2\gamma\dot{x} + \omega_0^2 x = F_0 \cos(\omega_{\text{ext}} t + \phi_0) \quad (3.16)$$

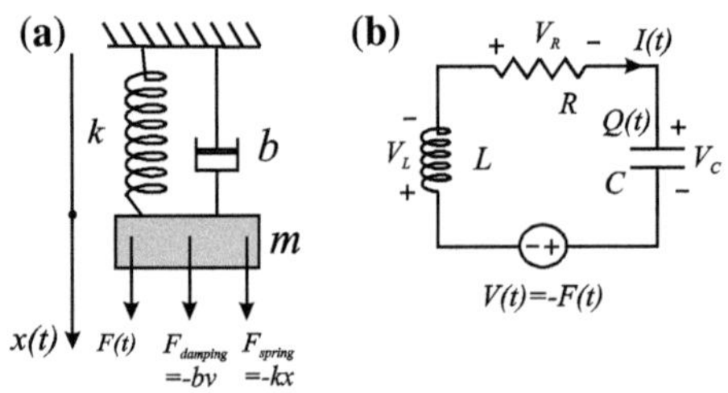

Fig. 3.3 Model of a forced linear oscillator. **a** Mechanical system and **b** electric circuit. The notation for the direction of forces and the direction of voltages across the elements are chosen to keep direct equivalence between the two systems

where $F_0 = \tilde{F}/m$ is the amplitude of the external force $F(t) = \tilde{F}\cos(\omega_{\text{ext}}t + \phi_0)$ normalised by the mass of the oscillator m. The solution of this differential equation is a superposition of a solution corresponding to the homogeneous equation (3.2) and of a particular solution of (3.16) [12, 32]:

$$x(t) = x_h(t) + x_{nh}(t) \tag{3.17}$$

where $x_h(t)$ is known from (3.8) or from (3.9).

The force on the right-hand side of Eq. (3.16) can be presented in the form $F(t) = \text{Re}[F_0 \exp(i\phi_0)\exp(i\omega_{\text{ext}}t)]$ and the quantity $\mathbb{F} = F_0 \exp(i\phi_0)$ is called the *complex amplitude* of the force or the *force phasor*. Because of the specific form this force, we will look for a particular solution of the non-homogeneous equation (3.16) in the form $x(t) = \text{Re}[A\exp(i\phi)\exp(i\omega_{\text{ext}}t)]$ where A and ϕ are the resulting amplitude and phase of forced oscillations. Correspondingly, the quantity $\mathbb{X} = A\exp(i\phi)$ is called the complex amplitude or the phasor of the displacement (or more generally, of a generic signal).

After substituting x and F into the original equation, we will obtain the equation linking two complex amplitudes:

$$(-\omega_{\text{ext}}^2 + 2i\gamma\omega_{\text{ext}} + \omega_0^2)\mathbb{X} = \mathbb{F} \tag{3.18}$$

We will skip the further steps of the derivation as they are straightforward. The resulting solution is

$$x_{nh}(t) = A\cos(\omega_{\text{ext}}t + \phi + \phi_0) \tag{3.19}$$

In this expression, the amplitude A of the forced oscillation is defined as

$$A = \frac{F_0}{\sqrt{(\omega_0^2 - \omega_{\text{ext}}^2)^2 + 4\gamma^2\omega_{\text{ext}}^2}} \tag{3.20}$$

while the phase ϕ of the forced oscillation is found from the following expressions:

$$\cos\phi = \frac{\omega_0^2 - \omega_{\text{ext}}^2}{\sqrt{(\omega_0^2 - \omega_{\text{ext}}^2)^2 + 4\gamma^2\omega_{\text{ext}}^2}}, \quad \sin\phi = \frac{-2\gamma\omega_{\text{ext}}}{\sqrt{(\omega_0^2 - \omega_{\text{ext}}^2)^2 + 4\gamma^2\omega_{\text{ext}}^2}} \tag{3.21}$$

Now we write the full solution of this system that represents the superposition of oscillations (3.8) and forced oscillations (3.19):

$$\begin{aligned} x(t) = x_h(t) + x_{nh}(t) = e^{-\gamma t}\left[C_1\cos\omega t + C_2\sin\omega t\right] + \\ + \frac{F_0}{\sqrt{(\omega_0^2 - \omega_{\text{ext}}^2)^2 + 4\gamma^2\omega_{\text{ext}}^2}}\cos(\omega_{\text{ext}}t + \phi + \phi_0) \end{aligned} \tag{3.22}$$

where the coefficients C_1 and C_2 are defined from the initial conditions

$$C_1 = x_0 - x_{nh}(0), \quad C_2 = \frac{v_0 - \dot{x}_{nh}(0) + \gamma(x_0 - x_{nh}(0))}{\omega} \tag{3.23}$$

and x_0 and v_0 denotes, as usual, the initial state of the oscillator at $t = 0$.

It is important to mention that after a sufficiently long time, the oscillator will display only the forced oscillations. Following (3.22), the eigen oscillation term is multiplied by $\exp(-\gamma t)$, and if time t is large, this term tends to zero. Without loss of generality, we may assume that (3.19) describes the steady-state regime of a forced linear oscillator.

Let us investigate the amplitude of forced oscillations A given by (3.20). It depends on the parameters of the external force (the force amplitude F_0 and the frequency ω_{ext}) and on the parameters of the oscillator itself (its natural frequency ω_0 and the dissipation factor γ).

It is obvious that for a particular oscillator with fixed ω_0 and γ, the dependence on the force amplitude F_0 is linear: as F_0 increases, the response of the oscillator becomes larger. However the dependence of A on the external frequency ω_{ext} is somewhat non-trivial (see Fig. 3.4). The amplitude A as a function of ω_{ext} displays an increase in the oscillator response to the external force in the case when the frequency ω_{ext} of the external force approached the natural frequency of the oscillator, i.e. $\omega_{\text{ext}} \approx \omega_0$. This increase is more visible for oscillators with higher Q-factors and less visible for oscillators with lower Q-factor.

This phenomenon is called *resonance*. On a qualitative level, it can be defined as an increase in the amplitude of forced oscillations when the frequency of the external force ω_{ext} is close to the natural frequency of the oscillator ω_0. For a conservative oscillator with $\gamma = 0$, the amplitude of forced oscillations will increase infinitely, as shown in Fig. 3.4 (see the line marked as $\gamma = 0$).

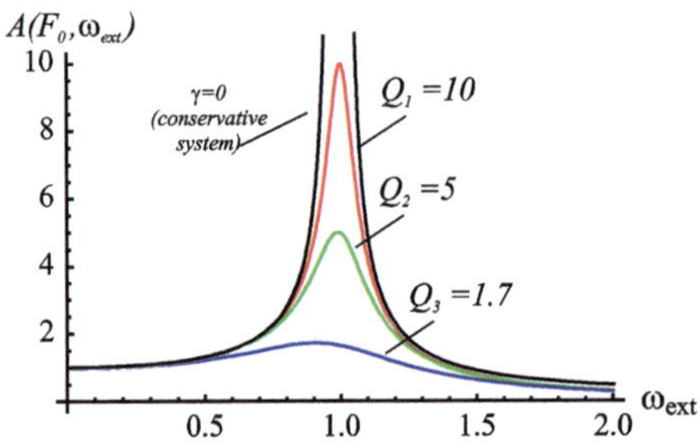

Fig. 3.4 The amplitude of the forced oscillations of a linear damped oscillator $A(F_0, \omega_{\text{ext}})$ as a function of the external frequency ω_{ext} for three different Q-factors. In the case of a conservative linear oscillator with $\gamma = 0$, the amplitude of forced oscillations increases infinitely. $\omega_0 = 1, F_0 = 1$

It is interesting to note that at $\omega_{\text{ext}} = \omega_0$ the phase shift of forced oscillation found from (3.21) is exactly $-\pi/2$:

$$\cos\phi = 0, \quad \sin\phi = -1 \quad \Longrightarrow \quad \phi = -\pi/2 \tag{3.24}$$

Sometimes, the case $\omega_{\text{ext}} = \omega_0$ is called *exact resonance* in the literature. On the other hand, as one can see from (3.20), the maximum oscillator displacement (i.e. the maximum of A) is reached when $\omega_{\text{ext}} = \sqrt{\omega_0^2 - 2\gamma^2}$.

The amplitude $A(F_0, \omega_{\text{ext}})$ is directly related to the transfer function of an oscillator. Such an approach, utilising transfer functions, may be more familiar to readers with a background in electric and electronic engineering. In order to obtain the transfer function of an oscillator, we use the Laplace method and substitute $x = X_0 \exp(st)$ into Eq. (3.16) (we use that $F = F_0 \exp(st)$), as a result we obtain

$$X_0 = H(s)F_0 = \frac{1}{s^2 + 2\gamma s + \omega_0^2} F_0 \tag{3.25}$$

Let us write the energy relation for a forced harmonic oscillator. Again, we will take Eq. (3.16) and multiply it by $v = \dot{x}$:

$$\frac{d}{dt}\left[\frac{mv^2}{2} + \frac{kx^2}{2}\right] = vF(t) - bv^2 \tag{3.26}$$

where $vF(t)$ expresses the instantaneous power of the external force $F(t) = mF_0 \cos(\omega_{\text{ext}} t + \phi_0)$. The above equation describes the energy conservation law in this system: the power of the external force is dissipated by the change of the energy of the oscillator and losses due to dissipation. If we find the average of (3.26) over the period $T = 2\pi/\omega_{\text{ext}}$ applying the following rule

$$\overline{F(t)} = \frac{1}{T} \int_t^{t+T} F(\tau) d\tau \tag{3.27}$$

we obtain that

$$\overline{bv^2(t)} = \overline{v(t)F(t)} \tag{3.28}$$

This law expresses the fact that the average power of the external force over the period of oscillations is equal to the power dissipated by the friction force. We can write the same expressions for the electric circuit. In this case, the role of the external force is played by an additional source of voltage $V(t)$

$$\frac{d}{dt}\left[\frac{LI^2}{2} + \frac{Q^2}{2C}\right] = IV(t) - RI^2 \tag{3.29}$$

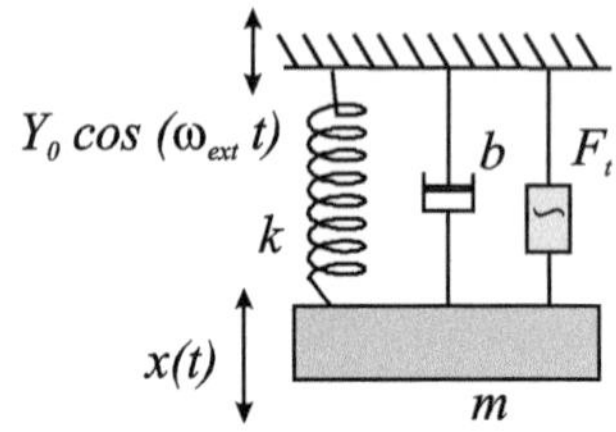

Fig. 3.5 A schematic view of the simplest vibration energy harvester (VEH), also called a kinetic energy harvester (KEH)

and

$$\overline{RI^2(t)} = \overline{I(t)V(t)} \tag{3.30}$$

The phenomenon of forced oscillations and resonance is widely used for macro- and micro-scale vibration energy harvesting [17, 18, 37]. A detailed discussion on the application of linear and nonlinear resonators for energy harvesting will be provided in Sect. 3.3. A schematic view of a vibration energy harvester (VEH), also called a kinetic energy harvester (KEH), is shown in Fig. 3.5. It consists of a mass m attached to a movable frame through the spring k. There is also a force F_t that acts on the oscillator as a dissipating force (it takes the energy of the system and transfers it from the mechanical domain to the electrical domain). The implementation of this force is discussed in Sect. 3.3.

In order to effectively transfer energy from the mechanical to the electrical domain, the resonator must display large amplitude vibrations, and in this way the kinetic energy of the resonator will be large. For this reason, resonance can be very useful for energy harvesting: by designing a resonator that operates near the resonance of external vibrations $\omega_0 \approx \omega_{\text{ext}}$, one ensures its large amplitude oscillations. There are many energy harvesting systems that employ linear resonators, and all of them operate in resonance.

3.1.3 Equilibrium Points and Stability. An Oscillator as a Dynamical System

In the previous section, we discussed free and forced oscillations of a conservative and damped ($\gamma > 0$) linear oscillator. However, we have not yet discussed other cases. What if damping is negative? Will the behaviour of such a system be different from the studied case? In order to obtain a more general view of the behaviour of the linear oscillator, let us present it in the form of a *dynamical system*:

$$\dot{\mathbf{x}} = \mathbf{F}(\mathbf{x}) \tag{3.31}$$

In this form, $\mathbf{x}(t)$ is the *state vector* containing the state variables, and the vector function $\mathbf{F}(\mathbf{x})$ defines the evolution of the system and also called the evolution operator.

Let us present Eq. (3.2) as a set of ordinary differential equations of the first order:

$$\begin{aligned} \dot{x} &= v \\ \dot{v} &= -2\gamma v - \omega_0^2 x \end{aligned} \tag{3.32}$$

From this form, we can introduce the state vector as $\mathbf{x} = (x, v)^T$ and the vector function $\mathbf{F} = (v, -2\gamma v - \omega_0 x)^T$

$$\begin{pmatrix} \dot{x} \\ \dot{v} \end{pmatrix} = \begin{pmatrix} v \\ -2\gamma v - \omega_0 x \end{pmatrix} \Leftrightarrow \begin{pmatrix} \dot{x} \\ \dot{v} \end{pmatrix} = \begin{pmatrix} 0 & 1 \\ -\omega_0^2 & -2\gamma \end{pmatrix} \begin{pmatrix} x \\ v \end{pmatrix} \tag{3.33}$$

From the transformation that has been carried out in the above equation, we can see that the linear oscillator belongs to the class of linear dynamical systems. Indeed, the vector function $\mathbf{F}$ can be presented in the form $\mathbf{F} = \mathbf{M}\,\mathbf{x}$ where the matrix $\mathbf{M} = \left(\begin{smallmatrix} 0 & 1 \\ -\omega_0^2 & -2\gamma \end{smallmatrix}\right)$ is constant matrix. It depends only on the parameters γ and ω_0 and does not depend on the variables x and v. Therefore, we can present the system in the form

$$\dot{\mathbf{x}} = \mathbf{M} \cdot \mathbf{x} \tag{3.34}$$

and hence it is a linear system.

The key property of a dynamical system is that knowing its evolution operator, the state of the system at any time $t > t_0$ can be defined from the initial conditions $x_0 = x(t_0)$ and $v_0 = v(t_0)$.

From Eq. (3.32) it is obvious that $x_e = 0$ and $v_e = 0$ is an *equilibrium* point: the solution starting from this point will remain zero all time. It is important to understand the stability type of equilibrium points because it defines the behaviour of the system around these points. Indeed, what happens in the case if the initial conditions are very close to zero $x_0 \approx 0$ and $v_0 \approx 0$, but not exactly zero? Will the solution decay to the equilibrium $(0, 0)$ or not?

There are a number of definitions of stability, but the most commonly used ones are Lyapunov stability and asymptotic stability. The equilibrium of a dynamical system $\mathbf{x}_e(t)$ is Lyapunov stable if every solution starting from a δ-neighbourhood of the equilibrium $\|\mathbf{x}(t_0) - \mathbf{x}_e\| < \delta$ remains in a small neighbourhood of this equilibrium $\|\mathbf{x}(t) - \mathbf{x}_e\| < \varepsilon$ for all $t > t_0$ $(\delta, \varepsilon > 0)$. Note that the solution $\mathbf{x}(t)$ need not approach the equilibrium $\mathbf{x}_e$ in order for $\mathbf{x}_e$ to be considered stable in the context of Lyapunov stability. Asymptotic stability is a stronger definition. It requires that the solution must eventually approach the equilibrium: $\|\mathbf{x}(t) - \mathbf{x}_e\| \to 0$ if $t \to \infty$.

To investigate the behaviour of a dynamical system (3.31) around an equilibrium $\mathbf{x}_e$, the state variables are presented in the form $\mathbf{x} = \mathbf{x}_e + \tilde{\mathbf{x}}$ and substituted in the governing equation (3.31). The right-hand side of this equations is linearised $\mathbf{F}(\mathbf{x}_e + \tilde{\mathbf{x}}) \approx \mathbf{F}(\mathbf{x}_e) + \mathbf{J}\tilde{\mathbf{x}}$ where $\mathbf{J} = \partial\mathbf{F}/\partial\mathbf{x}$ is the Jacobian matrix. Since $\dot{\mathbf{x}}_e = \mathbf{F}(\mathbf{x}_e)$, one obtains a linear equation for the perturbed $\tilde{\mathbf{x}}$ in the form $\dot{\tilde{\mathbf{x}}} = \mathbf{J}\tilde{\mathbf{x}}$. The matrix $\mathbf{J}$ is analysed in term of its eigenvalues λ_i. The general rule is that if $\mathrm{Re}(\lambda_i) < 0$ then the equilibrium $\mathbf{x}_e$ is asymptotically stable and if $\mathrm{Re}(\lambda_i) > 0$, it is unstable.

Let us apply this methodology to the analysis of the equilibrium (0, 0) of a linear oscillator. We will write the displacement x and the velocity v in the form $x = \tilde{x}$ and $v = \tilde{v}$ where $\tilde{x}$ and $\tilde{v}$ are very small perturbations from the equilibrium point (0, 0). Now we substitute this into (3.32). It is easy to show that

$$\begin{pmatrix} \dot{\tilde{x}} \\ \dot{\tilde{v}} \end{pmatrix} = \begin{pmatrix} 0 & 1 \\ -\omega_0^2 & -2\gamma \end{pmatrix} \begin{pmatrix} \tilde{x} \\ \tilde{v} \end{pmatrix} = \mathbf{M} \begin{pmatrix} \tilde{x} \\ \tilde{v} \end{pmatrix} \tag{3.35}$$

The eigenvalues of the matrix $\mathbf{M}$ (the same matrix as in (3.34)) are found from the equation $\lambda(2\gamma + \lambda) + \omega_0^2 = 0$ and $\lambda_{1,2} = -\gamma \pm \sqrt{\gamma^2 - \omega_0^2}$.

We immediately see that depending on the sign of γ, the eigenvalues $\lambda_{1,2}$ can have positive or negative real parts and depending of the relation between γ and ω_0 they can be real or complex.

For the case of equilibrium in a conservative system (3.4) with $\gamma = 0$ and $\omega_0 > 0$ (the harmonic oscillator we started our chapter with), there are two purely imaginary eigenvalues and it is neither stable nor unstable according to this definition. The corresponding equilibrium is called the centre point. To be more accurate, the centre point is not asymptotically stable, but it is Lyapunov stable.

Another type of a conservative system that belongs to the family of linear oscillators is the inverted pendulum [3]:

$$\ddot{x} - \omega_0^2 x = 0 \tag{3.36}$$

The equilibrium of an inverted pendulum is characterised by two real eigenvalues (one positive and one negative) and thus is unstable. It is called a saddle point.

In the case of a linear oscillator with small (positive) damping ($\gamma > 0$ and $\gamma < \omega_0$), the equilibrium point (0, 0) is characterised by eigenvalues whose real part is negative $\mathrm{Re}(\lambda_{1,2}) < 0$, however $\lambda_{1,2}$ are complex conjugates. Such an equilibrium is called a stable focus. In the case of a linear oscillator with large damping ($\gamma > 0$ and $\gamma > \omega_0$), the equilibrium point is characterised by eigenvalues that are real and negative $\lambda_{1,2} < 0$. Such an equilibrium is called a stable node.

In the equation of the linear oscillator, the coefficient γ can be negative. For an oscillator with small negative damping ($\gamma < 0$ and $|\gamma| < \omega_0$), the equilibrium point (0, 0) is characterised by eigenvalues whose real part is positive $\mathrm{Re}(\lambda_{1,2}) > 0$, however $\lambda_{1,2}$ are complex conjugates. Such an equilibrium is called an unstable focus. In the case of a linear oscillator with large negative damping ($\gamma < 0$ and $|\gamma| > \omega_0$), the equilibrium point is characterised by eigenvalues that are real and positive $\lambda_{1,2} < 0$. Such an equilibrium is called an unstable node. The equilibrium types are summarised in Table 3.2.

Finally, we would like to explain the concept of a phase portrait. In Eqs. (3.2) and (3.32), x and $v = \dot{x}$ are functions of time. We can plot them as functions of time to show a temporal evolution as we do it Fig. 3.2. Alternatively, we can present v as a function of x parametrically in the (x, v)-plane. This approach to present the

Table 3.2 Equilibrium types of the general model of a linear oscillator

Eignevalues	Point type
$\gamma > 0, \gamma < \omega_0$	Stable focus
$\gamma > 0, \gamma > \omega_0$	Stable node
$\gamma < 0, \lvert\gamma\rvert < \omega_0$	Unstable focus
$\gamma < 0, \lvert\gamma\rvert > \omega_0$	Unstable node
$\gamma = 0, +\omega_0^2$	Centre point
$\gamma = 0, -\omega_0^2$	Saddle point

evolution of a dynamic system is very common in nonlinear dynamics and is called the phase portrait. In order to show the direction of x and v evolution in the plane (x, v), one uses arrows.

The phase portraits of a linear oscillator for different combination of γ and ω_0 around the equilibrium point (0, 0) are shown in Fig. 3.6. For example, let us consider a conservative oscillator with $\gamma = 0$ and $\omega_0^2 > 0$. The evolution of $x(t)$ and $v(t)$ with time is shown in Fig. 3.2a. Now we will plot this solution parametrically in the (x, v)-plane. Such a solution, starting from particular initial condition x_0 and v_0, is called a trajectory. We can see that trajectories corresponding to different initial conditions form ovals around the equilibrium point (centre-type equilibrium) in the origin of Fig. 3.6a. Another example is a damped linear oscillator. The case of $\gamma > 0$ and $\gamma < \omega_0$ is shown in Fig. 3.2b by the two lines that are marked Q_1 and Q_2. The corresponding phase portrait is shown in Fig. 3.6c and the equilibrium point in the centre is a stable focus. The case of $\gamma > 0$ and $\gamma > \omega_0$ is shown in Fig. 3.2b by the line marked by Q_3. The corresponding phase portrait is shown in Fig. 3.6e and the equilibrium point in the centre is a stable node.

3.2 Nonlinear Oscillators

3.2.1 Free Nonlinear Oscillator

The *nonlinear oscillator* is the next concept we are going to consider in this chapter [11, 15, 21]. It is an expansion of the linear oscillator to include the case of nonlinear forces $F_{damping}$ and F_{spring}.

At this stage, it would be very useful to discuss briefly the nature of mechanical forces in the context of energy conversion. In classical mechanics, a force describes an interaction between objects and it depends only on the position of the objects. Classical mechanics introduces a fundamental notion of *potential* (also called *conservative*) forces. The work done by potential forces depends only on the starting and ending points x_1 and x_2 of the trajectory of a body and is independent on the shape of

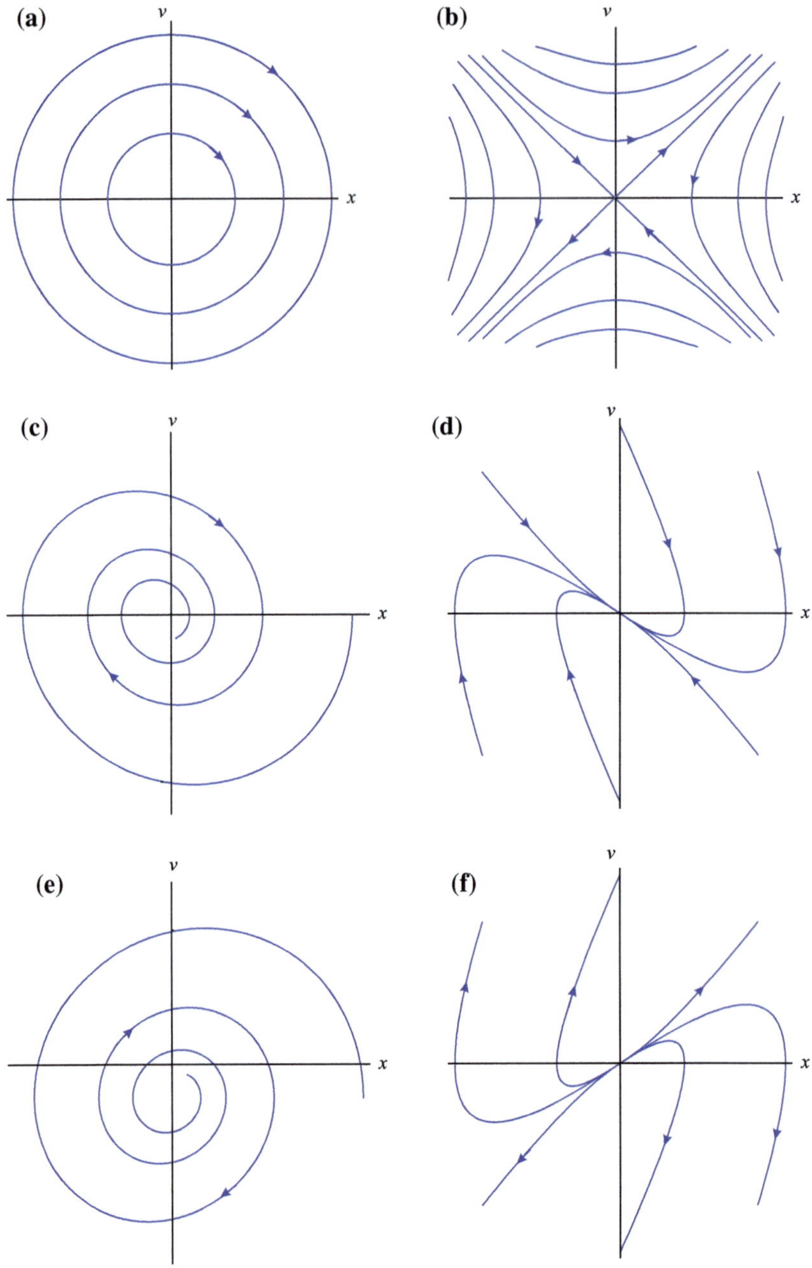

Fig. 3.6 Trajectories in the plane (x, v) spanned by the state variables x and $\dot{x}$ (phase portraits) around the six types of equilibrium points. **a** Centre point corresponding to a conservative linear oscillator (3.4). **b** Saddle point corresponding to a conservative inverted pendulum (3.36). **c** Stable focus, $\gamma > 0$ and $\gamma < \omega_0$. **d** Stable node, $\gamma > 0$ and $\gamma > \omega_0$. **e** Unstable focus, $\gamma < 0$ and $|\gamma| < \omega_0$. **f** Unstable node, $\gamma < 0$ and $|\gamma| > \omega_0$. The arrows show the direction of evolution along a trajectory. Different trajectories correspond to different initial conditions

the trajectory. For such forces, one can introduce a quantity called 'potential energy' that in a one-dimensional case is defined as the following integral:

$$W_p = -\int F(x)\mathrm{d}x, \tag{3.37}$$

where x is the position along the trajectory. It indeed follows from this definition that the work done by a potential force $F(x)$ on a body to move from x_1 to x_2 is $W_{x_1 \text{ to } x_2} = \Phi(x_2) - \Phi(x_1)$ (where Φ is the antiderivative of F) and is independent of the trajectory.

The potential forces have a very important property: they only depend on the position x and not on the velocity of the body. A force generated by a linear spring can be seen as a good example of a potential force. This force depends only on the compression or expansion of the spring $F_{spring} = -k\Delta x$, where Δx is the deformation of the spring. In a more general case, a potential force may be a nonlinear function of the objects's position.

In a isolated mechanical system with only potential forces, total energy is conserved. You may refer to the discussion about the linear oscillator (in particular, the derivation of Eq. (3.6)) as an illustration of this statement. For this reason, the potential forces are also called conservative.

However in practice, any mechanical system experiences the loss of energy, mainly due to different kinds of friction. For example, a body with an initial velocity v slides on a rough uniform horizontal surface. It will eventually lose its momentum and stop. This behaviour is modelled by the introduction of a force that acts on the body and that depends only on its velocity. For example, the damper that we have introduced in the model of a linear oscillator generates a linear friction force $F_{damping} = -bv$. In a more general case, a friction force can be a nonlinear function of the velocity.

Let us note that in the particular case of the air damping force we introduced, $b > 0$, and therefore energy decreases. In most real systems, damping forces reduce energy, and for this reason they are called *dissipative*. We must mention that there is the specific case of 'negative' dissipation (in the case of a linear oscillator, $b < 0$). It leads to the increase of the total energy of the system (such as in self-oscillators), however the discussion on self-oscillating systems it outside the scope of this book. Let us also note here that there is a collective term for such systems—non-conservative, since these forces modify the mechanical energy of the system.

Dissipative forces can also depend on both, the displacement and the velocity. In MEMS resonators, there are two known examples of such forces—a squeeze film damping force [31] and a transducer-generated force (used in the context of vibration energy harvesting, as explained in Sect. 3.3).

We will give an illustration of conservative and non-conservative forces using the example of oscillatory applications. Consider a case when a body whose position $x(t)$ varies periodically (the displacement is described by a sinusoidal function of time)

$$x(t) = X_0 \sin(\omega t). \tag{3.38}$$

is subjected to a nonlinear force $F(x(t), \dot{x}(t))$. In this case, the force $F(t)$ is also periodic, but, in the most general case, not sinusoidal. Often, only the first (fundamental) harmonic of the force is considered.

- In the case of a potential force $F = F(x(t))$, which can be nonlinear, the fundamental harmonic of the force has the same phase as the displacement (or with a π-phase shift), i.e. the force is also expressed as a sinusoidal function

$$F(t) = F_1 \sin(\omega t), \tag{3.39}$$

where F_1 is calculated as

$$F_1 = \frac{1}{\pi}\left(\int_{-1}^{1} F(-X_0\sqrt{1-y^2})dy - \int_{-1}^{1} F(X_0\sqrt{1-y^2})dy\right) \tag{3.40}$$

The derivation of this formula can be found in Appendix I. Note that if $F(x)$ is even, the first harmonic of the force is equal to zero. However, this does not happen in most of practical cases.

- In the case of a 'purely' dissipative force $F = F(\dot{x}(t))$ (dependent only on the velocity), its fundamental harmonic has the same phase as the velocity (or with a π-phase shift, i.e. it is expressed as cosine function

$$F(t) = F_1 \cos(\omega t), \tag{3.41}$$

where F_1 can be calculated as

$$F_1 = \frac{1}{\pi}\left(\int_{-1}^{1} F(\omega X_0\sqrt{1-y^2})dy - \int_{-1}^{1} F(-\omega X_0\sqrt{1-y^2})dy\right) \tag{3.42}$$

The derivation of this formula can be found in Appendix II. Note that if $F(\dot{x})$ is even, the first harmonic of the force is equal to zero.

- In the case of the force that depends on both the displacement and velocity $F(x, \dot{x})$, the fundamental harmonic of the force may have a phase shift with respect to both the velocity and displacement.

$$F = F_1^c \cos(\omega t) + F_1^s \sin(\omega t) = F_1 \cos(\omega t + \phi_1) \tag{3.43}$$

The corresponding sine and cosine components are calculated as

$$\left.\begin{array}{l} F_1^c = \frac{1}{\pi}\left(\int_{-1}^{1} F(X_0 y, \omega X_0\sqrt{1-y^2})dy - \int_{-1}^{1} F(X_0 y, -\omega X_0\sqrt{1-y^2})dy\right) \\ F_1^s = \frac{1}{\pi}\left(\int_{-1}^{1} F(X_0\sqrt{1-y^2}, \omega X_0 y)dy - \int_{-1}^{1} F(-X_0\sqrt{1-y^2}, \omega X_0 y)dy\right) \end{array}\right\} \tag{3.44}$$

The representation of the first harmonic of the force as the superposition of the sine and cosine components is sometimes interpreted as the superposition of potential and purely dissipative components (see the discussion in [31] on modeling of the squeeze film damping force). It should be noted though that it is only useful in the context of systems with periodically varying displacements and velocities.

In the context of this discussion, we are going to consider only those oscillators that are subjected to 'purely' dissipative (dependent of the velocity only) and potential forces. The specific case of a force depending on both x and $\dot{x}$ will appear later in the discussion on transducers and transducer forces (see Sect. 3.3).

Now let us consider an expansion of the model of a linear oscillator to the case of nonlinear potential forces. For example, Fig. 3.7 shows a mechanical oscillator with a nonlinear spring. In the similar model of a linear oscillator shown in Fig. 3.1, the spring follows Hooke's law so that $F_{spring} = -kx$. For a nonlinear spring however the force has a nonlinear dependance on the displacement. Therefore, in the most general case, a free nonlinear damped oscillator is described by the equation

$$\ddot{x} + 2\gamma\dot{x} + g(x) = 0 \tag{3.45}$$

where $g(x)$ is a nonlinear spring force of the displacement x the mass of oscillator m.

One of the most common models of nonlinear oscillators is obtained when we write a nonlinear spring force F_{spring} using its Taylor series and limit this series to the cubic term only, thus assuming that $F_{spring} = -(k_1x \pm k_3x^3)$. The odd order of this polynomial reflects the fact that the spring force is an odd function of the displacement, i.e. it always opposes to the deformation. In addition, following the note made in Sect. 3.1.3 on Eq. (3.36) that describes an oscillator near an unstable equilibrium, we can also consider that the linear term k_1x can appear with a minus sign. Substituting the force into (3.45), we obtain

$$\ddot{x} + 2\gamma\dot{x} \pm \omega_0^2 x + \kappa x^3 = 0 \tag{3.46}$$

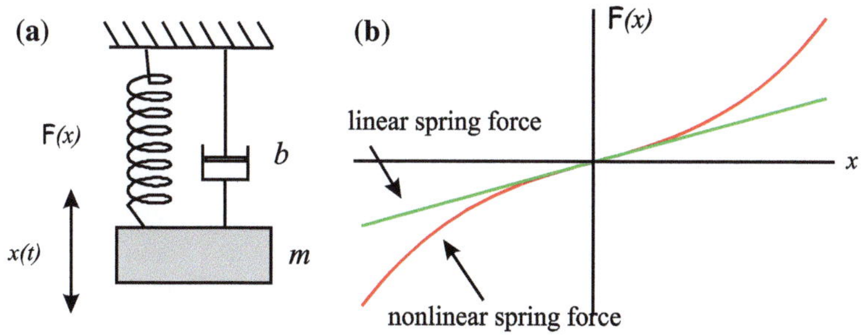

Fig. 3.7 **a** Simple nonlinear mechanical oscillator that is obtained from the linear oscillator in Fig. 3.1 by considering that the spring restoring force $F_{spring}(x)$ has a nonlinear dependance on the displacement x. **b** Example of nonlinear spring force in the form $F_{spring}(x) = k_1x + k_3x^3$. For the comparison, the linear spring force is shown together with the nonlinear one

where $\omega_0^2 = k_1/m$ and the coefficient $\kappa = \pm k_3/m$ can have any sign. Note that it is often accepted that the coefficient appearing before the linear term x is denoted as ω_0^2. Therefore if one wants to consider the sign of this term, one writes $\pm\omega_0^2 x$ (obviously we assume that $\omega_0^2 > 0$ since ω_0 is real). We do not have such a limitation on the term κ that can be both positive or negative.

This model is known as the *Duffing oscillator* [8, 13]. The Duffing type of nonlinearity often appears in micro-scale mechanical structures (MEMS) due to mechanical properties or it can be induced electrically [1, 7, 14, 31]. In micro-scale energy harvesting applications, there is a large number of examples of nonlinear MEMS resonators [5, 16, 23, 24, 33, 37] that are designed specifically to improve characteristics of harvesters.

The potential energy of a nonlinear oscillator is introduced in the standard way

$$\frac{dW_p}{dx} = m \cdot g(x) = -F_{spring} \tag{3.47}$$

For instance, we have already established that for a linear oscillator its potential energy associated with the spring force is $W_p = -kx^2/2$. Indeed, $F_{spring} = -kx$ and, as follows form the above equation, $W_p = -\int F(x)dx = \int kxdx = kx^2/2$.

In order to obtain the energy relation for a nonlinear oscillator, we will multiply Eq. (3.45) by $\dot{x}$:

$$\frac{d}{dt}\left[\frac{m\dot{x}^2}{2} + W_p\right] = -b\dot{x}^2 \tag{3.48}$$

where $b = 2\gamma\, m$ is a corresponding damping coefficient. This expression is similar to the case of the linear oscillator with the difference that W_p can now be some other function of the displacement, different from the parabolic $kx^2/2$ expression. For a nonlinear oscillator in the absence of friction $\gamma = 0$, its energy is conserved. While in the presence of friction $\gamma > 0$, energy dissipates.

The oscillations resulting from Eqs. (3.2), (3.4) or (3.45) can be seen as the coordinates of a particle that moves (oscillates) on a curved surface whose profile has the shape of the potential function W_p. If such a surface is ideal (no friction) it will correspond to the case $\gamma = 0$, and the particle slides (oscillates) on this surface without loss of energy. This analogy is shown in Fig. 3.8 where a particle of mass m moves in a potential well W_p. The displacement and the velocity of such a particle is the same as x and v we obtain from solving differential equations (3.2), (3.4) or (3.45).

Figure 3.8 shows the potential functions, W_p, of a linear and two nonlinear Duffing oscillators. For the linear oscillator, the potential function $W_p = kx^2/2$ is a parabola. For the nonlinear oscillator with $W_p = \alpha x^2 + \beta x^4$ ($\alpha = k_1/2$ and $\beta = k_3/4$), the potential function also resembles a parabola. Although it is qualitative similar to the linear case, the large-scale of the linear and nonlinear systems will be different. Only small-scale oscillations around the local minimum of W_p will be similar in both systems. For a nonlinear oscillator shown in Fig. 3.8c, the potential function is given

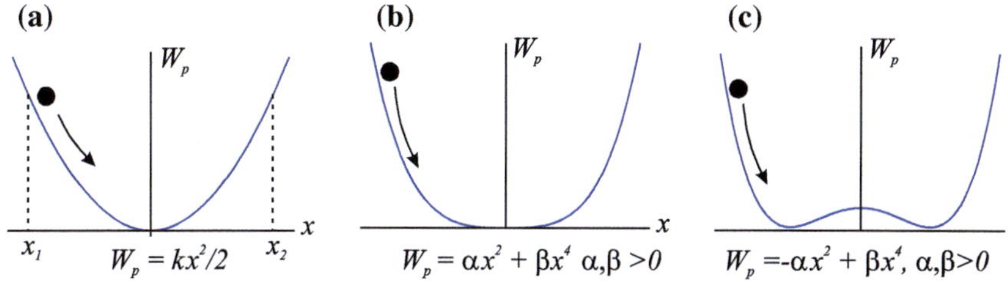

Fig. 3.8 Schematic view of the potential energy W_p of three different oscillators. **a** Linear oscillator with $g(x) = \omega_0^2 x$ and $W_p = kx^2/2$. **b** Nonlinear oscillator with $g(x) = \omega_0^2 x + \kappa x^3$ and $W_p = \alpha x^2 + \beta x^4$ where $\alpha = k_1/2$, $\beta = k_3/4$ and $\alpha, \beta > 0$. **c** Nonlinear oscillator with $g(x) = -\omega_0^2 x + \kappa x^3$ and $W_p = -\alpha x^2 + \beta x^4$ where $\alpha = k_1/2$, $\beta = k_3/4$ and $\alpha, \beta > 0$

by $W_p = -\alpha x^2 + \beta x^4$ ($\alpha = k_1/2$ and $\beta = k_3/4$). Such a potential well has two local minima and one local maximum.

Any local extremum of the potential function corresponds to an equilibrium point of Eq. (3.45). In a linear oscillator, its parabolic function W_p has only one extremum, a minimum. If the linear oscillator is conservative, i.e. $\gamma = 0$, this local minimum corresponds to a centre point, and if $\gamma > 0$ it corresponds to a stable focus or node. In a linear oscillator, such as an inverted pendulum (3.36), the potential function is $W_p = -kx^2/2$, and has only one extremum, a maximum. It corresponds to a saddle point. In a nonlinear oscillator from Fig. 3.8c, there are three extrema, two minima and one maximum. Each local minimum corresponds to a centre ($\gamma = 0$) or stable focus/node ($\gamma > 0$), while a maximum corresponds to a saddle point. Such an oscillator has two coexisting stable equilibria, and therefore is called *bistable*. Depending on initial conditions, the particle from Fig. 3.8c can oscillate either around one minimum of W_p or the other.

A number of fundamental consequences arise from the presence of the nonlinear function $g(x)$ in (3.45). For instance, we know that the frequency of free oscillations of a linear oscillator does not depend on the amplitude of oscillations or on the energy of the system. We say that these oscillations are *isochronous*. Indeed, the natural frequency $\omega_0^2 = k/m$ is a constant and is independent of the energy of the oscillator. For nonlinear oscillators, it is not even clear from Eqs. (3.45) or (3.46) what quantity must be considered as the natural frequency. It turns out that for a nonlinear oscillator, its frequency depends on the amplitude of oscillations (or its energy), i.e. it is not a fixed quantity. Such oscillations are called *anisochronous*.

Consider for simplicity the conservative case when $\gamma = 0$. Since the energy is conserved, Eq. (3.48) can be written as

$$\frac{m\dot{x}^2}{2} + W_p = W_0 \tag{3.49}$$

where W_0 is the full or initial energy of the oscillator. The velocity is expressed from the equation

$$\dot{x} = \pm\sqrt{\frac{2(W_0 - W_p)}{m}} \tag{3.50}$$

By separating the variables $\dot{x} = dx/dt$, we obtain $dt = dx/\sqrt{2m^{-1}(W_0 - W_p)}$. Let us assume that the turnoff points x_1 and x_2 bound the displacement of the system (see Fig. 3.8). We can find the time required for the mass m to make one full oscillation between x_1 and x_2 and back:

$$T = 2\int_{x_1}^{x_2} \frac{dx}{\sqrt{2m^{-1}(W_0 - W_p)}} \tag{3.51}$$

This time defines the period of oscillations (T) for an oscillator whose potential energy is given by the function W_p and whose full energy is W_0. Note that for symmetrical potential functions W_p, which account for most cases, $x_2 = -x_1 = A$ where A is the amplitude of oscillations.

It is easy to show that in the case of a linear oscillator with $W_p = kx^2/2$, the time calculated from (3.51) is $T = 2\pi/\sqrt{k/m} = 2\pi/\omega_0$ and is the period of natural oscillations. In the majority of other practical cases, including the Duffing oscillator, the result will depend on the full energy W_0 and as a consequence on the amplitude of oscillations A. For instance, for the Duffing oscillator

$$\omega = \frac{2\pi}{T} \approx \sqrt{\frac{k}{m}}\left[1 + \frac{3(k_3/k_1)A^2}{8}\right] \tag{3.52}$$

As is seen from the above formula, the larger the amplitude of oscillations A is, the larger their frequency ω is.

Finally, let us emphasise that there is no general analytical solution of Eq. (3.45). In order to understand and analyse the behaviour of a nonlinear oscillator, one can apply various perturbation techniques [19]. The two most commonly used methods in the context of weakly nonlinear oscillators are the Lindstedt—Poincaré method and the multiple scales method. Let us obtain the approximate solution of a weakly nonlinear Duffing oscillator using the multiple scale method.

Perturbation methods are applied when there is a 'small' parameter present in a nonlinear equation. In order to introduce a small parameter, we first must obtain a dimensionless system. In order to do so, let us introduce dimensionless time $\tau = \omega_0 t$ and dimensionless displacement $y = x/d$ where d is some characteristic scale of the displacement. Equation (3.46) will become

$$y'' + 2\beta y' + y + \alpha y^3 = 0 \tag{3.53}$$

where $\beta = \gamma/\omega_0 = b/(2\sqrt{km})$ is the dimensionless dissipation parameter and $\alpha = \kappa d^2/\omega_0^2$ is the nonlinear parameter. The prime denotes the derivative with respect to the dimensionless time τ. Moreover, following the approximation of a weakly nonlinear system, let us suggest that $\beta, \alpha \ll 1$. We will express the 'smallness' of these coefficients in (3.54) by presenting them in the form $\beta = \varepsilon\tilde{\beta}$ and $\alpha = \varepsilon\tilde{\alpha}$ where ε is an auxiliary 'small' parameter:

$$y'' + 2\varepsilon\tilde{\beta}y' + y + \varepsilon\tilde{\alpha}y^3 = 0 \tag{3.54}$$

Following the method of multiple scales, we introduce different time scales:

$$T_0 = t, \quad T_1 = \varepsilon t, \quad T_2 = \varepsilon t_2 \ldots \tag{3.55}$$

The variable $y(\tau)$ in (3.54) now can be presented as a series, and every term is this series is a function of these time scales

$$y(\tau) = y_0(T_0, T_1, T_2, \ldots) + \varepsilon y_1(T_0, T_1, T_2, \ldots) + \varepsilon^2 y_2(T_0, T_1, T_2, \ldots) + \ldots \tag{3.56}$$

The full derivative with respect to time τ can be expressed through partial derivatives with respect to times T_i:

$$\begin{aligned}
\frac{d}{d\tau} &= \frac{\partial}{\partial T_0}\frac{dT_0}{d\tau} + \frac{\partial}{\partial T_1}\frac{dT_1}{d\tau} + \frac{\partial}{\partial T_2}\frac{dT_2}{d\tau} + \ldots = \frac{\partial}{\partial T_0} + \varepsilon\frac{\partial}{\partial T_1} + \varepsilon^2\frac{\partial}{\partial T_2} + \ldots = \\
&= D_0 + \varepsilon D_1 + \varepsilon^2 D_2 \ldots \\
\frac{d^2}{d\tau^2} &= (D_0 + \varepsilon D_1 + \varepsilon^2 D_2 \ldots)^2 = D_0^2 + 2\varepsilon D_0 D_1 + \varepsilon^2(D_1^2 + 2D_0 D_2) + \ldots
\end{aligned} \tag{3.57}$$

where we introduced the operators $D_i = \partial/\partial T_i$. Using Eqs. (3.56) and (3.57), we rewrite the original Eq. (3.54):

$$\begin{aligned}
\left(D_0^2 + 2\varepsilon D_0 D_1 + \varepsilon^2(D_1^2 + 2D_0 D_2) + \ldots\right)(y_0 + \varepsilon y_1 + \varepsilon^2 y_2 + \ldots) + \\
+2\varepsilon\tilde{\beta}\left(D_0 + \varepsilon D_1 + \varepsilon^2 D_2 \ldots\right)(y_0 + \varepsilon y_1 + \varepsilon^2 y_2 + \ldots) + (y_0 + \varepsilon y_1 + \varepsilon^2 y_2 + \ldots) + \\
+\varepsilon\tilde{\alpha}(y_0 + \varepsilon y_1 + \varepsilon^2 y_2 + \ldots)^3 = 0
\end{aligned} \tag{3.58}$$

In the next step, we set every term of the ε^i order to zero:

$$\begin{aligned}
&D_0^2 y_0 + y_0 = 0 \\
&D_0^2 y_1 + y_1 = -2D_0 D_1 y_0 - 2\tilde{\beta} D_0 y_0 - \tilde{\alpha} y_0^3
\end{aligned} \tag{3.59}$$

From the first equation in (3.59) we obtain that

$$y_0 = A(T_1)e^{iT_0} + A^*(T_1)e^{-iT_0} = a(T_1)\cos(\tau + \varphi(T_1)) \tag{3.60}$$

where $A(T_1) = (a/2)\exp(i\varphi)$ is the slow amplitude of oscillations, A^* is the complex conjugate of A and φ is the phase of oscillations. Now we will use this in the second equation in (3.59), taking into account that $D_0 y_0 = iA(T_1)\exp(iT_0)$ and $D_1 y_0 = \dot{A}(T_1)\exp(iT_0)$, where $\dot{A}$ now denotes the derivative with respect to time scale T_1.

$$D_0^2 y_1 + y_1 = -2i\dot{A}e^{iT_0} + 2i\dot{A}^* e^{-iT_0} - 2i\tilde{\beta}Ae^{iT_0} + 2i\tilde{\beta}A^* e^{-iT_0} - \tilde{\alpha}(Ae^{iT_0} + A^* e^{-iT_0})^3 \tag{3.61}$$

Instead of one real-valued variable y, we just introduced a complex variable A (that contains in fact two variables, Re(A) and Im(A)). Therefore, we can suggest one additional equation for the variable A. In order to introduce such an equation, let us analyse the terms on the right-hand side of the above equation. The terms that are proportional to $\exp(iT_0)$ lead to resonance in a linear system: as discussed in Sect. 3.1.2, they will cause an infinite increase of the oscillations. These terms are also called *secular terms*. Since, this is not possible in a realistic system, we will choose an additional equation such that these terms are removed. As a result, we set

$$-2i\dot{A}e^{iT_0} - 2i\tilde{\beta}Ae^{iT_0} - 3\tilde{\alpha}A^2A^*e^{iT_0} = 0 \tag{3.62}$$

and, therefore, this is the equation to find the slow amplitude. (Implicitly we assume that the complex conjugate of (3.62) is also equal to zero.)

Expressing complex A through real a and φ, we obtain two equations that define the evolution of slow amplitude a and phase φ with time:

$$\begin{aligned} \dot{a} &= -\tilde{\beta}a \\ \dot{\varphi} &= -3\tilde{\alpha}a^2/8 \end{aligned} \tag{3.63}$$

These equations are called the truncated equations. Knowing the initial conditions $a(0) = a_0$ and $\varphi(0) = \varphi_0$, it is easy to solve these equations

$$\begin{aligned} a &= a_0 e^{-\tilde{\beta}T_1} \\ \dot{\varphi} &= \varphi_0 + \frac{3\tilde{\alpha}a_0^2}{16\tilde{\beta}}\left(1 - e^{-2\tilde{\beta}T_1}\right) \end{aligned} \tag{3.64}$$

Finally, in order to find y, we return to Eq. (3.61) after having excluded all resonant terms

$$D_0^2 y_1 + y_1 = -(\tilde{\alpha}a^3/4)\cos(3\tau + 3\varphi) \tag{3.65}$$

where we have taken into account that the only two terms left in the right-hand side are $A^3\exp(3iT_0) + (A^*)^3\exp(-3iT_0) = (a^3/4)\cos(3\tau + 3\varphi)$. The particular solution of Eq. (3.61) corresponding to a forced oscillation under a harmonic force

$F(t) = \tilde{\alpha} a^3 \cos(3\tau + 3\varphi)$ was found in Sect. 3.1.2. In order to use the solution (3.19), we set $F_0 = \tilde{\alpha} a^3/4$, $\omega_{ext} = 3$, $\gamma = 0$ and $\phi = -\pi$:

$$y_1 = \frac{\tilde{\alpha} a^3}{32} \cos(3\tau + 3\varphi) \tag{3.66}$$

Taking into account y_0 and y_1 and noting that $\varepsilon\tilde{\beta} = \beta$, $\varepsilon\tilde{\alpha} = \alpha$ and $\tilde{\alpha}/\tilde{\beta} = \alpha/\beta$, we can write

$$\begin{aligned} y(\tau) = y_0 + \varepsilon y_1 + \ldots \approx a_0 \cos(\tau + \varphi) + \frac{\alpha a^3}{32} \cos(3\tau + 3\varphi) = \\ = a_0 e^{-\beta\tau} \cos\left(\tau + \varphi_0 + \frac{3\alpha a_0^2}{16\beta}\left(1 - e^{-2\beta\tau}\right)\right) + \\ + \frac{\alpha a_0^3 e^{-3\beta\tau}}{32} \cos\left(3\tau + 3\varphi_0 + \frac{9\alpha a_0^2}{16\beta}\left(1 - e^{-2\beta\tau}\right)\right) \end{aligned} \tag{3.67}$$

Furthermore one can note that this solution does not depend on the auxiliary small parameter ε we introduced in Eq. (3.54) for our analysis. If a more accurate solution is required, one can continue and write the equation for y_2 and use the already found y_0 and y_1 to solve it.

From (3.67) we can see that the solution of a free, unforced Duffing oscillator contains not only the fundamental component $\cos(\tau + \varphi)$ but also the third harmonics $\cos(3\tau + 3\varphi)$, i.e. it is *anharmonic*. Figure 3.9 shows the evolution of the dimensionless displacement y of the normalised Duffing equation (3.54) as a function of time τ at $\beta = 0.05$ and $\alpha = 0.5$ obtained from a numerical simulation. For comparison, we also give the solution of a linear oscillator at the same dissipation $\beta = 0.05$ (and $\alpha = 0$ in this case). We also show the approximated solution (3.67) obtained from the multiple scale method. Although the parameter of nonlinearity α cannot be considered very small compared to unity, the approximated solution is still accurate.

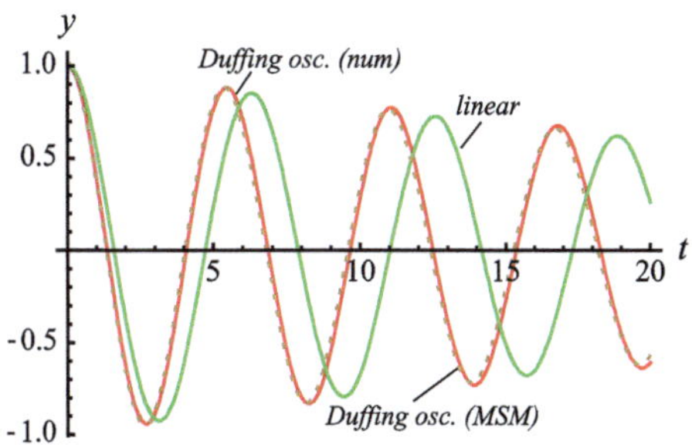

Fig. 3.9 Evolution of y as a function of time for a linear oscillator (3.2) and the Duffing oscillator (3.46) obtained from numerical simulation of the equation. The equations are normalised as shown for Eq. (3.54) $\beta = 0.05$ and $\alpha = 0.5$. For the comparison, the approximated solution (3.67) obtained from the multiple scales method (MSM) is shown by the dashed line

Note the change in the frequency of nonlinear oscillations due to cubic nonlinearity compared to the linear case: it is easily seen at the point when $y(\tau)$ crosses zero.

3.2.2 Forced Nonlinear Oscillator

We have already studied a linear oscillator under external harmonic actuation and established that if the frequency of the external force is close to the natural frequency of the oscillator, one observes resonance. In a nonlinear oscillator, as we know from the previous section, oscillations are anisochronous, i.e. the frequency of such oscillation is not a constant, but rather depends on the energy of the oscillator. However, resonance is also possible in nonlinear systems, but the resulting forced oscillations are more complex than in the linear case. We will consider a Duffing oscillator as an example to illustrate nonlinear resonance. We will write a forced nonlinear Duffing oscillator in the following dimensionless form

$$\ddot{y} + 2\beta\dot{y} + y + \alpha y^3 = f_0 \cos((1+\sigma)\tau + \phi_0) \tag{3.68}$$

where we remind the reader that α can be positive or negative. The term $\cos[(1+\sigma)\tau + \phi_0]$ appeared after the normalisation procedure. Indeed, $\cos(\omega_{\text{ext}}t + \phi_0) = \cos(\omega_{\text{ext}}(\omega_0^{-1}\tau) + \phi_0) = \cos(\Omega\tau + \phi_0)$. We will present the dimensionless external frequency in the form $\Omega = 1 + \sigma$ where σ is a frequency mismatch between the dimensionless natural frequency (equal to unity after normalisation) and the dimensionless external frequency Ω. As in the previous section, following the procedure of introducing a small parameter ε, we will assume that $\beta = \varepsilon\tilde{\beta}$, $\alpha = \varepsilon\tilde{\alpha}$, $\sigma = \varepsilon\tilde{\sigma}$ and $f_0 = \varepsilon\tilde{f}_0$:

$$\ddot{y} + 2\varepsilon\tilde{\beta}\dot{y} + y + \varepsilon\tilde{\alpha}y^3 = \varepsilon\tilde{f}_0 \cos((1+\varepsilon\tilde{\sigma})\tau + \phi_0) \tag{3.69}$$

We will apply the multiple scale method as shown in the previous section. In order to do so, we will use (3.56) and (3.57). Equation (3.69) will transform into

$$\begin{aligned} \left(D_0^2 + 2\varepsilon D_0 D_1 + \varepsilon^2(D_1^2 + 2D_0 D_2) + \ldots\right)(y_0 + \varepsilon y_1 + \varepsilon^2 y_2 + \ldots) + \\ + 2\varepsilon\tilde{\beta}\left(D_0 + \varepsilon D_1 + \varepsilon^2 D_2 \ldots\right)(y_0 + \varepsilon y_1 + \varepsilon^2 y_2 + \ldots) + (y_0 + \varepsilon y_1 + \varepsilon^2 y_2 + \ldots) + \\ + \varepsilon\tilde{\alpha}(y_0 + \varepsilon y_1 + \varepsilon^2 y_2 + \ldots)^3 = \varepsilon\tilde{f}_0 \cos((1+\varepsilon\tilde{\sigma})\tau + \phi_0) \end{aligned} \tag{3.70}$$

The term of the external force can also be presented using the complex form: $\tilde{f}_0 \cos((1+\varepsilon\tilde{\sigma})\tau + \phi_0) = (\tilde{f}_0/2)\exp[i(1+\varepsilon\tilde{\sigma})\tau + \phi_0] + c.c.$

Collecting the terms corresponding to different orders of ε, we obtain:

$$
\begin{aligned}
&D_0^2 y_0 + y_0 = 0 \\
&D_0^2 y_1 + y_1 = -2D_0 D_1 y_0 - 2\tilde{\beta} D_0 y_0 - \tilde{\alpha} y_0^3 + \left((\tilde{f}_0/2)\exp[(1+\varepsilon\tilde{\sigma})T_0 + \phi_0] + c.c.\right)
\end{aligned} \tag{3.71}
$$

The solution has the form $y_0 = A(T_1)\exp(iT_0)$ where the complex slow amplitude $A = (a/2)\exp(i\varphi)$ can be written through a real slow amplitude a and real slow phase φ. In the second equation of the set (3.71), we set all secular terms to zero in order to avoid a linear resonance in a lossless system. This yields the equation for A:

$$
-2i\dot{A}e^{iT_0} - 2i\tilde{\beta}Ae^{iT_0} - 3\tilde{\alpha}A^2A^*e^{iT_0} + (\tilde{f}_0/2)e^{i\varepsilon\tilde{\sigma}T_0+\phi_0}e^{iT_0} = 0 \tag{3.72}
$$

Note that the term $\exp(i\varepsilon\tilde{\sigma}T_0 + \phi_0) = \exp(i\tilde{\sigma}T_1 + \phi_0)$ is in fact a slowly varying term, as $T_1 = \varepsilon T_0$. Now using the representation of the complex A through real a and φ, we write to equations

$$
\begin{aligned}
\dot{a} &= -\tilde{\beta}a + \frac{\tilde{f}_0}{2}\sin(\tilde{\sigma}T_1 + \phi_0 - \varphi) \\
a\dot{\varphi} &= \frac{3\tilde{\alpha}a^3}{8} - \frac{\tilde{f}_0}{2}\cos(\tilde{\sigma}T_1 + \phi_0 - \varphi)
\end{aligned} \tag{3.73}
$$

where we remind the reader that the overdot denotes the derivative with respect to the slow time scale T_1. Finally, we will introduce a new variable phase $\psi = \tilde{\sigma}T_1 + \phi_0 - \varphi$. Since $\dot{\psi} = \tilde{\sigma} - \dot{\phi}$, Eq. (3.73) will take the form

$$
\begin{aligned}
\dot{a} &= -\tilde{\beta}a + \frac{\tilde{f}_0}{2}\sin\psi \\
\dot{\psi} &= \tilde{\sigma} - \frac{3\tilde{\alpha}a^2}{8} + \frac{\tilde{f}_0}{2a}\cos\psi
\end{aligned} \tag{3.74}
$$

One can compare the set of equations (3.74) to the truncated equations (3.63) of a free oscillator. The external driving is presented in the form of terms proportional to $\tilde{f}_0$. From this equation, we will find the amplitude a_0 and phase ψ_0 of steady-state forced oscillations by assuming that $a_0, \psi_0 = \text{const}$, i.e. $\dot{a}_0, \dot{\psi}_0 = 0$:

$$
\begin{aligned}
\frac{\tilde{f}_0}{2}\sin\psi_0 &= \tilde{\beta}a_0 \\
\frac{\tilde{f}_0}{2a}\cos\psi_0 &= -\tilde{\sigma} + \frac{3\tilde{\alpha}a_0^2}{8}
\end{aligned} \tag{3.75}
$$

Since $\cos^2\psi_0 + \sin^2\psi_0 = 1$, we will obtain an equation that allows us to find the slow real amplitude a_0:

$$\tilde{\beta}^2 a_0^2 + \left(\tilde{\sigma} a_0 - \frac{3\tilde{\alpha} a_0^3}{8}\right)^2 = \frac{\tilde{f}_0^2}{4} \tag{3.76}$$

Note again that since $\tilde{\beta} = \beta/\varepsilon, \tilde{\sigma} = \sigma/\varepsilon, \tilde{\alpha} = \alpha/\varepsilon$ and $\tilde{f}_0 = f_0/\varepsilon$, the above equation can be rewritten in terms of the original parameters β, σ, α and f_0 and is independent of the auxiliary parameter ε we introduced in (3.69). Therefore,

$$y_0(\tau) = a_0 \cos(\tau + \varphi_0) = a_0 \cos((1 + \varepsilon\tilde{\sigma})\tau + \phi_0 - \psi_0) \tag{3.77}$$

and we can find y_1 from the equation:

$$D_0^2 y_1 + y_1 = -(\tilde{\alpha} a_0^3/4)\cos(3(1 + \varepsilon\tilde{\sigma})\tau + 3\phi_0 - 3\psi_0) \tag{3.78}$$

The particular solution of this equation can be found from (3.19) where we set $F_0 = \tilde{\alpha} a^3/4$, $\omega_{ext} = 3$, $\gamma = 0$ and $\phi = -\pi$:

$$y_1 = \frac{\tilde{\alpha} a_0^3}{32} \cos(3(1 + \varepsilon\tilde{\sigma})\tau + 3\phi_0 - 3\psi_0) \tag{3.79}$$

Taking into account y_0 and y_1 and noting that $\varepsilon\tilde{\beta} = \beta$, $\varepsilon\tilde{\alpha} = \alpha$, $\tilde{\alpha}/\tilde{\beta} = \alpha/\beta$ and $1 + \varepsilon\tilde{\sigma} = \Omega$, we can write a steady-state solution in the form

$$y(\tau) \approx a_0 \cos(\Omega\tau + \phi_0 - \psi_0) + \alpha_0 \frac{a^3}{32} \cos(3\Omega\tau + 3\phi_0 - 3\psi_0) \tag{3.80}$$

where a is the solution of (3.76) and ψ is found from (3.75).

The amplitude of steady-state forced oscillations in the Duffing oscillator a_0 as a function of the frequency mismatch $\sigma = \omega_{\text{ext}}/\omega_0 - 1$ obtained from (3.76) is shown in Fig. 3.10. Three cases are shown: $\alpha = 0$ (corresponds to the resonance in a linear damped oscillator), $\alpha < 0$ (nonlinear, spring softening) and $\alpha > 0$ (nonlinear, spring hardening). All of them are calculated for the same amplitude of the external force $f_0 = 0.2$. In contrast to the linear case shown by the grey line, resonance in nonlinear oscillators leads to *bistability* and *hysteresis*.

From this figure, one can see that for certain values of the frequency mismatch σ there are three overlapping branches of the graph. This means that there are three solutions for a_0 found from (3.76). Further analysis of these solutions shows that two of them are stable (shown by the solid line) while the third is not stable (this branch is shown by the dashed line). Depending on the initial conditions $x(t_0)$ and $v(t_0)$, the oscillator can display forced vibrations with a large amplitude $a_{0,1}$ or with a smaller amplitude $a_{0,2}$. This phenomenon is called multistability.

Now imagine that we can slowly change the frequency of the external oscillations ω_{ext}, for example over the range from some $-\omega^*_{\text{ext}}$ to $+\omega^*_{\text{ext}}$ (this is called a frequency sweep). If one sweeps the frequency of the external force in the direction shown by the arrows on the resonance curves, one will observe hysteresis. For example, let us consider the case $\alpha > 0$. Increasing the frequency ω_{ext}, we will observe gradual increase of the oscillation amplitude (arrow moving right in the figure). When the peak of the resonance curve is reached at some $\omega_{\text{ext},1}$, we will observe a sharp jump as we will switch to the lower stable branch of the curve. Decreasing the frequency ω_{ext}, we also will observe a gradual increase of the forces oscillations. At another frequency $\omega_{\text{ext},2} < \omega_{\text{ext},1}$, we observe a jump to a higher stable branch of oscillations. This phenomenon is known as the hysteresis of the nonlinear resonance curve (Fig. 3.10).

Figure 3.11 shows the transformation of the resonance curve $a_0(\sigma)$ if the amplitude of the external force f_0 increases. If f_0 is small enough, the system behaves similarly to a linear oscillator and does not display bistability or hysteresis. However, as the amplitude f_0 increases, the amplitude of forced oscillations also increases. The segment with bistability and hysteresis increases correspondingly. As is clearly seen from this figure, nonlinear oscillators can respond to external vibrations (maintain resonance, i.e. large amplitude oscillations) for a wide frequency range of ω_{ext}, which can be significantly larger than the 'resonance' frequency range of a linear oscillator. This property has been exploited in vibration energy harvesters by designing and fabricating MEMS resonators with nonlinear springs [2, 23, 24]. It has been shown in these studies that such systems respond more effectively to ambient wideband or noise-like vibrations.

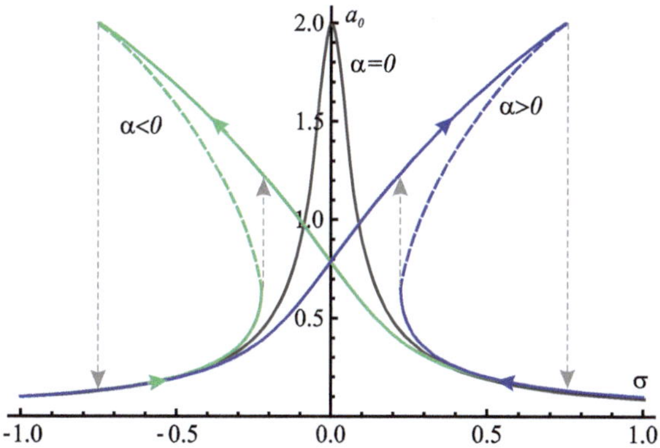

Fig. 3.10 Resonance in linear and nonlinear oscillators. The amplitude a_0 of steady-state forced oscillations as a function of the normalised frequency mismatch $\sigma = \omega_{\text{ext}}/\omega_0 - 1$. The grey line corresponds to the linear case when $\alpha = 0$ (see also Fig. 3.4), while the blue and green lines correspond to nonlinear cases ($\alpha > 0$ and $\alpha < 0$). For the nonlinear cases one can see multistability and hysteresis: there is a segment of the resonance curve $a_0(\sigma)$ where two stable solutions of a_0 and one unstable a_0 (shown by the *dashed line*) coexist

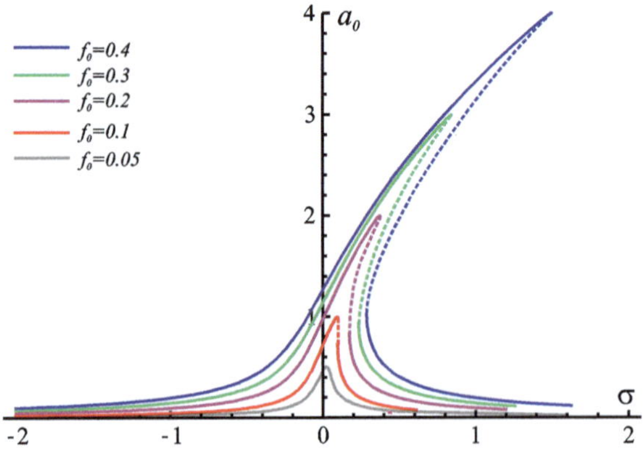

Fig. 3.11 Resonance in a nonlinear Duffing oscillators with $\alpha > 0$. The larger the amplitude of the external force f_0, the larger the amplitude of forced oscillations. The segment of the resonance curve $a_0(\sigma)$ with bistability and hysteresis (shown by the *dashed lines*) increases correspondingly

3.3 Resonators and Kinetic Energy Harvesting

The last two sections presented a few issues related with the fundamental properties and analysis of resonators. This section explains how mechanical resonators can be used for conversion of kinetic energy towards a non-mechanical domain. Conversion towards the electrical domain will be cited as an example, although the presentation of this section is valid for any conversion of kinetic energy.

3.3.1 Architecture of a Kinetic Energy Harvester (KEH)

A general structure of a kinetic energy harvester is given in Fig. 3.12a [17, 30, 36]. The harvester is placed into the reference frame Ox attached to the external vibration structure (e.g. a car). The main element of such a harvester is a mobile mass, sometimes called a 'proof mass' or 'seismic mass'. It should be understood that, in general, the energy of external vibrations is very high compared to what can be stored in the harvester. The role of the mobile mass is to capture *a small part* of external kinetic energy into the harvester. In order to achieve this, the mass must be mobile in the frame Ox. The energy transfer is done through an apparent force acting on the mass, which is, in fact, due to the non-inertial nature of the reference frame Ox. Indeed, according to theoretical mechanics [34], a non-inertial reference frame moving with acceleration a_{ext} can be considered as inertial if a force $-ma_{ext}$ is applied to every mass m of the system. In this way, the source of mechanical energy inside a harvester is given by the force $-ma_{ext}$ applied to the mobile mass. With this, the reference frame Ox should be considered as inertial, and the Newtonian laws valid for inertial systems can be used for the system analysis.

In order to convert mechanical energy that the mass receives from the force $-ma_{ext}$, a transducer is used. The role of the transducer is the generation of a force called 'transducer force' F_t. The nature and origin of this force depends on the domain

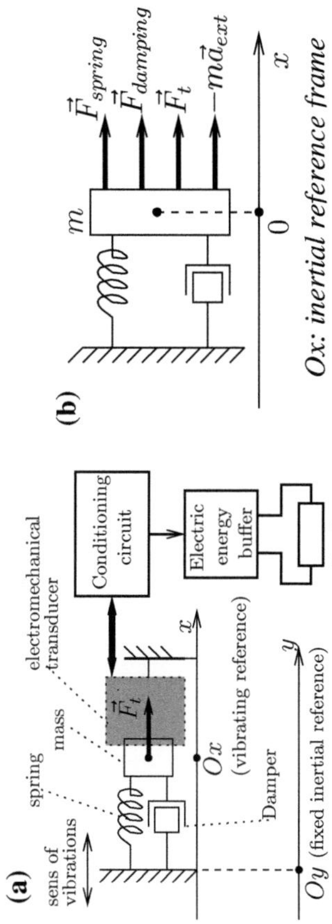

Fig. 3.12 Structure of a kinetic energy harvester: **a** general structure, **b** mechanical model of the harvester

into which the energy should be converted. In our study, we are interested in the generation of electricity. Hence, the force F_t is electrical, and it should reduce the kinetic energy of the mobile mass while increasing the electrical energy of the force F_t generator. The F_t generator should be designed to maximise the power of the energy conversion. The transducer is composed of the physical device interfacing the mechanical and electrical domains, and the electronic networks connected to the physical device from the electrical side. These networks (circuits) impact the value of the generated force F_t, and should be considered part of the system related to the transducer. In practice, the mobile mass should be mechanically attached/suspended to the harvester frame related to the Ox reference frame. These suspensions are usually implemented as springs (linear or nonlinear), which can be of zero stiffness (in the case of a freely moving mass [9]). In addition, parasitic losses are usually present (friction, air damping). These losses are modelled by a damper, which can also be linear or nonlinear.

The equivalent model of the harvester in the mechanical domain is given in Fig. 3.12b: in this diagram, all forces acting on the mass in the new reference frame $0x$ are shown explicitly. Note that this figure defines the conventional positive directions of the forces.

The mathematical model of such a system is given by the equation given by the second Newtonian law:

$$- ma_{ext} + F_{spring}(x) + F_{damping}(x, \dot{x}) + F_t(x, \dot{x}, \theta) = m\ddot{x} \tag{3.81}$$

In this equation, $F_{spring}(x)$ is an elastic force related with the suspensions of the mobile mass. It depends only on the mobile mass position [34]. In the case of a linear spring, it is equal to $-kx$, but the expression is more complex in practice (nonlinear springs as in Sect. 3.2.1, stoppers [31], etc.). The force $F_{damping}(x, \dot{x})$ represents the damping force; it depends on the mobile mass velocity and in general, on x. In the case of linear damping it is equal to $-b\dot{x}$ [34], but in practice, the model for damping is more complex (e.g. nonlinear squeeze film damping, cf. [31]). Because of the difficulties in modeling the high amplitude squeeze film damping, in most studies this force is considered linear.

The transducer force depends, generally, on the position, the velocity and on some set of parameters θ related with the construction of the transducer itself.

In this way, a KEH can be considered as a resonator characterised by its three components (inertial, elastic and losses), submitted to an external force and associated with a transducer generating the force F_t.

The harvested power is equal to the rate at which the energy is extracted from the mechanical domain by the force F_t

$$P = -\frac{1}{\Delta t}\int_t^{t+\Delta t} F_t\dot{x}\mathrm{d}t = -\overline{F_t\dot{x}}. \tag{3.82}$$

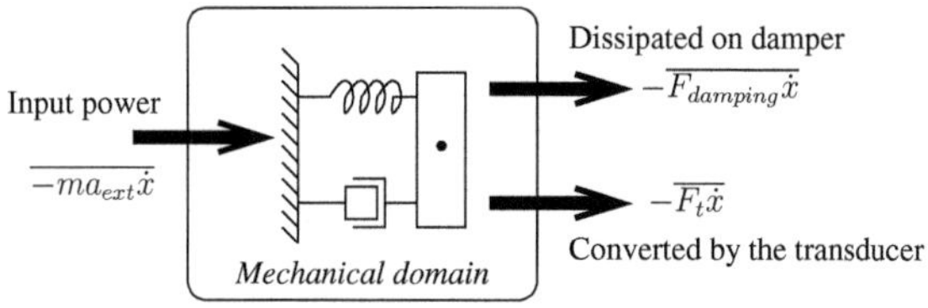

Fig. 3.13 Diagram illustrating the energy flow in a KEH

where Δt is the time interval at which the system is considered. The minus sign means that we are interested in the negative work done by this force.

The energy transformations in a KEH are characterised by the power, which, in general, means the rate of energy evolution in a (sub)system. We highlight three zones of the system, between which there is meaningful energy exchange: the external vibrating frame (mechanical domain), the mechanical domain of the kinetic energy harvester (energy of mass and spring) and the non-mechanical domain into which the damper and the transducer convert the energy of the harvester (cf. Fig. 3.13). It should be noted that at steady-state operation, the time average of the input energy flow is equal to the time average of the output flow. Indeed, non-equality of these flows would mean one of two impossible (absurde) behaviours: (i) the mechanical energy of the resonator (mass and spring) increases without limits, which is in contradiction with a limited geometry of the system, and with mandatory limitations on the mobile mass motion magnitude, (ii) the mechanical energy of the resonator decreases to zero, so the mass does not move and there is no energy flow, but this is in contradiction with the hypothesis of steady-state operation.

3.3.2 Design and Optimization of KEH

The goal the design of a KEH is generally formulated with relation to the application. Typically, the application specifies the desirable size of the system, the parameters of external vibrations and the desired harvested power (which should generally be as high as possible). Moreover, for different reasons related to the context of a particular project, the technology is fixed as well, hence constraining the geometry of the key resonator. Given these constraints, usually present in practice, two questions are of importance

- What is the theoretical limit of the maximum power the system can convert.
- Given a resonator, what should be the transducer dynamics maximising the converted power.

Presently, exhaustive answers to these questions exist only for the case of resonators excited by vibrations having sinusoidal acceleration a_{ext}:

$$a_{ext} = A_{ext}\cos(\omega t). \tag{3.83}$$

We propose here an analysis of a kinetic energy harvester submitted to sinusoidal vibrations. In this section, we calculate the maximum power which can be extracted and converted from external vibrations given the mobile mass, the system size and the parameters of external vibrations. Then, we show what should be the force generated by the transducer in order to maximise the power of the energy conversion.

3.3.3 Study of a General Case

The kinetic energy harvester is described by the second Newtonian law (3.81). To calculate the instantaneous power injected into the mechanical domain by each force, we multiply both sides of the equation by $\dot{x}$ and integrate them over t

$$-ma_{ext}\dot{x} + F_{sprig}(x)\dot{x} + F_{damping}(x,\dot{x})\dot{x} + F_t(x,\dot{x},\theta)\dot{x} = m\ddot{x}\dot{x} \left| \frac{1}{\Delta t}\int_t^{t+\Delta t} \cdot \, \mathrm{d}t \right. \quad (3.84)$$

It is reasonable to assume that if the input force is sinusoidal, the displacement of the mobile mass $x(t)$ is periodic, with the period of the external vibrations $T = 2\pi/\omega$, so that $x(t) = x(t+T)$. This assumption excludes the cases of complex system behaviour, when the motion can be chaotic. In the case of a periodic motion of the mobile mass, the average power can be calculated over the period, i.e. taking $\Delta t = T/(2\pi\omega)$ in Eq. (3.84). At the same time, x is periodic and can be represented by a sum of its harmonics with frequencies 0, ω, 2ω, etc.

$$x(t) = \sum_{n=1}^{\infty} [a_n \cos(n\omega t) + c_n \sin(n\omega t)] \quad (3.85)$$

Eq. (3.84) represents the power balance of the system. The term related to $-ma_{ext}$ represents the source of mechanical energy and is positive. The terms related to F_{spring} and to $m\ddot{x}$ represent the energy flow on the elastic spring and on the mass, which is zero in the context of periodic motion. The terms related to $F_{damping}$ and F_t represent the energy extracted from the mechanical domain: the first term is the parasitic dissipation, the latter term represents the wanted energy transfer into the electrical domain. These two terms are negative if they extract energy from the mechanical domain.

Maximum Energy that can be Extracted from External Vibrations

We now calculate the maximum power that can be provided by the input vibrations, i.e. the maximum achievable value of the power term related with $-ma_{ext}$. We consider the expression which gives power injected into the mechanical domain by the external force

$$P_{ext} = \frac{1}{T}\int_t^{t+T} (-ma_{ext})\dot{x}\mathrm{d}t \tag{3.86}$$

a_{ext} is expressed by (3.83), and as a consequence only the ω harmonic of x contributes to P_{ext}. According to (3.85), the first harmonic of $\dot{x}(t)$ is

$$\dot{x}_1(t) = -a_1\omega\sin(\omega t) + c_1\omega\cos(\omega t). \tag{3.87}$$

Thus we obtain the power

$$P_{ext} = \frac{1}{T}mA_{ext}\int_t^{t+T} \cos(\omega t)[-a_1\omega\sin(\omega t) + c_1\omega\cos(\omega t)] = \frac{1}{2}mA_{ext}c_1\omega\mathrm{d}t \tag{3.88}$$

From this equation, it can be seen that the power given by the external force is unlimited if no constraints are applied on the amplitude of the first harmonic. However, practical (geometrical) considerations usually apply a limit on the maximum value of the function $x(t)$, that we call X_{lim}. It can be shown that the maximum amplitude c_1 of the sinusoidal component of the first harmonic of a periodic function $x(t)$ such as $|x(t)| \leq X_{lim}$ corresponds to the case when $x(t)$ is an odd rectangular function with amplitude X_{lim}

$$x(t) = \begin{cases} X_{lim}, & t\%T < T/2, \\ -X_{lim}, & t\%T \geq T/2, \end{cases}, \tag{3.89}$$

where % denotes the modulo operator. This is proven by considering the formula calculating c_1

$$c_1 = \frac{2}{T}\int_0^T x(t)\sin(\omega t)\mathrm{d}t \tag{3.90}$$

The interval of the integration T can be split into two halves at which the sinus function keeps its sign

$$c_1 = \frac{2}{T}\int_0^{T/2} x(t)\sin(\omega t)\mathrm{d}t - \frac{2}{T}\int_{T/2}^{T} x(t)|\sin(\omega t)|\mathrm{d}t \tag{3.91}$$

From the latter formula, it is clear that for $0 < t < T/2, x(t)$ should be positive and maximised, and for $T/2 < t < T$, $x(t)$ should be negative and minimised. Because of the limit on extreme values of $x(t)$, Eq. (3.89) is obtained. For such $x(t)$, c_1

is $4/\pi \cdot X_{lim}$. This trajectory of the mobile mass guarantees the extraction of the maximum power from the external force. This maximum power is given by

$$P_{ext_{max}} = \frac{2}{\pi} m A_{ext} X_{lim} \omega. \tag{3.92}$$

This expression states the absolute upper limit of power what can be extracted from external sinusoidal vibrations, given the four parameters of Eq. (3.92). In practice, this value of extracted power cannot be obtained. Indeed, the velocity corresponding to the optimal $x(t)$ is a train of Dirac functions

$$v(t) = X_{lim}(\delta(t + nT) - \delta(t + nT - T/2)), \quad n \in \mathbb{N} \tag{3.93}$$

In this formula, the Delta function has a dimension of one over the time (1/sec). Such a velocity corresponding to an infinite kinetic energy of the mobile mass during the transition between the levels $-X_{lim}$ and X_{lim} has no physical sense. Moreover, the formula (3.92) bounds the *extracted* energy, and only a part of it can be converted into electricity, because of unavoidable losses (cf. the next section).

Maximum Power Which can be Converted by the Transducer

As mentioned in Sect. 3.3.1, only a part of the energy extracted from the external vibrations is converted into electricity if the resonator is subject to mechanical losses. In this section, we calculate the maximum power which can be converted by the transducer (P_t), given the geometry of the resonator and the parameters of the external vibrations, under constraints of maximum amplitude of the mobile mass displacement X_{lim}. We start from the Eq. (3.84), and we consider the power extracted by the transducer

$$\begin{aligned} P_t &= -\frac{1}{T} \int_t^{t+T} F_t \dot{x} \ \mathrm{d}t = \\ &= -\frac{1}{T} \int_t^{t+T} \left[m a_{ext} - F_{spring}(x) - F_{damping}(x, \dot{x}) + m\ddot{x} \right] \dot{x} \mathrm{d}t \end{aligned} \tag{3.94}$$

The minus sign before the integral means that we express the power extracted from the mechanical domain.

It can be shown that the terms $m\ddot{x}$ and $F_{spring}(x)$ do not contribute to the power P_t, i.e.

$$\int_t^{t+T} F_{spring}(x)\dot{x} \ \mathrm{d}t = 0 \tag{3.95}$$

and

$$\int_t^{t+T} m\ddot{x}\dot{x} \ \mathrm{d}t = 0 \tag{3.96}$$

for any $x(t)$ with period T. Physically, it means that in steady-state mode, the average energy of the mass and of the spring remains constant.

In this way, only the terms ma_{ext} and $F_{damping}$ contribute with non-zero power. So, we can write

$$P_t = -\frac{1}{T}\int_t^{t+T} \left[ma_{ext}\dot{x} - F_{damping}(x,\dot{x})\dot{x}\right]\mathrm{d}t \tag{3.97}$$

We should now look for $x(t)$ maximising this expression, where $x(t)$ is defined by (3.85).

The integral (3.97) has two components including $\dot{x}$. In the first one, only the fundamental sinusoidal harmonic of $x(t)$ has a non-zero contribution to the value of the integral, since a_{ext} is cosinusoidal (cf. (3.83)). The first term is equal to

$$-\frac{1}{T}\int_t^{t+T} ma_{ext}\dot{x}\mathrm{d}t = -\frac{1}{2}mc_1A_{ext}\omega \tag{3.98}$$

The second term of the integral can be calculated if the function $F_{damping}(x,\dot{x})$ is known. In this study, we consider the most common case, where the damping is linear and is given by

$$F_{damping}(x,\dot{x}) = -b\dot{x}. \tag{3.99}$$

The second term of (3.97) becomes:

$$-\frac{1}{T}\int_t^{t+T} b\dot{x}^2\mathrm{d}t = -\frac{1}{2}b\sum_{n=1}^{\infty}\left[(a_nn\omega)^2 + (c_nn\omega)^2\right] \tag{3.100}$$

Therefore, for the power absorbed by the transducer, we have

$$P_t = -\frac{1}{2}mc_1A_{ext}\omega - \frac{1}{2}b\sum_{n=1}^{\infty}\left[(a_nn\omega)^2 + (c_nn\omega)^2\right],$$
$$\max_{t\in[0,T)}[x(t)] \le X_{lim} \tag{3.101}$$

We need to find a trajectory given by coefficients a_n and c_n to maximise (3.101). This optimization problem does not have an analytical solution, but can easily be solved with a numerical tool.

One particular case allows an exact solution in closed form. If there is no limitation on the maximum value of $x(t)$, the optimization problem is trivial: all coefficients except c_1 should be zero (the displacement is sinusoidal), and the problem is reduced to an optimisation of a function with one parameter c_1. It can be shown that in this case, the optimal c_1 is

$$c_1 = -0.5mA_{ext}/(b\omega), \tag{3.102}$$

the optimal mobile mass displacement is

$$x_{opt}(t) = -\frac{mA_{ext}}{2b\omega}\sin(\omega t), \tag{3.103}$$

and the maximum power is

$$P_{t\,max} = \frac{1}{8}\frac{(mA_{ext})^2}{b}. \tag{3.104}$$

The corresponding optimal transducer force can be found if the obtained optimal $x(t)$ is submitted into the Eq. (3.105). We have for $F_t(t)$

$$F_t(t) = ma_{ext} + m\ddot{x} - F_{sprig}(x) - F_{damping}(x, \dot{x}) \tag{3.105}$$

This equation provides the optimal evolution in time of the force of transducer. In general, this force should be nonsinusoidal, in order to balance the harmonics generated by the nonlinear spring. We consider a particular common case, when the elastic force is linear and is expressed as

$$F_{spring}(x) = -kx. \tag{3.106}$$

We obtain for optimal F_t

$$\begin{aligned} F_t(t) = ma_{ext} + m\ddot{x} + b\dot{x} + kx = \\ mA_{ext}\cos(\omega t) - mx_1\omega^2\sin(\omega t) + bx_1\omega\cos(\omega t) + kx_1\sin(\omega t) = \\ (mA_{ext} + b\omega x_1)\cos(\omega t) - x_1(m\omega^2 - k)\sin(\omega t) = \\ \frac{1}{2}mA_{ext}\cos(\omega t) + \frac{mA_{ext}}{2b}\left(m\omega - \frac{k}{\omega}\right)\sin(\omega t) \end{aligned} \tag{3.107}$$

3.3.4 Case of a Narrow Band Resonator

If the resonator is of high quality (narrow frequency band), and if the frequency of the external sinusoidal vibration is in the resonator passband, its motion can be considered sinusoidal, as higher harmonics of nonlinear nonsinusoidal forces are filtered out by the resonator frequency characteristic [19]. In this case, the general expression for the mobile mass displacement is

$$x(t) = x_1(t) = a_1 \cos(\omega t) + c_1 \sin(\omega t). \tag{3.108}$$

This case can be analysed rigorously, with simple mathematical tools. The analysis is presented in this section.

Maximum Power Which can be Extracted from the External Force

The expression P_{ext} defined in Eqs. (3.86) and (3.88) is maximised if $c_1 = X_{lim}$, $a_1 = 0$, i.e. when

$$x(t) = X_{lim} \sin(\omega t). \tag{3.109}$$

The maximum power that can be extracted is

$$P^1_{ext_{max}} = \frac{1}{2} m A_{ext} X_{lim} \omega. \tag{3.110}$$

Obviously, $P^1_{ext_{max}}$ is below the absolute limit given by (3.92).

Maximum Power Which can be Converted by the Transducer

In order to find the maximum power which can be extracted by a transducer, we consider again the expression (3.97) providing the power converted by the transducer in the context of a periodic motion of the mobile mass. We submit (3.108) into it, and we are looking for a_1 and c_1 maximising the converted power. Considering linear damping (Eq. 3.99) and submitting (3.108) into (3.97), we have

$$P_t = -\frac{1}{2} m A_{ext} \omega c_1 - \frac{1}{2} b \omega^2 (a_1^2 + c_1^2) \tag{3.111}$$

This expression is maximised when a_1 is zero, and when $c_1 = -mA_{ext}/(2\omega b)$ (the global maximum of a quadratic function with a negative quadratic coefficient). This is the same result as that obtained in Sect. 3.3.4 (Eq. 3.102). Now, we consider two cases.

(a) X_{lim} is less than the absolute value of Eq. (3.102). In this case, the limitation on the maximum displacement is not restrictive. The optimal system behaviour is described in Sect. 3.3.4.

(b) However, when X_{lim} is less than $mA_{ext}/(2\omega X)$, the valid interval for c_1 is $-X_{lim} \leq c_1 \leq 0$ (the optimal a_1 is still zero), and this interval corresponds to the right (increasing) branch of the parabolic function $P_t(c_1)$ (3.111). Hence, the optimal value of c_1 is $-X_{lim}$, and, for the optimal power, we get

$$P_t = \frac{1}{2} m A_{ext} \omega X_{lim} - \frac{1}{2} b (\omega X_{lim})^2. \tag{3.112}$$

We let the reader calculate the transducer force F_t guaranteeing this power of conversion.

This result can be summarised in one equation, showing the maximum power a transducer can extract from the mechanical domain

$$P_t = \begin{cases} \dfrac{1}{8} \dfrac{(mA_{ext})^2}{b}, & \dfrac{mA_{ext}}{2b\omega} < X_{lim}, \\ \dfrac{1}{2} \omega X_{lim} m A_{ext} - \dfrac{1}{2} b (\omega X_{lim})^2, & \dfrac{mA_{ext}}{2b\omega} \geq X_{lim}, \end{cases} \tag{3.113}$$

The corresponding optimal transducer force is given by

$$F_t(t) = \begin{cases} \dfrac{1}{2} m A_{ext} \cos(\omega t) + \dfrac{mA_{ext}}{2b} \left(m\omega - \dfrac{k}{\omega} \right) \sin(\omega t), & \dfrac{mA_{ext}}{2b\omega} < X_{lim}, \\ (mA_{ext} - b\omega X_{lim}) \cos(\omega t) + X_{lim} \omega \left(m\omega - \dfrac{k}{\omega} \right) \sin(\omega t), & \dfrac{mA_{ext}}{2b\omega} \geq X_{lim}, \end{cases} \tag{3.114}$$

3.3.5 Conclusion

The above analysis presents the ideal behaviour of the transducer in the context of kinetic energy harvesting and defines the upper limit of the performance of a harvester composed of a realistic resonator (with losses) and some ideal transducer. Unfortunately, physical constraints and limitations of realistic transducer devices make the transducer behaviour deviate from the idealistic scenario. Nevertheless, an awareness of the upper bound of performances of kinetic energy harvesters may help in design of the KEH systems, and may provide an objective measure of the efficiency of the designed and implemented KEH.

Appendix I

In this appendix, we calculate the fundamental harmonic of the force generated by a nonlinear spring submitted to sinusoidal deformation. Let the spring be characterised by the relationship between the force F_{spring} and the deformation x

$$F_{spring} = F(x), \tag{3.115}$$

and x is given by

$$x = X\sin(\omega t). \tag{3.116}$$

The complex amplitude of the fundamental harmonic of the force $\dot{F}_{spring}$ is given by

$$\dot{F}_{spring} = \frac{\omega}{\pi}\int_0^T f[X\sin(\omega t)]e^{-j\omega t}\mathrm{d}t \tag{3.117}$$

(Do not confuse $\dot{F}_{spring}$ used in this section with the notation for the derivative.)

We first calculate the real part of this integral

$$Re(\dot{F}_{spring}) = \frac{\omega}{\pi}\int_0^T f[X\sin(\omega t)]\cos(\omega t)\mathrm{d}t = \\ \frac{1}{\pi}\int_0^T f[X\sin(\omega t)]\mathrm{d}\sin(\omega t) = \frac{1}{\pi}\oint_y f[Xy]\mathrm{d}y \tag{3.118}$$

Since the force of a spring is potential, this integral on a closed path is zero, so that

$$Re(\dot{F}_{spring}) = 0. \tag{3.119}$$

Now we calculate the imaginary part

$$Im(\dot{F}_{spring}) = -\frac{\omega}{\pi}\int_0^T f[X\sin(\omega t)]\sin(\omega t)\mathrm{d}t = \\ \frac{1}{\pi}\int_0^T f[X\sin(\omega t)]\mathrm{d}\cos(\omega t) = \\ \frac{1}{\pi}\int_1^{-1} f[X\sqrt{1-y^2}]\mathrm{d}y + \frac{1}{\pi}\int_{-1}^{1} f[-X\sqrt{1-y^2}]\mathrm{d}y \tag{3.120}$$

For the complex amplitude of the spring force, we get

$$\dot{F}_{spring} = j\left[\frac{1}{\pi}\int_{1}^{-1} f[X\sqrt{1-y^2}]\mathrm{d}y + \frac{1}{\pi}\int_{-1}^{1} f[-X\sqrt{1-y^2}]\mathrm{d}y\right] \tag{3.121}$$

and as a consequence, the force in the time domain is given by

$$\dot{F}_{spring} = \left[\frac{1}{\pi}\int_{1}^{-1} f[X\sqrt{1-y^2}]\mathrm{d}y + \frac{1}{\pi}\int_{-1}^{1} f[-X\sqrt{1-y^2}]\mathrm{d}y\right]\sin(\omega t). \tag{3.122}$$

Appendix II

In this appendix, we calculate the fundamental harmonic of the force generated by a nonlinear damper submitted to sinusoidal deformation. Let the damper characterised by the relationship between the force F_{damper} and the deformation x

$$F_{damper} = F(\dot{x}), \tag{3.123}$$

and x is given by

$$x = X\sin(\omega t). \tag{3.124}$$

The complex amplitude of the fundamental harmonic of the force F_{damper} is given by

$$\dot{F}_{damper} = \frac{\omega}{\pi}\int_{0}^{T} f[X\omega\cos(\omega t)]e^{-j\omega t}\mathrm{d}t \tag{3.125}$$

We first calculate the real part of this integral

$$\begin{aligned} Re(\dot{F}_{damper}) &= \frac{\omega}{\pi}\int_{0}^{T} f[X\omega\cos(\omega t)]\cos(\omega t)\mathrm{d}t = \\ &\frac{1}{\pi}\int_{0}^{T} f[X\omega\cos(\omega t)]\mathrm{d}\sin(\omega t) = \\ &\frac{1}{\pi}\int_{T/4}^{5T/4} f[X\omega\cos(\omega t)]\mathrm{d}\sin(\omega t) = \\ \frac{1}{\pi}\int_{1}^{-1} f[X\omega\sqrt{1-y^2}]\mathrm{d}y &+ \frac{1}{\pi}\int_{-1}^{1} f[-X\omega\sqrt{1-y^2}]\mathrm{d}y \end{aligned} \tag{3.126}$$

Now we calculate the imaginary part

$$
\begin{aligned}
Im(\dot{F}_{damper}) = -\frac{\omega}{\pi}\int_0^T f[X\omega\cos(\omega t)]\sin(\omega t)\mathrm{d}t = \\
\frac{1}{\pi}\int_0^T f[X\omega\cos(\omega t)]\mathrm{d}\cos(\omega t) = \\
\frac{1}{\pi}\oint_y f[X\omega y]\mathrm{d}y
\end{aligned}
\tag{3.127}
$$

According to the Green-Riemann theorem [29], this integral is zero.

The expression of the reaction force of a damper is

$$
\dot{F}_{damper} = \left[\frac{1}{\pi}\int_1^{-1} f[X\omega\sqrt{1-y^2}]\mathrm{d}y + \frac{1}{\pi}\int_{-1}^{1} f[-X\omega\sqrt{1-y^2}]\mathrm{d}y\right]\cos(\omega t).
\tag{3.128}
$$

References

1. Adams, S. G., Bertsch, F. M., Shaw, K. A., & MacDonald, N. C. (1998). Independent tuning of linear and nonlinear stiffness coefficients [actuators]. *Journal of Microelectromechanical Systems*, *7*(2), 172–180.
2. Amri, M., Basset, P., Cottone, F., Galayko, D., Najar, F., & Bourouina, T. (2011). Novel nonlinear spring design for wideband vibration energy harvesters. In *Proceedings of the power MEMS* (pp. 15–18).
3. Andronov, A. A. (1987). *Theory of oscillators* (Vol. 4). Courier Dover Publications.
4. Beeby, S., Torah, R., Tudor, M., Glynne-Jones, P., O'Donnell, T., Saha, C., et al. (2007). A micro electromagnetic generator for vibration energy harvesting. *Journal of Micromechanics and Microengineering*, *17*, 1257.
5. Beeby, S. P., Tudor, M. J., & White, N. M. (2009). Energy harvesting vibration sources for microsystems applications. *Measurement Science and Technology*, *17*, R175–R195.
6. Benzi, R., Sutera, A., & Vulpiani, A. (1981). The mechanism of stochastic resonance. *Journal of Physics A: Mathematical and General*, *14*(11), L453.
7. DeMartini, B. E., Rhoads, J. F., Turner, K. L., Shaw, S. W., & Moehlis, J. (2007). Linear and nonlinear tuning of parametrically excited mems oscillators. *Journal of Microelectromechanical Systems*, *16*(2), 310–318.
8. Duffing, G. (1918). *Erzwungene Schwingungen bei veränderlicher Eigenfrequenz und ihre technische Bedeutung*. 41–42. R, Vieweg and Sohn.
9. Galchev, T., Aktakka, E. E., & Najafi, K. (2012). A piezoelectric parametric frequency increased generator for harvesting low-frequency vibrations. *Journal of Microelectromechanical Systems*, *21*(6), 1311–1320.
10. Gammaitoni, L., Neri, I., & Vocca, H. (2009). Nonlinear oscillators for vibration energy harvesting. *Applied Physics Letters*, *94*, 164,102.
11. Guckenheimer, J., & Holmes, P. (1983). *Nonlinear oscillations, dynamical systems, and bifurcations of vector fields* (Vol. 42). New York: Springer.
12. Hairer, E., Nørsett, S. P., & Wanner, G. (1991). *Solving ordinary differential equations* (Vol. 2). Springer.
13. Holmes, P., & Rand, D. (1976). The bifurcations of Duffing's equation: An application of catastrophe theory. *Journal of Sound and Vibration*, *44*(2), 237–253.
14. Kaajakari, V. (2009). *Practical MEMS: Design of microsystems, accelerometers, gyroscopes, RF MEMS, optical MEMS, and microfluidic systems*. Las Vegas, NV: Small Gear Publishing.

15. Kuznetsov, A. P., Kuznetsov, S. P., & Ryskin, N. (2002). *Nonlinear oscillations*. Moscow: Fizmatlit.
16. Li, H., Preidikman, S., Balachandran, B., & Mote, C, Jr. (2006). Nonlinear free and forced oscillations of piezoelectric microresonators. *Journal of Micromechanics and Microengineering*, *16*(2), 356.
17. Meninger, S., Mur-Miranda, J., Amirtharajah, R., Chandrakasan, A., & Lang, J. (2001). Vibration-to-electric energy conversion. *IEEE Transactions on Very Large Scale Integration (VLSI) Systems*, *9*(1), 64–76.
18. Mitcheson, P., Yeatman, E., Rao, G., Holmes, A., & Green, T. (2008). Energy harvesting from human and machine motion for wireless electronic devices. *Proceedings of the IEEE*, *96*(9), 1457–1486.
19. Nayfeh, A. (1993). *Introduction to perturbation techniques*. Wiley.
20. Nayfeh, A. H., & Balachandran, B. (2008). *Applied nonlinear dynamics* (Vol. 24). Wiley-VCH.
21. Nayfeh, A. H., & Mook, D. T. (2008). *Nonlinear oscillations*. Wiley.
22. Nayfeh, A. H., Younis, M. I., & Abdel-Rahman, E. M. (2005). Reduced-order models for mems applications. *Nonlinear dynamics*, *41*(1–3), 211–236.
23. Nguyen, D., Halvorsen, E., Jensen, G., & Vogl, A. (2010). Fabrication and characterization of a wideband mems energy harvester utilizing nonlinear springs. *Journal of Micromechanics and Microengineering*, *20*(12), 125,009.
24. Nguyen, S. D., & Halvorsen, E. (2011). Nonlinear springs for bandwidth-tolerant vibration energy harvesting. *Journal of Microelectromechanical Systems*, *20*, 1225–1227.
25. Nicolis, C. (1981). Solar variability and stochastic effects on climate. In *Physics of Solar Variations* (pp. 473–478). Springer.
26. Nicolis, C. (1982). Stochastic aspects of climatic transitions response to a periodic forcing. *Tellus*, *34*(1), 1–9.
27. Pelesko, J. A., & Bernstein, D. H. (2002). *Modeling MEMS and NEMS*. CRC Press.
28. Rabinovich, M. I. (1989). *Oscillations and waves: In linear and nonlinear systems* (Vol. 50). Taylor and Francis.
29. Riley, K., Hobson, P., & Bence, S. (2006). *Mathematical methods for physics and engineering: a comprehensive guide*. Cambridge University Press. http://books.google.com.ua/books?id=Mq1nlEKhNcsC.
30. Roundy, S., Wright, P., & Pister, K. (2002) Micro-electrostatic vibration-to-electricity converters. In *Proceedings of 2002 ASME international mechanical engineering congress*.
31. Senturia, S. D. (2001). *Microsystem design* (Vol. 3). Boston: Kluwer Academic Publishers.
32. Tang, K. T. (2007). *Mathematical methods for engineers and scientists*. Springer.
33. Tang, L., Yang, Y., & Soh, C. K. (2010). Toward broadband vibration-based energy harvesting. *Journal of Intelligent Material Systems and Structures*, *21*(18), 1867–1897.
34. Taylor, J. R. (2005). *Classical mechanics*. University Science Books.
35. Trubetskov, D. I., & Rozhnev, A. G. (2001). *Linear oscillations and waves*. Moscow: Fizmatlit.
36. Williams, C., & Yates, R. (1996). Analysis of a micro-electric generator for microsystems. *Sensors and Actuators A*, *52*, 8–11.
37. Zhu, D., Tudor, M. J., & Beeby, S. P. (2010). Strategies for increasing the operating frequency range of vibration energy harvesters: A review. *Measurements and Science Technology*, *21*, 022,001.

Chapter 4
Transducers for Energy Harvesting

E. Blokhina, A. El Aroudi and D. Galayko

4.1 Capacitive Transducers

This section presents the basic information describing capacitive transducers used as converters of mechanical energy into electricity [6, 16].

The widespread use of capacitive transducers has become possible thanks to the miniaturization of electronic systems. Indeed, capacitive transducers are inefficient at microscale, for reasons which will be explained later in this section. Capacitive transducers are mainly implemented with MEMS silicon technologies which are compatible with the requirements of batch fabrication.

Capacitive transducers are used as sensors/actuators for the transfer of information between the mechanical and electrical domains. For information processing, the functions describing relations between mechanical and electrical quantities should be linear. For that reason, the preferable mode of operation of the transducer/actuator is a small-signal mode, where the magnitude of dynamic quantities is small enough to negate the nonlinear distortions.

The energy conversion, however, sets very different constraints. Not only is the linearity of the conversion not important, but in many cases nonlinear behavior of electrical and mechanical devices is unavoidable or even desirable. The energy conversion devices operate at *large amplitude* mode, and linearized small-signal mode is not adequate for modeling of the behavior.

E. Blokhina
University College Dublin, Dublin, Ireland
e-mail: elena.blokhina@ucd.ie

A. El Aroudi
University Rovira i Virgili, Tarragona, Spain
e-mail: abdelali.elaroudi@urv.cat

D. Galayko (✉)
UMPC — Sorbonne Universities, Paris, France
e-mail: dimitri.galayko@lip6.fr

E. Blokhina et al. (eds.), *Nonlinearity in Energy Harvesting Systems*,
DOI 10.1007/978-3-319-20355-3_4

As will be shown, a capacitive transducer is an intrinsically nonlinear device. Moreover, in the mode of the energy conversion, the transducer is associated with conditioning electronics having a time-variant operation. Nonlinear models are required in the study and design of energy conversion systems with use of capacitive transducers.

4.1.1 Presentation of a Capacitive Transducer

A capacitive transducer is a physical capacitor whose geometry can change in time and so, to vary the electrical capacitance. Although a capacitor can be of any geometrical shape (spheric, cylindric,...), in practice, the most common shape is a parallel plate capacitor, whose geometry is given in Fig. 4.1. Such a capacitor is constituted from a pair of parallel conductive planes (electrodes) spaced by some distance, called "gap." A dielectric material can be present between the planes. The capacitance C_t of such a transducer is

$$C_t = \varepsilon_0 \varepsilon_r \frac{S}{d}, \tag{4.1}$$

where d is the distance between the planes (gap), S is the overlapping area of the planes, ε_0 is the permittivity of the vacuum (a fundamental constant, equal to $8.85 \cdot 10^{-12}\,\mathrm{Fm}^{-1}$), and ε_r is the dielectric constant of the material between the electrodes.

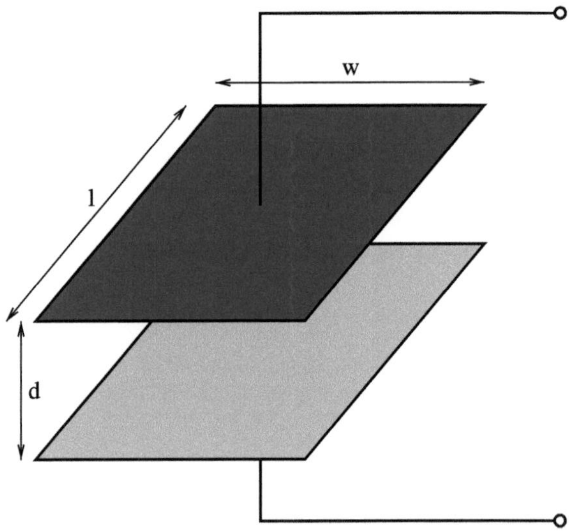

Fig. 4.1 Geometry of a parallel plate capacitor

The capacitance of a parallel plate capacitor is a function of three parameters, and a variation of each of them produces a variation of the capacitance

$$C_t(t) = \varepsilon_0\varepsilon_r(t)\frac{S(t)}{d(t)}. \tag{4.2}$$

Most existing variable capacitors operate in air or in vacuum, so that $\varepsilon_r \approx 1$. However, there are "exotic" cases where the variation of the capacitance is produced by a motion of the dielectric material separating the electrodes(cf. Fig. 4.2), in particular, in fluidic devices.

A variable capacitor is usually obtained when one electrode of the capacitor moves with regard to the other. To simplify the analysis, it is usually considered that one electrode of the capacitor is fixed, and the other moves. This is the most common configuration in energy harvesters (and will generally be assumed in this book), although there are many other applications of capacitive transducers where both electrodes are mobile [7]. In principle, the motion can be in any direction, but in the majority of devices there are only two possible and exclusive kinds of motion: (i) electrodes move in their plane, or (ii) electrodes move along the axis normal to their planes. The choice of the particular motion mode is obtained by implementation of a particular geometry of capacitor, so that all undesirable directions of motion are blocked. We now consider two cases of variable capacitor geometry.

(1) *The parallel motion of electrodes*. Such a capacitor is called an "area overlap capacitor" (Fig. 4.3). In this case the distance between the electrodes is kept constant, and the capacitance varies according to

$$C_t(t) = \varepsilon_0\frac{S(t)}{d}. \tag{4.3}$$

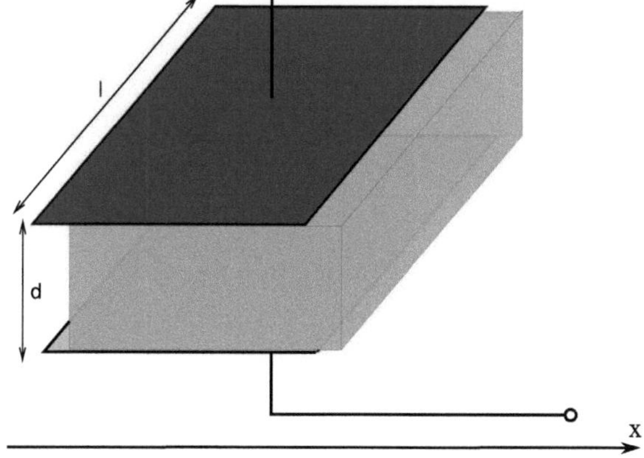

Fig. 4.2 Principle of a capacitive transducer with movable dielectric

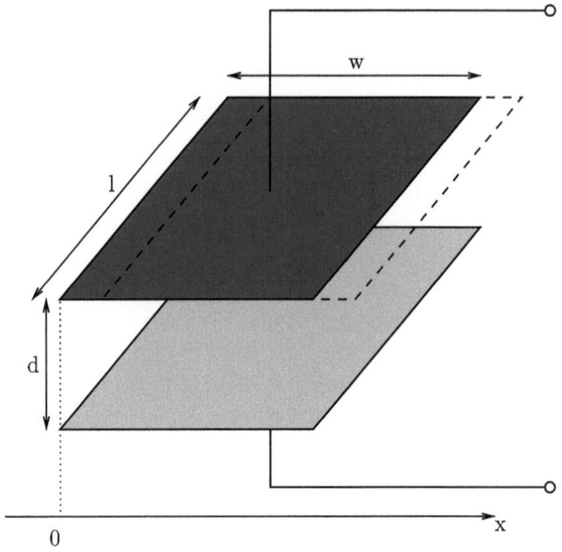

Fig. 4.3 Geometry of a transducer with parallel motion of electrodes

The variation of the overlapping area can be related to the relative displacement of the electrodes by a function $S(x)$, where $S(\cdot)$ depends on the geometry of the transducer. If the transducer electrodes have a rectangular shape (the most common in energy harvesters), and the mobile electrode moves in parallel with one of its sides, the function S is given by

$$S(x) = l(x_0 \pm x), \tag{4.4}$$

where l is the length of the side perpendicular to the motion, x_0 is the length of the overlapping rectangular area at rest. The parameter x_0 and the sign of x depend on the choice of the reference frame. For the structure given in Fig. 4.3, $x_0 = w$, and $S(x)$ is expressed as

$$S(x) = l(w - |x|). \tag{4.5}$$

In general, the relationship between the capacitance of a parallel plate transducer and the position of the mobile electrode (x) can be expressed as

$$C_t(x) = \varepsilon_0 \frac{l(x_0 \pm x)}{d}. \tag{4.6}$$

It should be noted that the function (4.4) may become zero. It is very important to remember, that the model of a parallel plate capacitor is only valid when the dimensions of the overlapping area are much greater than the gap. If the overlapping area goes to zero, there is residual capacitance, which is not accounted for anymore by (4.1), but which is non zero. A typical plot of capacitance for a rectangular area overlap transducer versus position of the mobile electrode is given in Fig. 4.4. As can be seen, the relation between C_t and x is linear only when the overlapping area is large. Even if no overlap exists and the plates are separated far each from other (formally, it corresponds to a negative overlapping area), the capacitance is still not zero. Ignoring this point may lead to completely wrong results in modeling and simulation.

(2) *Perpendicular motion of the electrodes.* A transducer whose mobile electrode moves following the direction normal to the plane is called a "gap closing variable capacitor" (Fig. 4.5). Its capacitance changes according to:

$$C_t(t) = \varepsilon_0 \frac{S}{d(t)} \tag{4.7}$$

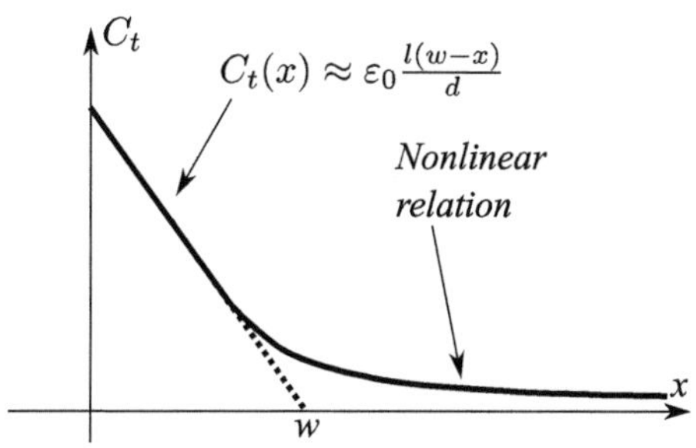

Fig. 4.4 The relationship between the capacitance of an area overlap transducer and the position of the movable electrode (Fig. 4.3). For the case when the overlapping area is large, $C_t(x)$ is linear, otherwise, when the overlapping area goes to zero and becomes negative, the characteristic is nonlinear

Fig. 4.5 Geometry of a capacitive transducer with gap-closing geometry

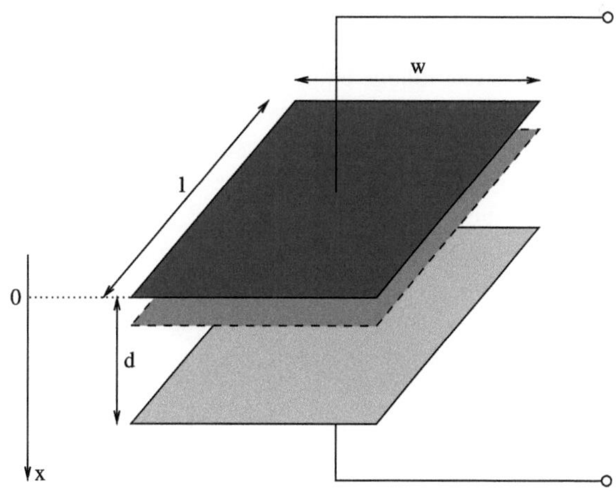

The transducer gap is linearly related to x

$$d(t) = d_0 \pm x(t). \tag{4.8}$$

Here d_0 is the initial gap of transducer at rest ($x = 0$). d_0 and the sign before x depend on the choice of the reference frame.

The capacitance is expressed over x as

$$C_t(x) = \varepsilon_0 \frac{S}{d_0 \pm x}. \tag{4.9}$$

It should be noted that $C_t(x)$ tends toward infinity when the denominator becomes zero. In practice, this corresponds to a strong increase in the transducer capacitance when the electrodes become separated by very small distance. In this case the attracting force approaches infinity as well, resulting in a strong risk of instability. For this reason stoppers are added in practical devices, to prevent the reduction of gap between the electrodes below some predefined value.

An important conclusion of this subsection is that the capacitance is a geometrical parameter, and the function $C_t(x)$ depends only on the geometry of the transducer.

4.1.2 Electrical Operation of a Variable Capacitor

From an electrical point of view, a capacitor is a system of two electrodes separated by a dielectric or by a vacuum. A capacitor behaves as an electrical element when its two electrodes have different electrical charges. If electrodes 1 and 2 have charges Q_1 and Q_2, the voltage between the electrodes is given by

$$V_{12} = \phi_1 - \phi_2 = \frac{Q_1 - Q_2}{2C}. \tag{4.10}$$

Here V_{12} is the voltage equal to a potential difference between the electrodes ($\phi_1 - \phi_2$). Note that the electrode having the highest charge has the highest potential.

In electronics, a capacitor is usually considered as an electrically neutral device, so that its electrodes have the same absolute charge, but of opposite sign. This charge is called "the charge of the capacitor." Its sign is defined by the sign of the charge in one of the electrodes chosen arbitrarily, called "positive electrode" and labeled by character "+" (cf. Fig. 4.6). This figure demonstrates also the definition of the conventional positive voltage of the capacitor (the arrow points toward the positive electrode). The equation describing the capacitor is written as

$$Q = CV \tag{4.11}$$

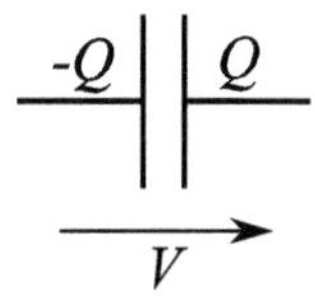

Fig. 4.6 Definition of the charge of the capacitor

The neutrality of a capacitor is a very important hypothesis for electrical circuit analysis. In particular, it allows the use of Kirchhoff laws for the circuit analysis (the current law). In practice, the neutrality of capacitors is ensured by existing DC (direct current) paths to the ground, thanks to different electrical devices or thanks to leakages.

It can be shown that a capacitor stores energy. This is the energy of the electrical field existing between the electrodes of the capacitance. Since this field is created by the charges on the capacitor, the stored energy W depends directly on the quantity of charges stored by the capacitance

$$W = \frac{Q^2}{2C} \tag{4.12}$$

This is a potential energy, and it is always positive.

4.1.3 Forces in a Capacitive Transducer

The existence of mechanical forces created by the electrical field of a capacitor allows the use of a capacitor as a potential electromechanical transducer. These forces are applied to all parts of a capacitor.[1] To calculate the mechanical force applied to some part of the capacitor along an axis x, the following mental experiment should be made:

- The considered part of the system should be made freely movable along the axe x,
- The considered part should be moved along this axis by infinitesimal distance $\mathrm{d}x$,
- the capacitance variation $\mathrm{d}C$ should be measured.

The force along the axe x is then calculated as

$$F_x = \frac{1}{2} V^2 \frac{\mathrm{d}C}{\mathrm{d}x}, \tag{4.13}$$

where V is the voltage of the capacitor.

[1] A force is a notion from mechanics, but sometimes in the literature the forces created by electrical phenomena are called "electrical forces." Their action on mechanical system are described by usual laws of mechanics.

As a consequence, an electrostatic force is only applied to the parts whose geometric position impacts the value of the capacitance: the electrodes and the dielectric between the plates. The force is oriented in the direction of *increasing* capacitance.

For the geometries of transducers considered in the previous subsections, the forces F_t are given by the following expressions. For an area overlap capacitor whose capacitance is given by (4.6)

$$F_t = \pm\frac{1}{2}V_t^2\varepsilon_0\frac{l}{d}. \tag{4.14}$$

For a gap-closing transducer whose capacitance is given by (4.9)

$$F_t = \mp\frac{1}{2}V_t^2\varepsilon_0\frac{S}{(d_0 \pm x)^2}. \tag{4.15}$$

Note that in the case where the transducer gap goes to zero, the force becomes infinite.

These expressions explain why a capacitive transducer is only efficient at the microscale. Indeed, the transducer force does not scale with the device dimensions (since both S and $(d_0 - x)^2$ scale quadratically), whereas mechanical forces are at least proportional to the dimensions for the springs. For the masses, the inertia force is proportional to the cube of linear dimensions. It means that electrostatic forces are too weak to be useful at the macroscale.

These expressions highlight the dependence of the transducer force on the voltage. The ability to modulate the transducer force by the applied voltage is a powerful tool for electrical synthesis of mechanical behavior. However, the transducer force is unilateral, i.e., cannot change sign, because of the square dependence of the voltage. This creates some problems both for sensor/actuator implementation [16] and for the energy conversion. One possible solution is the use of a differential capacitive transducer. Differential structures are often used for sensing/actuating applications, but they are too complex to be widespread in energy harvesting devices.

4.1.4 Energy Conversion with a Capacitive Transducer

The work of the force generated by the capacitive transducer represents the energy transferred between the mechanical and the electrical domains.

Suppose that during a time interval $[t_1, t_2]$ the capacitance of transducer C_t changes monotonically. The work produced by the force of transducer during a time interval $[t_1, t_2]$ is given by

$$A = \int_{t_1}^{t_2} F_t v dt = \int_{t_1}^{t_2} \frac{1}{2}V_t^2\frac{\mathrm{d}C_t}{\mathrm{d}x}\frac{\mathrm{d}x}{\mathrm{d}t}dt = \int_{C_1}^{C_2} \frac{1}{2}V_t^2 \mathrm{d}C_t, \tag{4.16}$$

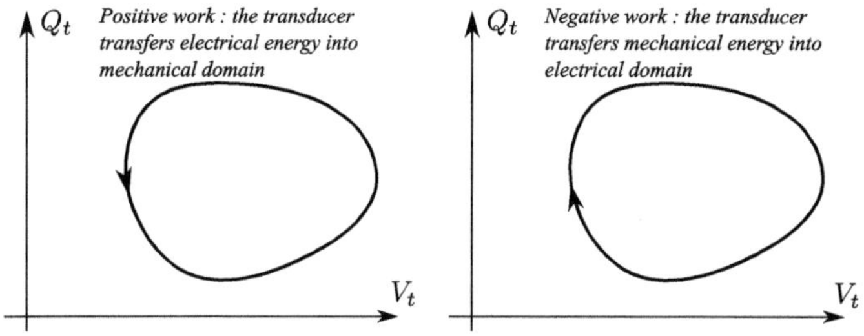

Fig. 4.7 Example of a cyclic operation of a capacitive transducer plotted in the charge-voltage plane

where $C_1 = C_t(t_1)$ and $C_2 = C_t(t_2)$, and v is the instantaneous velocity of the mobile electrode.

Since $V_t^2 \geq 0$, the work is positive when the capacitance increases, and is negative when the capacitance decreases. Mechanical energy is converted into electricity when this work is negative. By consequence, for the energy harvesting applications, the voltage on the transducer should be minimized when the capacitance increases, and maximized when the capacitance decreases.

The quantity C_t evolving in time can be seen as a path defined in the plane (V_t, Q_t), given by the relation $C_t = Q_t/V_t$, and the integral can be seen as a path integral. This allows an application of the formalism of vector calculus. Given $\mathrm{d}C_t = \mathrm{d}Q_t/V_t - (Q_t/V_t^2)\mathrm{d}V_t$, the above formula can be written as

$$A = \frac{1}{2}\int_{\Gamma} [V_t \mathrm{d}Q_t - Q_t \mathrm{d}V_t], \tag{4.17}$$

where Γ is the path which the transducer state follows in the plane QV, between the times t_1 and t_2, and $\int_{\Gamma}$ is the path integral.

If the curve Γ forms a cycle so that the $C_t(t_1) = C_t(t_2)$, the work A is written as

$$A = \frac{1}{2}\oint_{\Gamma} [V_t \mathrm{d}Q_t - Q_t \mathrm{d}V_t]. \tag{4.18}$$

According to Green's theorem, this formula expresses the area enclosed by the curve Γ, if the path is positively oriented in the plane (V_t, Q_t), i.e., the path is counter clockwise in the QV plane in Fig. 4.7. (cf. proof in the appendix I). When the path is negatively oriented (clockwise), Eq. (4.18) is negative, and the area enclosed by the path represents the energy converted into electricity during the cycle of C_t variation. The line representing the state of the transducer in the (V_t, Q_t) plane is called "QV diagram," and is a very elegant representation widely used for the analysis of the energy conversion achieved by capacitive transducers.

4.1.5 Optimization of the Operation of a Capacitive Transducer

The goal when designing energy harvesters is to maximize the area of the cycle QV corresponding to the converted energy. Let us consider a transducer whose capacitance variation is between C_{min} and C_{max}. In the QV plane, all possible states of this transducer are limited by loci $Q_t = C_{min} V_t$ and $Q_t = C_{max} V_t$ (Fig. 4.8). This defines an open segment with infinite area on the QV plane, if there is no limit on the voltage on the transducer. In practice, the voltage is always limited by the technology, let us say, by a value V_{max}. In this way, an ideal QV cycle is a triangle, formed by the lines $C = C_{max}$, $C = C_{min}$, $V_t = V_{max}$ (triangle OMN, Fig. 4.8). Its area is given by

$$\Delta W = \frac{1}{2} V_{max}^2 (C_{max} - C_{min}). \tag{4.19}$$

Such a QV cycle is called "constant voltage QV cycle" [11]. This term emphasizes the fact that the energy conversion is achieved when the voltage on the transducer is constant.

This formula provides an opportunity to estimate the maximal energy and power that can be generated by a capacitive transducer in a realistic context. We take the value for C_{min} and C_{max} from a state-of-the art MEMS capacitive transducer [1] (40 pF and 140 pF respectively) and 50 V for V_{max} (the limit for the 0.35 μm technology of Austrian Microsystem). We obtain 125 nJ per cycle. And with the frequency of the capacitance variation at 100 hertz, it corresponds to 12 μW of converted power. This figure should be seen as the order of magnitude of the maximal convertible power with capacitive transducers at microscale. This value can be changed if one assumes different hypotheses on the frequency, transducer parameters, and the maximal voltages.

As follows from the formula (4.16), the operation of a capacitive transducer is fully controlled by the voltage waveform $V_t(t)$ applied to its electrodes as the transducer capacitance C_t varies. Indeed, for each value of C_t, as far as V_t is defined, the electrical state of the transducer is uniquely defined through the formula $Q_t = CV_t$. In this

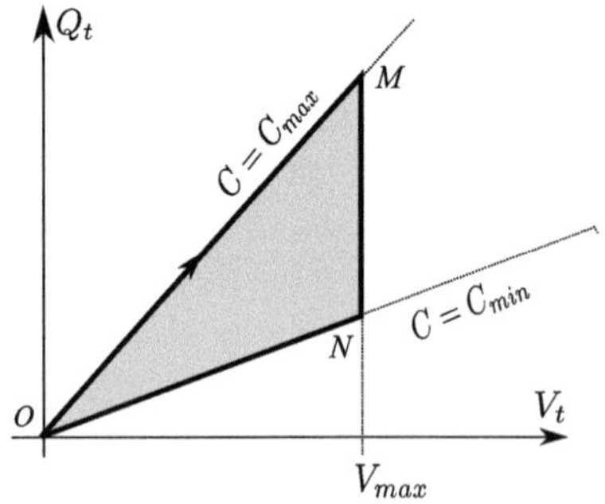

Fig. 4.8 Ideal QV cycle giving the following system parameters and limitations: the transducer capacitance varies from C_{min} to C_{max}, the maximal allowable voltage is V_{max}

way, by generating an appropriate voltage waveform on the transducer, it is possible to "synthesize" any desirable QV cycle. This is one of the roles of the *conditioning circuit*: definition of a dynamic biasing required for the energy conversion by the transducer. In that way, a capacitive vibration energy harvester is composed of a mechanical device, of a capacitive transducer, and of a conditioning circuit which sets dynamically the voltage V_t and the charge Q_t of the transducer, cf. Fig. 1.4

However, the QV cycles implemented practically are often different from the optimal cycle given in Fig. 4.8. The first reason is the difficulty to generate the optimal QV cycle at a reasonable energy cost. Second, the optimization of the converted energy is only one of two roles of the conditioning circuit. The other role is the optimal transmission of the converted energy toward the storage or load device. The compromise between the efficiency of these two functions results in suboptimal power conversion. Design of conditioning circuits will be discussed in Chap. 8.

4.1.6 Electromechanical Coupling

In the previous subsections, we assumed a defined variation of the transducer capacitance, between C_{min} and C_{max}. This hypothesis is non realistic, and that can be highlighted in the following mental experiment. Imagine a transducer attached to a given resonator submitted to some external vibrations. For some conditions, the capacitance of the transducer varies between C_{min} and C_{max}. According to (3.92), there is an upper bound of the power $P_{ext_{max}}$, that the system is able to convert from the mechanical to the electrical domain. Suppose that the triangular cycle of Fig. 4.8 is used. According to the Eq. (4.19), the energy converted by the transducer can have any large value, if the voltage V_{max} is not limited. There is an apparent contradiction, which is solved by the consideration of the electromechanical coupling. Indeed, assuming a given variation of the transducer capacitance is equivalent to assume a given motion of the mobile mass. However, the energy conversion is done through an application to the mass of the transducer's force, which is proportional to the square of the voltage. If the voltage is high, this force is large, and the motion of the mobile mass is likely to be affected by the process of the energy conversion. By consequence, the capacitance variation of the transducer is affected, therefore enforcing the fundamental limit given by (3.92).

This situation explains the difficulty in analysis and design of capacitive vibration energy harvesters. In order to analyze the energy conversion of the transducer, the capacitance variation (and hence, the motion of the mobile mass) should be known, but the mechanical dynamics of the system are strongly affected by the electrical operation of the transducer, especially when the energy conversion is to be maximized. More insight into the methods allowing analysis and design of kinetic energy harvesters with capacitive conversion will be presented in Chaps. 8 and 9.

4.2 Piezoelectric Transducers

4.2.1 Piezoelectric Mechanism

This section describes briefly energy harvesting using the piezoelectric mechanism. Piezoelectric energy harvesting is the process of acquiring the energy surrounding a vibrating system and converting it into usable electrical energy using a piezoelectric transducer. Piezoelectricity means electrical energy that results from mechanical pressure. It is generated by the accumulation of electric charge in certain solid materials when they are mechanically pressed. The piezoelectric effect was discovered in 1880–1981 by the French Curie brothers [2, 3] in naturally occurring crystals. However, it was only in 1950 s that started to be used for industrial applications. Since then man-made materials have been also demonstrated to exhibit piezoelectric effects which have been increasingly used and can be regarded as a mature technology since it is being exploited in various applications such as medical [17], transportation, [20] and cell phone battery chargers among others. In the transportation industry, for example, piezoelectric elements are used, among others, as knock sensors for detecting irregular combustion, for ultrasonic distance sensors for parking, fuel injection systems, active vibration reduction, and for energy harvesting [13]. Depending of what type of physical effect is used, piezoelectric can be designed to operate as sensors, actuators, or transducers [18]. The first case makes use of the direct piezoelectric effect and the mechanical energy is transformed into an electrical energy which is manifested as voltage signal between the surfaces of the piezoelectric material (Fig. 4.9).

Actuators takes advantage of the reversible process of piezoelectricity in the sense that when a voltage is applied to a piezoelectric material this will deform (Fig. 4.10).

The deformation in the reverse process is usually very slight and proportional to the voltage applied, and so the reverse effect finds application in precise movement detection on the microscale. The piezoelectric sensor converts mechanical energy into electrical energy, and the actuator converts electrical energy into mechanical energy. Finally, in transducers both effects are used within the same device. Therefore, a transducer may be used as an actuator.

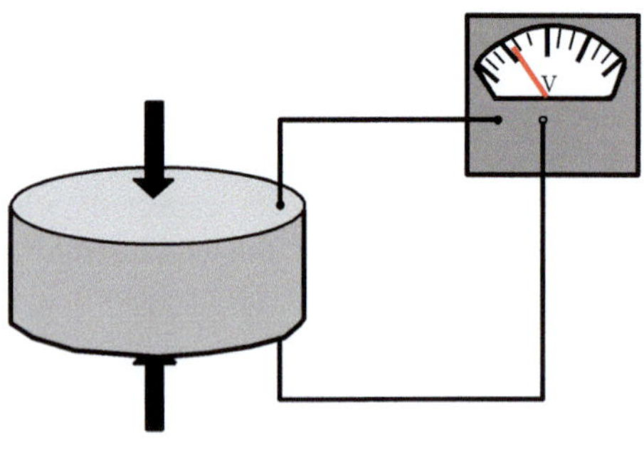

Fig. 4.9 Direct piezoelectric effect

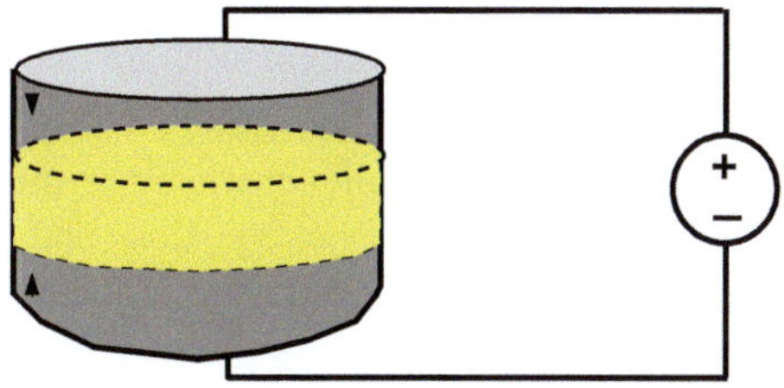

Fig. 4.10 Reverse piezoelectric effect

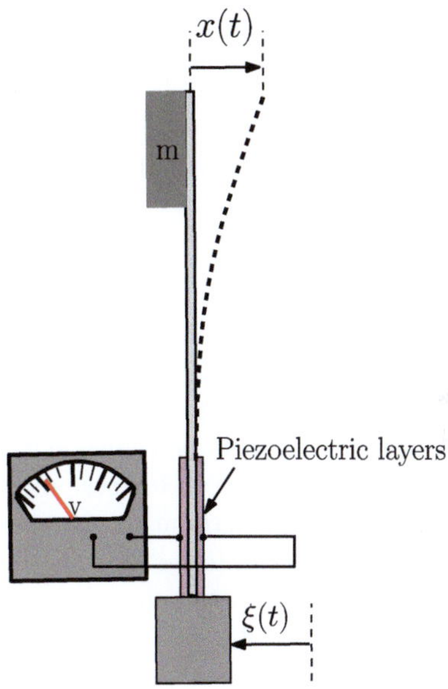

Fig. 4.11 A cantilever beam-based energy harvester

4.2.2 Energy Harvesting Using Piezoelectric Transducers

The use of piezoelectric effect in energy harvesting applications has been investigated since the beginning of the 1990s and since then it became an emerging technology. When a body, to which a piezoelectric material is attached, moves, the last one vibrates and produces electricity. The piezoelectric energy harvesting produces relatively higher voltage and power density levels than electromagnetic and electrostatic harvesters. One of the most effective methods of implementing an energy harvester system using piezoelectric materials is to use mechanical vibration to apply a strain energy to it. Energy harvesting from mechanical vibration usually uses ambient vibration around the harvester as an energy source, and then converts it into electrical energy, in order to power other devices ranging from digital electronics to wireless transmitters. The piezoelectric energy harvester is typically a cantilever beam structure with piezoelectric layers attached on the beam and a mass at its free end to amplify strains resulted from a given external force $\xi(t)$ (Fig. 4.11).

Beams and cantilevers can be considered as elastic systems and are usually modeled by distributed-parameter models represented by partial differential equations with specified boundary conditions. The motion of the beam can be described by the so called Euler–Bernoulli equation. For small oscillations, the response can be adequately described by linear equations and boundary conditions. Energy harvesters can be modeled by a reduced-order single degree of freedom spring–mass systems. However, in general, the governing equations, boundary conditions, or both are nonlinear.

The schematic diagram representation of the model is shown in Fig. 4.12. It consists of a typical spring–mass-damper system with a mass m, a total damping b, and an external force $\xi(t)$ as source of vibration. By developing the Euler–Bernoulli equation and performing a model reduction the following mass–spring dynamic model is obtained for the considered transducer,

$$m\ddot{x} = -\frac{\partial W_p}{\partial x} - b\dot{x} + \xi(t) \tag{4.20}$$

where x is the displacement, m represents the effective mass of the layer, and b stands for the damping factor. Different kinds of external vibrational source $\xi(t)$ can be considered. In some cases, these sources are governed by stochastic laws and their parameters can only be known in terms of statistical estimators such as mean values and variances [4, 10]. However, there are other applications where these sources can be considered deterministic signals such as in rotating machines and in vehicle and aircraft tires [20]. In the first case the idealized excitation sinusoidal term will only represent an approximation of the real case.

In order to take into account the dynamic coupling of the piezoelectric device, an extra differential equation describing the output voltage v_o (applied to the load) must be added to the system in (4.20). Considering resistive load R is connected at the output of the transducer and applying Kirchoff's voltage law for the electrical subsystem and the second Newton law for the mechanical part, the following coupled governing differential equation is obtained

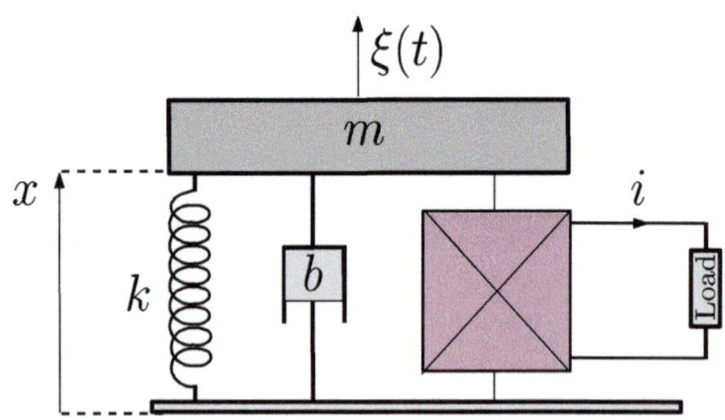

Fig. 4.12 Schematic diagram the spring-mass-damper model for a piezoelectric transducer

$$m\ddot{x} + \frac{\partial W_p}{\partial x} + b\dot{x} = \Gamma_v v_o + \xi(t) \tag{4.21}$$

$$\dot{v}_o + \frac{1}{RC} v_o = \alpha \dot{x} \tag{4.22}$$

where Γ_v is the piezoelectric coupling parameter in the mechanical part of the system, R is the resistive load, C is the equivalent capacitance of the piezoelectric layers, and α is the piezoelectric coupling parameter in the electrical part of the system. In the linear case, the potential energy is given by

$$W_p(x) = \frac{1}{2} k x^2 \tag{4.23}$$

Therefore, the model described in (4.21)–(4.22) becomes as follows:

$$m\ddot{x} + b\dot{x} + kx = \Gamma_v v_o + \xi \tag{4.24}$$

$$\dot{v}_o + \frac{1}{RC} v_o = \alpha \dot{x}. \tag{4.25}$$

A state-space model can be obtained from the previous mathematical representation for numerical simulation and a study of the harvested voltage v_o and the corresponding harvested power $P = v_{o,\text{rms}}/R$ can be analyzed in terms of the excitation level using simple linear theory.

For applications where the vibration frequency is known, linear piezoelectric harvesters can be efficient. However, like in other kind of energy harvesting technologies, the main drawback of a linear piezoelectric vibration energy harvester is a narrow bandwidth implying a tight tuning of the linear resonant harvester to match the vibrating source frequency when this is uncertain or time varying. The next section gives more details on the shortcoming of linear resonators operating as energy harvesters and presents some existing alternatives.

As mentioned previously, in most of the reported studies, the energy harvesters are designed as linear resonators by matching the resonant frequency of the harvester with that of the external excitation to extract maximum power. This maximum power extraction depends on the quality factor (Q factor) of the linear resonator. However, it is viable only when the excitation frequency is known a priori. Moreover, a maximum energy extraction with a high Q factor will paradoxically imply a limited and narrow frequency range within which energy can be harvested. The performance of these systems is therefore rapidly degraded if the excitation frequency is far from the resonant one and they are efficient only when an optimum design is implemented by tuning the resonance frequency to match with the ambient source vibrations frequency. However, in environments where no single dominant frequency exists, these performances can be lowered significantly as the excitation frequency moves away from the designed frequency [9]. Some solutions have been reported recently to remedy these problems. Among them, resonance tuning and frequency up-conversion techniques [14, 19]. These methods can overcome the above mentioned problems at the expense of making their implementation a challenging task. For instance, resonance tuning implies the change of the mass of the harvester while frequency-up conversion would imply the use of an array of resonators which would increase the size and cost of the harvester and making it not a suitable choice for small self-powered portable devices.

Traditionally, nonlinearities are to be avoided in device design. However, recently these nonlinearities have been shown to have potential to allow designers to take advantages of nonlinear behavior in certain applications [9], where the performance of energy harvesters is enhanced by inducing a bistable potential well through introducing suitable polynomial nonlinearities inducing a double well potential effect which makes the harvester efficient in a broad frequency range including low frequencies. Using this approach, rather than resonance frequency tuning, the nonlinearity of the system is exploited to improve the performances of the energy harvester within a wide frequency range outperforming, in this way, classical resonant energy harvesters [9, 21]. These techniques have been demonstrated to work both at the microscale [8] and nano-scale [10]. Polynomial nonlinearity is not the only way to enhance the performances of the harvesters at low frequencies. Other alternative inducing similar double well effect is in [22].

Double well potential in piezoelectric transducers can be induced by placing permanent magnets in the proximity of the proof mass (Fig. 4.13) forming the double well beam system studied in [12] for the first time.

It consists of a cantilever beam hung vertically with the free end attached by two magnets as shown in Fig. 4.13. The harvester is realized with a piezoelectric beam, in which magnetic effect induces double well potential. On the free end of the beam

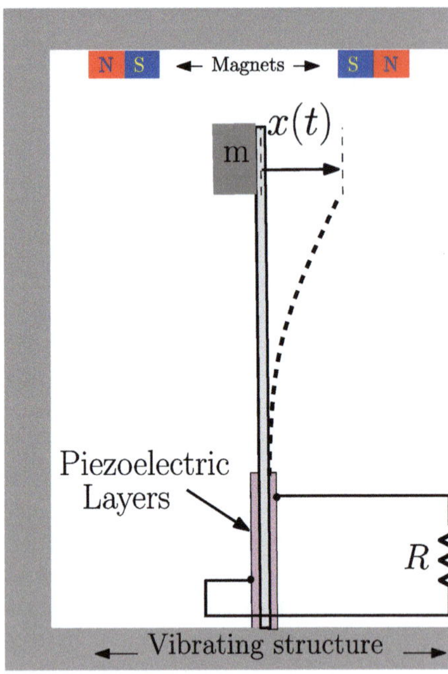

Fig. 4.13 An example of a piezoelectric cantilever beam energy harvester system arranged such that it has two different equilibrium point in the absence of excitation force by placing two magnets in the proximity of the free end

two magnets has been added. In the presence of vibration the structure oscillates making the piezoelectric beam to generate a voltage. The magnetic filed makes the potential energy to be nonharmonic and the equation of motion of the harvesting piezoelectric beam to be nonlinear The resulting potential energy can be expressed as follows:

$$W_p(x) = \frac{1}{2}k_1x^2 + \frac{1}{4}k_2x^4 \tag{4.26}$$

and the resulting equation of motion becomes as follows: [5]

$$m\ddot{x} + b\dot{x} + k_1x + k_2x^3 = \Gamma_v v_o + \xi, \tag{4.27}$$

$$\dot{v}_o + \frac{1}{RC}v_o = \alpha\dot{x}. \tag{4.28}$$

The voltage produced from the piezoelectric layers is an irregular AC signal which is then rectified by a diode bridge AC-DC rectifier. A filtering capacitor C is also placed in parallel with the load as shown in Fig. 4.14.

Let us consider the linear case for simplicity. The equivalent circuit representation of the piezoelectric harvester is depicted in Fig. 4.15. There, the mass m has been replaced by the inductance L_{eq}, the damping coefficient b has been replaced by a resistor with resistance R_{eq} and finally, the spring has been modeled by a capacitor

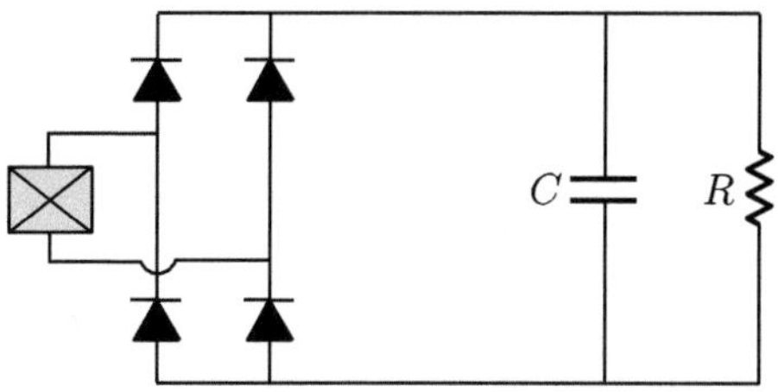

Fig. 4.14 Schematic diagram of the energy harvester based on a piezoelectric transducer

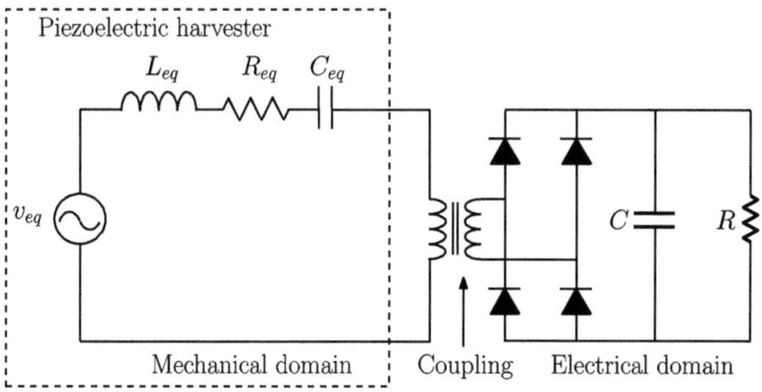

Fig. 4.15 Equivalent circuit diagram of the energy harvester based on a piezoelectric transducer

with capacitance C_{eq}. This model can be used for simulations, design, and analysis. The transformer represents the coupling effect. Introducing nonlinear effect can also be done by replacing the linear equivalent capacitance C_{eq} by a nonlinear one.

Appendix I

In this section, we present the demonstration of the fact that the area of a charge-voltage cycle performed by a variable capacitance is numerically equal to the electrical energy generated or absorbed by the capacitance, depending on the cycle direction.

The demonstration starts from the formula (4.18) expressing the work achieved by the capacitive transducer in the mechanical domain

$$A = \frac{1}{2} \oint_{\Gamma} [V \mathrm{d}Q - Q \mathrm{d}V] . \tag{4.29}$$

The Green theorem states that for a positively oriented, piecewise smooth, simple closed curve Γ in a right-handed plane (V, Q), for a region D bounded by Γ and for functions $L(V, Q)$, $M(V, Q)$ defined on an open region containing D and having continuous partial derivatives, the following equality is true [15]:

$$\oint_{\Gamma} (L \mathrm{d}V + M \mathrm{d}Q) = \iint_{D} \left[\frac{\partial M}{\partial V} - \frac{\partial L}{\partial Q} \right] \mathrm{d}V \mathrm{d}Q \tag{4.30}$$

Applying this theorem to Eq. (4.29), we get

$$A = \frac{1}{2} \oint_{\Gamma} [-Q \mathrm{d}V - V \mathrm{d}Q] = \frac{1}{2} \iint_{D} [1 + 1] \mathrm{d}V \mathrm{d}Q = \iint_{D} \mathrm{dVdQ}. \tag{4.31}$$

The last double integral expresses the area of the domain D enclosed by the curve. For a positively oriented (counterclockwise) path, A is positive: that means that the energy is transferred from the electrical into the mechanical domain. Conversely, for a negatively inverted (clockwise) path, the transducer's force work is negative, and the elctrical energy is converted from the mechanical energy.

References

1. Basset, P., Galayko, D., Cottone, F., Guillemet, R., Blokhina, E., Marty, F., & Bourouina, T. (2014). Electrostatic vibration energy harvester with combined effect of electrical nonlinearities and mechanical impact. *Journal of Micromechanics and Microengineering*, *24*(3), 035,001.

2. Curie, J., & Curie, P. (1880). Development, via compression, of electric polarization in hemihedral crystals with inclined faces. *Bulletin de la Societe de Minerologique de France*, *3*, 90–93.
3. Curie, J., & Curie, P. (1881). Contractions and expansions produced by voltages in hemihedral crystals with inclined faces. *Comptes Rendus*, *93*, 1137–1140.
4. El Aroudi, A., Lopez-Suarez, M., Alarcon, E., Rurali, R. & Abadal, G. (2013). Nonlinear dynamics in a graphene nanostructured device for energy harvesting. In *IEEE International Symposium on Circuits and Systems (ISCAS)*, pp. 2727–2730.
5. Erturk, A., & Inman, D. (2011). Broadband piezoelectric power generation on high-energy orbits of the bistable duffing oscillator with electromechanical coupling. *Journal of Sound and Vibration*, *330*(10), 2339–2353.
6. Fedder, G. K. (1994). Simulation of microelectromechanical systems. Ph.D. thesis, University of California at Berkeley.
7. Galayko, D., Kaiser, A., Legrand, B., Buchaillot, L., Collard, D., & Combi, C. (2005). Tunable passband t-filter with electrostatically-driven polysilicon micromechanical resonators. *Sensors and Actuators A: Physical*, *117*(1), 115–120.
8. Gammaitoni, L., Neri, I., & Vocca, H. (2009). Nonlinear oscillators for vibration energy harvesting. *Applied Physics Letters*, *94*, 164,102.
9. Gammaitoni, L., Travasso, F., Orfei, F., Vocca, H., & Neri, I. (2011). Vibration energy harvesting: Linear and nonlinear oscillator approaches. INTECH Open Access Publisher.
10. López-Suárez, M., Rurali, R., Gammaitoni, L., & Abadal, G. (2011). Nanostructured graphene for energy harvesting. *Physical Review B*, *84*(16), 161,401.
11. Meninger, S., Mur-Miranda, J., Amirtharajah, R., Chandrakasan, A., & Lang, J. (2001). Vibration-to-electric energy conversion. *IEEE Transactions on Very Large Scale Integration (VLSI) Systems*, *9*(1), 64–76.
12. Moon, F., & Holmes, P. J. (1979). A magnetoelastic strange attractor. *Journal of Sound and Vibration*, *65*(2), 275–296.
13. Nuffer, J., & Bein, T. (2006). Applications of piezoelectric materials in transportation industry. In: *Global Symposium on Innovative Solutions for the Advancement of the Transport Industry*, San Sebastian, Spain.
14. Ramlan, R., Brennan, M., Mace, B., & Kovacic, I. (2010). Potential benefits of a non-linear stiffness in an energy harvesting device. *Nonlinear Dynamics*, *59*(4), 545–558.
15. Riley, K., Hobson, P., & Bence, S. (2006). Mathematical Methods for Physics and Engineering: A Comprehensive Guide. Cambridge University Press. http://books.google.com.ua/books?id=Mq1nlEKhNcsC.
16. Senturia, S. D. (2001). Microsystem design, vol. 3. Kluwer academic publishers Boston.
17. Smith, W. A. (1986). Composite piezoelectric materials for medical ultrasonic imaging transducers—a review. In *Sixth IEEE International Symposium on on Applications of Ferroelectrics*, pp. 249–256.
18. Sodano, H. A., Inman, D. J., & Park, G. (2004). A review of power harvesting from vibration using piezoelectric materials. *Shock and Vibration Digest*, *36*(3), 197–206.
19. Tang, L., Yang, Y., & Soh, C. K. (2010). Toward broadband vibration-based energy harvesting. *Journal of Intelligent Material Systems and Structures*, *21*(18), 1867–1897.
20. Toh, T. T., Bansal, A., Hong, G., Mitcheson, P. D., Holmes, A. S., & Yeatman, E. M. (2007). Energy harvesting from rotating structures. Technical Digest PowerMEMS 2007, Freiburg, Germany, 28–29 November 2007 pp. 327–330.
21. Trigona, C., Dumas, N., Latorre, L., Andò, B., Baglio, S., & Nouet, P. (2011). Exploiting benefits of a periodically-forced nonlinear oscillator for energy harvesting from ambient vibrations. *Procedia engineering*, *25*, 819–822.
22. Vocca, H., Neri, I., Travasso, F., & Gammaitoni, L. (2012). Kinetic energy harvesting with bistable oscillators. *Applied Energy*, *97*, 771–776.

Chapter 5
Nonlinear Structural Mechanics of Micro-and Nanosystems

Hassen M. Ouakad

In the nanoscale, when trying to fabricate straight clamped-clamped single-walled CNT transistors, the outcome comes as nonperfectly straight beam with an initial curvature, also called slack. Hence, we can see that understanding the dynamics of arches serve both the micro- and nanoworlds.

The dynamic behavior of CNTs also is greatly needed. Researchers have several reported unexplained and unjustified phenomena for electrically actuated slacked CNTs. These include multiple resonances, frequency crossing, frequency avoiding of crossing (veering), undistinguished resonances whether they are out-of-plane or in-plane, unexplained low values of quality factor, etc. A robust model of these complicated structures can reveal many of their dynamical related issues.

In the first part of this chapter, we summarize the main contributions in the area of modeling the structural mechanics of carbon nanotubes used as NEMS devices. We also address the need to add the slack effect when modeling CNTs. Then, we introduce general concepts (essentially reduced-order modeling and perturbation technique) in modeling the nonlinear structural mechanics problems of beams under several loading conditions (mainly electric actuation). We also include derivation of the equations of motion of MEMS arches along with discussing some conditions for applicability of the continuum theory in modeling the mechanical behavior of CNTs.

In the second part of this chapter, we present an investigation into modeling and analyzing the nonlinear structural mechanics of electrically actuated carbon nanotube resonators. We investigate in details the nonlinear structural mechanics of such devices including the effect of their initial curvature (level of slack). We present a framework and a platform to properly understand the dynamics of these complicated systems by explaining and revealing the meaning of their various detected resonance

H.M. Ouakad (✉)
Mechanical Engineering Department, King Fahd University of Petroleum and Minerals, Dhahran, Kingdom of Saudi Arabia
e-mail: houakad@kfupm.edu.sa

E. Blokhina et al. (eds.), *Nonlinearity in Energy Harvesting Systems*,
DOI 10.1007/978-3-319-20355-3_5

frequencies. Other topics that are covered include study of the possibility of energy exchange among the detected modes in slacked carbon nanotubes.

5.1 Literature Review

Since their discovery by Iijima [1] in 1991, CNTs attracted the interest of the nano scale world, especially the NEMS community [2, 3], due to their unique and distinguished mechanical and electrical properties. However, people how worked on investigating the dynamics of these tiny structures encountered serious obstacles such as their inherent nonlinear behavior even for very small dynamic loading [3], their low quality factors [4], detection of unexpected resonances, and unexplained patterns of the dependence of their natural frequencies on the gate voltage [5].

As an example for the structural related issues, it was shown that clamped-clamped CNT resonators are fabricated with some level of curvature (slack) when using the so-called Chemical Vapor Deposition (CVD) process [5]. This slack effect has shown to lead to various scenarios such as mode crossings, mode avoided crossings, multiple resonances, frequency crossing, frequency avoiding of crossing (veering), undistinguished resonances whether they are out-of-plane or in-plane, unexplained low values of quality factor, ...etc [5, 6].

Estimating accurately the natural and resonant frequencies of CNTs has been the center of research attention over the past few years [7]. This is because predicting precisely the resonance frequencies forms the basis of utilizing CNTs as resonant sensors. In addition, relating the measured resonance frequencies to the predicted from theory has been proposed as an effective way to extract the mechanical properties of CNTs, such as Young's modulus [8, 9].

Clamped-clamped straight CNTs have been under extensive researches in the NEMS community. Most of the modeling work has especially focused on simulating their static response (linear and nonlinear) and free vibration (natural frequencies). Dequesnes et al. [10] used molecular dynamics and linear beam theory with electrostatic force to investigate pull-in and static behavior of CNT switches. The effect of van der Walls forces was studied and found negligible for gaps above 4 nm. Sapmaz et al. [11] investigated the static behavior and free vibrations of CNTs for various DC voltages using nonlinear beam equations for clamped-clamped beam assuming the DC load as constant. Dequesnes et al. [12] investigated pull-in and natural frequencies of clamped-clamped (including mid-plane stretching) and cantilever CNTs using molecular dynamics and continuum models. They concluded that nonlinear continuum model yields good match with molecular dynamics model.

Lefèvre et al. [13] measured the deflection versus DC for a clamped-clamped CNT using AFM to extract Young's modulus and simulated the static behavior of the CNTs using nonlinear beam theory. Ke and Espinosa [14] and Pugno et al. [15] conducted a nonlinear analysis for the static response of a doubly clamped CNT using a nonlinear elastic beam equation. Postma et al. [3] used a nonlinear model of clamped-clamped CNT along with a Galerkin procedure to descritize the beam partial differential equation. They concluded that CNTs are almost useless in the linear regime even for small electric load and this due to the presence of thermal random noise. Witkamp et al. [16] and Poot et al. [17] presented modeling and testing of clamped-clamped CNTs. They used a beam model, including mid-plane stretching, to show the variations of natural frequencies of the CNT versus DC load. Peng et al. [18] used linear Euler–Bernoulli beam model to calculate the resonance frequency and quality factor.

Mecular dynamics (MD) analysis represented, in the past, the most used computational method in studying the physical as well as the mechanical behaviors of CNTs [12, 19–28]. In fact, MD can be accurately used to simulate the physical–chemical properties of CNTs at the atomic scale. But still being the most precise method in that fact, MD is of limited capabilities computationally point of view (maximum of 109 atoms [29] and 10–16 s as time step [30]. Basically in this method, the Newton's second law is applied for each considered atom forming the CNT (i.e., the acceleration term of each atom is equal to the spatial gradient of the empirical potential energy of the CNT). Then, the evaluation of that empirical potential energy is based on several methods such as: the force field method, the bond order method, and semiempirical method [31].

Several investigations have been conducted to simulate the response of CNTs using molecular dynamics and their results were compared to results obtained using continuum mechanics theories, such as beams and shells. All the investigations concluded that continuum mechanics serve the purpose of modeling and simulating CNTs both accurately and efficiently from a computational point of view. For example, Yakobson et al. [32] studied using a molecular dynamics model the large deformation of CNTs. They estimated the buckling of CNTs using continuum theory, a beam model, to the predictions of the molecular mechanics simulations. They concluded that this behavior of CNTs can be well described by a continuum model. Based on scaling analysis, Harik [33, 34] tackled the buckling problem of CNTs and proposed three nondimensional numbers related to the CNT geometric parameters to check the validity of the beam assumption for modeling the mechanics of CNTs. Liu and Chen [35] mentioned that investigating the global responses of CNTs such as deformations, effective stiffness, or load transfer can be done using continuum mechanics both effectively and efficiently. Pantano et al. [36] used finite element FE approaches to model a single-walled carbon nanotube SWCNT and a multi-walled carbon nanotube MWCNT. The CNT walls were modeled as thin shells while the inter-wall interactions were modeled as pressures. The pressures were defined as functions of separation distance. Pantano et al. [36] validated their model by comparing the FE results with the molecular mechanics simulations and the experimental data. They found good agreement among all the results. Arroyo and Belytschko [37] and Arroyo [38] also developed a FE model for MWNTs. They developed a membrane wall model directly using a Tersoff-Brenner potential and a modified Cauchy–Born rule. They were able to reproduce local buckling, kinking, and rippling effects, which are nearly identical to the deformed states of the parent molecular simulation, by using fine meshes for a variety of loadings including compression, torsion, and bending. They concluded that the continuum/finite element calculations are surprisingly accurate compared to the atomistic calculations [38]. Dequesnes et al. [10, 12] employed a nonlinear beam theory in which they used the molecular dynamics simulations to extract the beam material properties (the beam bending (EI) and the beam stretching (EA) constants). They finally showed good agreement when comparing the beam model to the molecular dynamic simulations except for slight difference very close to pull-in. It was shown by Garg [31] that for the purpose of investigating the global behavior and response of CNTs, the continuum mechanics can be safely used under certain conditions.

In relatively more recent works, Wang et al. [39] remarked that atomistic and molecular methods are limited to CNTs with small number of atoms, due to the high cost of computation, and are therefore restrained to the study of localized effects on small portion of the CNT. In order to simulate the mechanical behavior of large-sized atomic CNTs, Wang et al. [39] proposed the use of continuum models. Sears and Batra [40] showed that continuum models predict both global and local responses for buckling of SWCNT by comparing their continuum predictions for bending and buckling to atomistic simulations. They employed a simple Euler beam to model the cantilever bending, and they ended up by showing that the continuum SWCNT strain energy was found to match that of the molecular simulation very well. Also, the buckling of a SWCNT was studied for tubes of different length to study and it was found to predict the critical buckling strain of the carbon nanotube for the whole length scale compared to molecular simulations.

Motivated by the 2-D problem of CNTs, Conley et al. [41] proposed a model for a CNT accounting for both the in- and out-of-plane motions. They reported the onset possibility of the non-planar motions in straight CNT resonators that they attributed to the symmetry of the device, as well as, to the nanotube stretching as it deforms.

Recently, Elishakoff and Pentaras [42] derived analytical expressions for the fundamental natural frequencies of double-walled carbon nanotubes under various boundary conditions (simply supported and doubly clamped) using two different decritization schemes, the Bubnov–Galerkin and the Petrov–Galerkin methods. They used a linear beam model and showed a possibility to quickly evaluate the natural frequencies of such systems. In another investigation [43], they adopted a simplified Bresse-Timoshenko beam model to evaluate the natural frequencies of simply supported CNTs taking into account the shear deformation as well as the rotary inertia effects. They showed that the model yields excellent results compared to the full Bresse-Timoshenko theory as well as the Euler–Bernoulli beam model. Lately, Georgantzinos et al. [44] proposed a linear spring-based model with lumped masses to describe and evaluate the vibration characteristics of a single-walled CNT. They observed that the aspect ratio has a significant role and influence on the basic modes of vibration of the nanotubes. Hawwa and Al-Qahtani [45] adopted an elastic continuum approach for modeling the primary resonance of a double-walled carbon nanotube under a linear harmonic excitation using the Galerkin approach. They showed several transitions from quasiperiodic to chaotic behavior accompanied with some nonlinear jump phenomena and nonlinear bifurcations leading to chaos.

Cantilever CNTs received less interest in modeling their nonlinear behavior. Among the few works on this field, Liu et al. [35] used a nonlinear beam model (with cubic geometric nonlinearity) for a cantilever CNT. Ke and Espinosa [46, 47] presented the modeling of the static behavior of cantilever and doubly clamped CNTs. Ke et al. [48] studied theoretically and experimentally the static behavior of cantilever CNTs while considering the effect of the charge concentration, van der Wall forces, and the large kinematics (geometric nonlinearities). Isacsson et al. [49] investigated the dynamic response of a three-terminal cantilever CNT resonator structure. They used a perturbation theory to descritize the beam equation of motion and then determine analytically the frequency response of the system. Isacsson and Kinaret [50] modeled the parametric excitation of an array of cantilever CNTs excited by DC and AC excitations.

The above theoretical investigations were motivated especially by the several reported phenomenon of previously conducted experimental works [51–62].

Most of the previously mentioned literature work models clamped-clamped electrically actuated CNTs to be perfectly straight. However, due to their fabrication process using chemical vapor deposition (CVD), fabricating perfectly straight CNTs with controlled geometry and orientation is very difficult. Indeed, many studies have indicated that clamped-clamped CNTs are fabricated with some level of curvature (slack). Kang et al. [63] studied the interatomic interaction between a CNT bridge and the substrate underneath it used as a NEMS memory device. They showed that the value of the CNT bridge slack is very important for the operation of the NEMS memory device as a nonvolatile memory. Gibson et al. [7] stressed the importance of modeling the slack of CNTs and indicated that no consistent model has been presented so far to address this issue despite knowing the strong effect of the curvature of curved beams on modal frequencies.

Among those who reported experimental investigations showing the importance of slack on estimating accurately the natural frequencies of CNTs, Sazonova et al. [4] and Sazanova [5] tested slacked clamped-clamped CNT to a DC and AC load and characterized experimentally their free-vibration response. Comparing their experimental data with the numerical results of [6], they reported some unexplained phenomena [5]. Some of those experimentally observed phenomena are [5]: the sublinear variation of the frequencies, the avoided crossings (veering), the abundance of resonances, and the negative variation of the frequencies for certain carbon nanotubes. Üstünel et al. [6] were among the very few who attempted to investigate theoretically the effect of slack on a CNT oscillator. They based their work on a one-dimensional elastic continuum model by assuming the electrostatic force to be uniformly distributed along the length of the CNT and ignoring the nonlinear elastic effects. Then, they derived approximate analytical expressions of the natural frequencies of the CNT for several behavior regimes of the CNT depending on the gate voltage value. They identified three zones for the response: bending dominated, catenary dominated, and elastic dominated. They related this to the impact of slack on the natural frequencies. They reported that the bending regime can be described by a buckled doubly clamped beam model, the catenary regime is modeled as a simple string under variable tension, and the elastic regime is similar to the hanging chain under constant tension. Another group [64] attempted to model the CNT using a finite element method. They also reported discrepancy among their theoretically predicted and experimentally measured resonance frequencies of electrostatically actuated CNTs and cited slack as a possible reason of this. Mayoof and Hawwa [65] proposed a nonlinear curved beam model to describe the nonlinear dynamics of a slacked single-walled carbon nanotube under linear harmonic excitation. They showed several scenarios of chaotic behavior for the CNT.

5.2 Background

In this section, we introduce general concepts in modeling the nonlinear structural mechanics problems of beams and then the case where electrostatic forces are used as actuation method. We also include derivation of the frequency equation using the so-called method-of-multiples scales. Finally, we present some conditions for applicability of the continuum theory in modeling the mechanical behavior of CNTs.

5.2.1 Beams

Beams are perhaps the most commonly used structural component in the MEMS and NEMS field. Micro and nanobeams form the backbone of a wide range of devices including resonators, resonant sensors, actuators, filters, atomic force microscope probes, and RF switches. They are also used as spring elements with other microstructures and MEMS components, such as comb-drive actuators.

(a) Equation of motion

Here, the linear equation of motion of a beam in bending is derived, The derivation here follows Hamilton's principle, which is variational mechanics energy-based approach. This presents an alternative technique to the Newtonian or vectorial method used in the derivations [66, 67]. Hamilton's principle is considered very powerful for deriving the equation of motion, along with the associated boundary conditions, of complicated distributed-parameter systems of multiple bodies and complicated boundary conditions. As an example, we will derive the equation of motion and associated boundary conditions governing the motion of an initially curved beam (arch beam) using the Hamilton's principle.

We consider a clamped-clamped shallow arch, Fig. 5.1, of initial shape $w_0(x)$, width b, thickness h, length L, modulus of elasticity E, cross sectional is $A = bh$, moment of inertia I, mass density ρ, and subjected to a constant axial force of magnitude N, as shown in Fig. 5.2. The axial displacement is denoted by $u(x, t)$ and the transverse displacement is denoted by $w(x, t)$ measured from the initial curvature $w_0(x)$. The beam is modeled according to the Euler–Bernoulli beam theory and in which the planes of the cross sections remain planes after deformation.

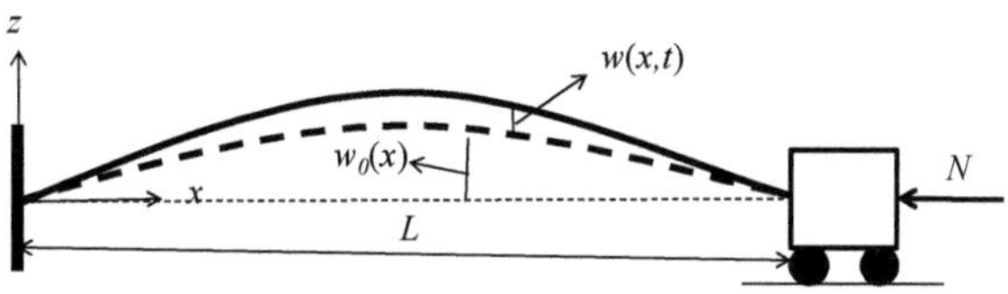

Fig. 5.1 Schematic of a clamped-clamped arch beam under a compressive axial load

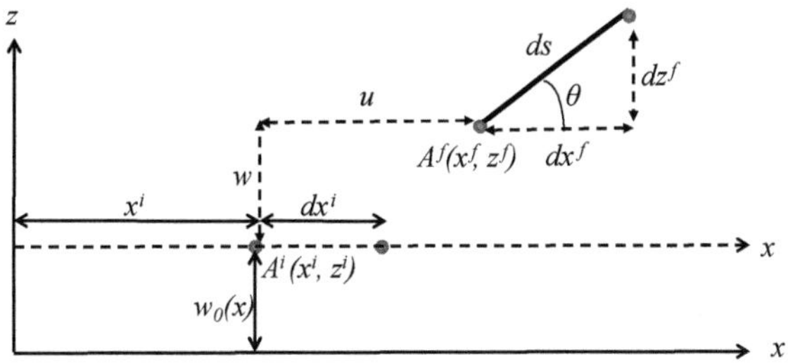

Fig. 5.2 Segment of the beam after deformation

(i) The total axial strain

We first determine the geometrical relations of the beam points using Fig. 5.2. Figure 5.2 describes how a point A^i, with coordinates x^i and z^i in the axial and transverse directions respectively, moves to a point A^f with coordinates denoted by x^f and z^f after a deformation.

The geometrical relations are determined from Fig. 5.2 which contains a differential element located at the point A^i with length dx^i. The coordinates of A^f and the element length of the deformed configuration are determined respectively as follows, where the "′" denotes the derivative with respect to x:

$$\begin{cases} x^f = x^i + u = x + u, \\ z^f = z^i + w = w_0 + w, \end{cases} \tag{5.1}$$

$$\begin{aligned} ds = \sqrt{\left(dx^f\right)^2 + \left(dz^f\right)^2} &= \sqrt{(dx + du)^2 + (dw + dw_0)^2} \\ &= \sqrt{(1 + u')^2 + \left(w' + w_0'\right)^2} dx, \end{aligned} \tag{5.2}$$

Now, under the assumptions of a shallow arch theory (the shallow arch approximation), in which the slope of the initial rise of the curved beam is considered smaller compared to unity [68], i.e., $\left(w_0'\right)^2 << 1$, Eq. (2.2) reduces to

$$ds = \sqrt{1 + 2u' + u'^2 + w'^2 + 2w'w_0'} dx \tag{5.3}$$

From the deformed element length, Eq. (2.3), we determine the strain of the deformed element and the stretch ratio for a small deformation, respectively, as

$$\varepsilon = \frac{ds - dx}{dx} = \sqrt{1 + 2u' + u'^2 + w'^2 + 2w'w_0'} - 1, \tag{5.4}$$

$$\lambda = \frac{ds}{dx} = \sqrt{1 + 2u' + u'^2 + w'^2 + 2w'w_0'} \tag{5.5}$$

The rotation angle, shown in Fig. 5.2, is determined as

$$\begin{cases} \sin(\theta) = \dfrac{dy^1}{ds} = \dfrac{w_0' + w'}{\lambda}, \\ \cos(\theta) = \dfrac{dx^1}{ds} = \dfrac{1 + u'}{\lambda}, \end{cases} \tag{5.6}$$

Differentiating Eq. (2.6), and using the shallow arch approximation, we get

$$\theta' = \frac{\left(1 + u'\right)\left(w_0'' + w''\right) - u''\left(w_0' + w'\right)}{\lambda^2} \tag{5.7}$$

The curvature of the arch mid-plane is given by

$$\begin{aligned} \kappa = \tfrac{d\theta}{ds} = \theta' \frac{dx}{ds} &= \frac{\left(1 + u'\right)\left(w_0'' + w''\right) - u''\left(w_0' + w'\right)}{\lambda^{3/2}} \\ &= \frac{\left(1 + u'\right)\left(w_0'' + w''\right) - u''\left(w_0' + w'\right)}{\left(1 + 2u' + u'^2 + w'^2 + 2w'w_0'\right)^{3/2}} \end{aligned} \tag{5.8}$$

Expanding Eqs. (2.4) and (2.8) up to the quadratic terms using a Taylor series expansion for small u' and w', we get

$$\varepsilon \approx u' + \frac{w'^2}{2} + w'w_0' + \cdots \tag{5.9}$$

$$\kappa \approx \left(w_0'' + w''\right) - u''\left(w_0' + w'\right) - 2\left(w_0'' + w''\right)u' + \cdots \tag{5.10}$$

Now, the total axial strain at a point of distance z from the arch mid-plane line is measured to be

$$\varepsilon_T \approx \varepsilon - zw'' + \cdots \tag{5.11}$$

(ii) The kinetic and potential energies

Next, we develop expressions for the kinetic and potential energies of the arch beam. The potential energy due to the beam elastic deformation and the stretching of its mid-plane is given by

$$
\begin{aligned}
V &= \tfrac{1}{2}\int_0^l \int_{-h/2}^{h/2}\int_{-b/2}^{b/2}\left(E\varepsilon_T^2\right) dydzdx,\\
&= \frac{E}{2}\int_0^l \int_{-h/2}^{h/2}\int_{-b/2}^{b/2}\left\{\left(u'+\frac{w'^2}{2}+w'w_0'\right)^2 - zw''\left(u'+\frac{w'^2}{2}+w'w_0'\right)+z^2w''^2\right\} dydzdx,\\
&= \frac{EA}{2}\int_0^l \left(u'+\frac{w'^2}{2}+w'w_0'\right)^2 dx + \frac{EI}{2}\int_0^l w''^2 dx,
\end{aligned}
\tag{5.12}
$$

where $I = bh^3/12$.

The kinetic energy of the arch while neglecting the axial inertia term is given by

$$
T = \frac{\rho A}{2}\int_0^L \dot{w}^2 dx, \tag{5.13}
$$

where the dot denotes the partial derivative with respect to the time variable t.

(iii) The extended Hamilton principle

The Hamilton's principle is an important variational method in deriving the equation of motion and the associated boundary conditions of continuous systems. In such systems, the state is described by using continuous functions of space and time. The extended Hamilton Principle for such bodies is given by

$$
\int_{t_1}^{t_2} \delta L\, dt = \int_{t_1}^{t_2} \left(\delta T - \delta V + \delta W_e\right) dt = 0, \tag{5.14}
$$

where L is the Lagrangian, T is the kinetic energy, V is the elastic energy, W_e is the nonconservative work done by external loads on the system, and t_1, t_2 are the initial and final times, respectively.

The variation of the potential energy is obtained by integrating by parts over time Eq. (5.12) as follows:

$$
\int_{t_1}^{t_2} \delta V\, dt = \int_{t_1}^{t_2}\left\{
\begin{aligned}
& EA\left[\left(u'+\frac{w'^2}{2}+w'w_0'\right)\delta u\right]_0^L - EA\int_0^L \left(u'+\frac{w'^2}{2}+w'w_0'\right)' \delta u dx +\\
& +EA\left[\left(u'+\frac{w'^2}{2}+w'w_0'\right)w'\delta w\right]_0^L - EA\int_0^L \left(\left(u'+\frac{w'^2}{2}+w'w_0'\right)w'\right)' \delta w dx +\\
& +EA\left[\left(u'+\frac{w'^2}{2}+w'w_0'\right)w_0'\delta w\right]_0^L - EA\int_0^L \left(\left(u'+\frac{w'^2}{2}+w'w_0'\right)w_0'\right)' \delta w dx +\\
& +EI\left[w''\delta w'\right]_0^L - EI\left[w'''\delta w\right]_0^L + EI\int_0^L w''''\delta w dx
\end{aligned}
\right\} dt
\tag{5.15}
$$

The variation of the kinetic energy is obtained by integrating by parts over time Eq. (5.13) as follows:

$$\int_{t_1}^{t_2} \delta T \; dt = \frac{\rho A}{2}\int_{t_1}^{t_2} \delta\left[\int_0^L \left(\frac{\partial w}{\partial t}\right)^2 dx\right] dt = \rho A \int_0^L \int_{t_1}^{t_2} \left[\frac{\partial w}{\partial t}\left(\frac{\partial}{\partial t}\delta w\right)\right] dt\, dx,$$
$$= \rho A \cancel{\left[\frac{\partial w}{\partial t}\delta w\right]_{t_1}^{t_2}}^{t_1=t_2} - \rho A \int_0^L \frac{\partial^2 w}{\partial t^2}\delta w\, dx = -\rho A \int_0^L \frac{\partial^2 w}{\partial t^2}\delta w\, dx, \tag{5.16}$$

Finally, the variation of the nonconservative forces is given as follows:

$$\int_{t_1}^{t_2} \delta W_e \, dt = \int_{t_1}^{t_2} \left(F\left(x,t\right)\delta w - c\dot{w}\right) dt, \tag{5.17}$$

where F is a distributed load in the transverse direction and c is the viscous damping coefficient.

Plugging then Eqs. (5.15)–(5.17) into Eq. (5.14), and then collecting the terms with the arbitrary coefficient "δu", we get

$$\left(u' + \frac{w'^2}{2} + w'w_0'\right)' = 0 \tag{5.18}$$

Integrating Eq. (5.18) over the beam domain, we obtain the beam's axial elongation as

$$u\left(L,t\right) - u\left(0,t\right) = \left(u' + \frac{w'^2}{2} + w'w_0'\right)L - \int_0^L \left(\frac{w'^2}{2} + w'w_0'\right)dx, \tag{5.19}$$

where $u(L,t)$ and $u(0,t)$ are the axial displacements at the ends of the beam.

From Fig. 5.2, we have a fixed end at $x = 0$ and compressive axial load acting at the other end, hence

$$u\left(L,t\right) = -\frac{NL}{EA} \tag{5.20}$$

We finally obtain from Eqs. (5.18) and (5.20) that

$$u' + \frac{w'^2}{2} + w'w_0' = -\frac{N}{EA} + \frac{1}{L}\int_0^L \left(\frac{w'^2}{2} + w'w_0'\right)dx \tag{5.21}$$

Plugging now Eqs. (5.15)–(5.17) into Eq. (5.14), and then collecting the terms with the arbitrary coefficient "δw", we get

$$\rho A\ddot{w} + EIw'''' + c\dot{w} = EA\left(u' + \frac{w'^2}{2} + w'w'_0\right)' \left(w' + w'_0\right) + \\ + EA\left(u' + \frac{w'^2}{2} + w'w'_0\right)\left(w'' + w''_0\right) + F(x,t) \quad (5.22)$$

Plugging Eqs. (5.18)–(5.21) into Eq. (5.22), we get the nonlinear equation of motion of the shallow arch [69]

$$\rho A\ddot{w} + EIw'''' + c\dot{w} = \left(-N + \frac{EA}{2L}\int_0^L (w'^2 + 2w'w'_0)dx\right)(w'' + w''_0) + F(x,t), \quad (5.23)$$

(b) Reduced-order modeling

If the equation characterizing the behavior of a beam is nonlinear or if the beam has some irregular geometrical and material properties, closed-form solutions may not be accessible. In such cases, numerical methods (such as Rayleigh–Ritz method, finite element method, and the weighted-residual method) need to be utilized to simulate the behavior of the beam. However, the previously mentioned methods can be inconvenient or even computationally very expensive, especially for multi-physics, nonlinear, and time-dependent problems, such as those commonly encountered in MEMS and NEMS. Hence, we propose to discuss a powerful technique, the Galerkin method, which is computationally efficient, capable of handling any systems, and suits nonlinear beams problems. Beam equations of motion cab be discretized using the Galerkin technique to yield a reduced-order model (ROM) [70]. To derive a ROM from a distributed-parameter system, one can either work with the Lagrangian [71], or work with the partial differential equations, boundary conditions, and orthogonality conditions of the beam [72]. We decide here to present the latter approach. We consider a general system with an equation and boundary conditions expressed respectively as

$$M(w(x,t)) = f(x,t), \quad (5.24)$$

$$BC_1(w) = w_1, \qquad BC_2(w) = w_2, \quad (5.25)$$

where $w(x,t)$ is the dependent variable in space x and time t, M is a differential operator in space and time, which can be linear or nonlinear, $f(x,t)$ is the forcing term, BC_1 and BC_2 are boundary operators and w_1 and w_2 are non-time-varying boundary conditions. In the Galerkin discretization technique, we seek an approximate solution of the above system in the form of

$$w(x,t) = \phi_0(x) + \sum_{i=1}^{n} u_i(t)\phi_i(x), \tag{5.26}$$

where $\phi_0(x)$ is a space function that satisfies the nonzero boundary conditions (w_1 and w_2 different of zero). If the boundary conditions are homogeneous ($w_1 = w_2 = 0$) then $\phi_0(x) = 0$. The time-varying functions $u_i(t)$ are to be determined for the sake to get the solution $w(x,t)$. The functions $\phi_i(x)$ are space-varying trial functions that should satisfy the following conditions [73]

- The homogenous form of all the boundary conditions of the problem ($w_1 = w_2 = 0$)
- Differentiable as the order of the differential equation of the system.
- Form and belong to a complete set of linearly independent functions.

To proceed, Eq. (5.26) is substituted into Eq. (5.24) yielding the following equation

$$M\left[\phi_0(x) + \sum_{i}^{n} u_i(t)\phi_i(x)\right] = f(x,t) \tag{5.27}$$

Then, multiplying Eq. (5.27) by ϕ_j, and integrating the equation over the domain of the problem yields

$$\int_{\Gamma} \phi_j(x)\left\{M\left[\phi_0(x) + \sum_{i}^{n} u_i(t)\phi_i(x)\right]\right\}dx = \int_{\Gamma} \phi_j(x) f(x,t)\,dx, \tag{5.28}$$

Once evaluating the integrals of Eq. (5.38), we get n differential equations in time of the $u_i(t)$ functions, which can be integrated numerically using Runge–Kutta techniques. Then the results are substituted back into Eq. (5.26) to yield the total response of the beam. The number n of necessary modes to ensure convergence needs to be examined.

The reduction of a distributed-parameter system of infinite degree of freedom or a partial differential equation in space and time into a number of ordinary differential equations in time (a finite-degree-of-freedom-system) is considered significant advantage from a computational point of view. Indeed, this is one of the major benefits of using the Galerkin method over finite element approaches. Thus, the Galerkin method is considered a powerful technique to generate reduced-order models. It is worth to mention that for static problems ($f(x,t) = f(x)$), the Galerkin discretization yields to a system of algebraic equations of unknown constant coefficients, which can be solved using Newton's Raphson methods.

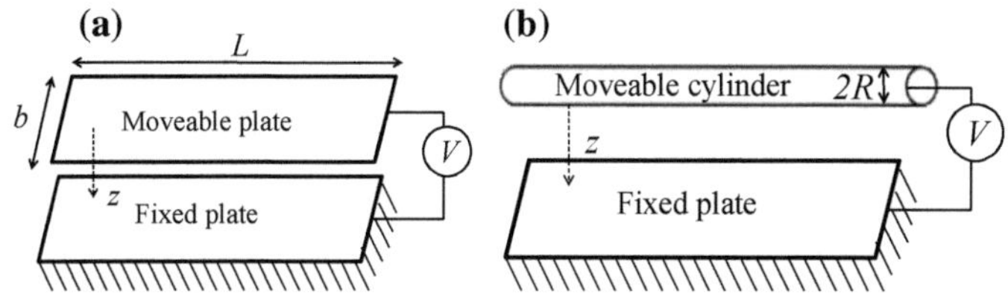

Fig. 5.3 Schematic of (**a**) a parallel-plates capacitor, and (**b**) a cylinder-plate capacitor

5.2.2 Electric Actuation

(a) About the actuation technique

Electrostatic loading is the most common actuation method in MEMS. The electrostatic actuation scheme is based on the simple parallel-plate capacitors. Those kinds of capacitors require a voltage source and they are characterized by having very low power consumption and being one of the fastest actuation methods. Electrostatic actuation depends on the attractive force between the two parallel plates of a capacitor, which is nonlinearly proportional to the gap width that separates both of them.

We will derive here the expression of the electrostatic force for two different cases: two parallel plates and two parallel cylinder and plate. We first consider the parallel-plate capacitor shown in Fig. 5.3. Here, we assume under the parallel-plate theory conditions that the electric field lines between the two considered plates are perpendicular to the plates even near edges (no fringing field effect near the edge of the plates).

Then, we assume that each capacitor shown in Fig. 5.3 is driven by a voltage source of load V. The electrical charge and the potential energy stored in each capacitor are given by the following expressions [74], respectively,

$$Q = C(z)\,V, \tag{5.29}$$

$$E = \frac{V^2}{2} C(z) \tag{5.30}$$

The attractive electrostatic force between each of the two electrodes of both the capacitors in Fig. 5.3 can be obtained as [74]

$$F_e = \frac{-\partial E(z)}{\partial z} = \frac{V^2}{2}\frac{\partial C(z)}{\partial z} \tag{5.31}$$

Now, we will derive the analytical expressions of $C(z)$ depending on the geometries of the considered electrodes.

Considering the case of a capacitor formed by two parallel rectangular plates, Fig. 5.3a, the capacitance can be expressed as [74]

$$C(z) = \frac{\varepsilon_o L b}{z}, \tag{5.32}$$

where $\varepsilon_o = 8.85 \times 10^{-12}\left(C^2/Nm^2\right)$ is the air permittivity. We substitute Eq. (5.23) into Eq. (5.31) and get the attractive electrostatic force between the two electrodes of Fig. 5.3a

$$F_e = \frac{-\varepsilon_o b L V^2}{2z^2} \tag{5.33}$$

Now, we consider the case of a capacitor formed by parallel cylinder and rectangular plate, Fig. 5.3b. The capacitance can be expressed as [74]

$$C(z) = \frac{2\pi \varepsilon_o L}{\cosh^{-1}(1 + z/R)} \tag{5.34}$$

Substituting Eq. (5.34) into Eq. (5.31), we get the attractive electrostatic force between the two electrodes of Fig. 5.3b

$$F_e = \frac{-\pi \varepsilon_o L V^2}{\sqrt{z(z+2R)}\left(\cosh^{-1}(1 + z/R)\right)^2} \tag{5.35}$$

We can notice, from both Eqs. (5.33) and (5.35), the inherent nonlinear dependence of the electrostatic force on the moveable electrodes (the planar electrode in Fig. 5.3a and the cylindrical electrode in Fig. 5.3b) displacement.

(b) The pull-in instability

In "parallel-plate" electrostatic actuation, Fig. 5.4a, a DC electrostatic load (V_{DC}) is applied between the lower and upper electrode, which is typically a flexible structure, such as a rectangular micro or nanobeam or a carbon nanotube. If V_{DC} is small, the structure stays in the deflected position, at which the elastic restoring force of the structure is in equilibrium with the opposing electrostatic force. There is an upper limit for V_{DC}, beyond which the mechanical restoring force of the structure can no longer resist the opposing electrostatic force. This leads to a sudden collapse of the structure, which is known as the pull-in instability [75–77]. Typically, the DC load tends to soften the actuated structure, which decreases its linear natural frequency [78].

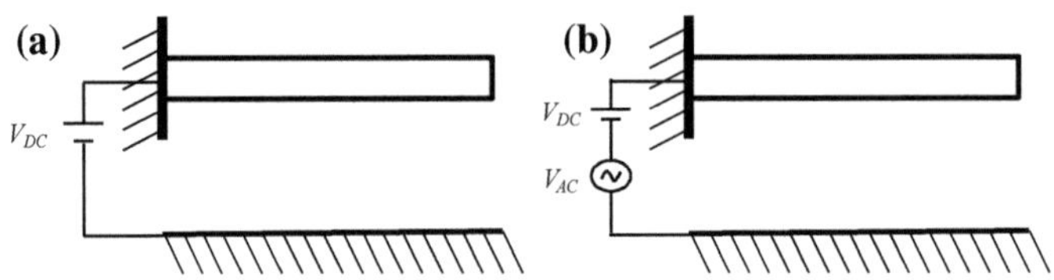

Fig. 5.4 Schematic of a parallel-plate electrostatic actuation with (**a**) DC load only, and (**b**) DC and AC harmonic load

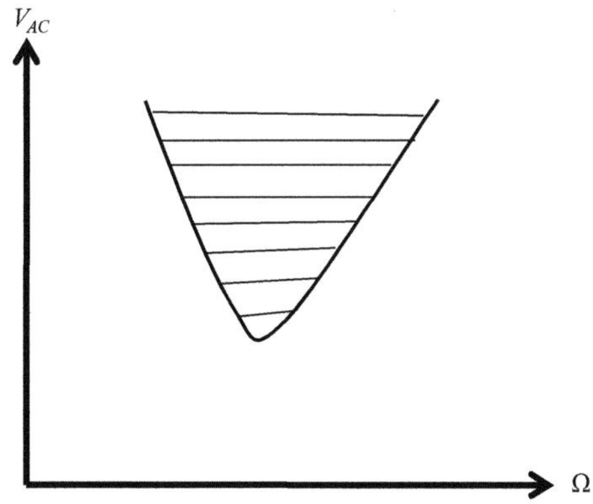

Fig. 5.5 The pull-in band (escape phenomenon) of an electrically actuated system

In addition to the DC electrostatic load of amplitude V_{DC}, resonant sensors and resonators are actuated by an AC harmonic load of amplitude V_{AC}, Fig. 5.4b. Here, V_{DC} deflects the movable electrode slightly and V_{AC} vibrates the electrode around the new deflected position. If V_{AC} is smaller than the dynamic pull-in limit and the structure is inherently nonlinear, such as the case of doubly clamped beams, then it can exhibit either softening or hardening behavior (depending on whether the electrostatic nonlinearity of quadratic nature dominates the geometric nonlinearity of cubic nature or vice versa). This effect of V_{AC} leads to a further shift in the resonance frequency, also called nonlinear resonance frequency [79–81]. While attempting to exceed the noise level of some structures, such as the thermal noises in CNTs [3, 51, 54], the V_{AC} amplitude might be raised to be high enough to trigger a dynamic pull-in instability [80, 82–84]. This instability, which characterizes the instability of the structure due to dynamics considerations, is called "dynamic pull-in".

For a certain DC and AC loads, the upper electrode can oscillate in a stable state, and varying the forcing frequency can lead to the dynamic pull-in phenomenon also referred to an escape from a potential well. Figure 5.5 shows a schematic of the escape band of the electrically actuated system of Fig. 5.4ba. The figure shows that for any dynamic load V_{AC} with frequency Ω in the shaded area then the escape phenomenon occurs definitely (inevitable escape) and for a frequency Ω lying in the nonshaded line, the system can oscillate in a stable state or escape from the potential well depending on its initial conditions (fractal behavior) [85, 86].

5.2.3 Perturbation Series and the Method of Multiple Scales

In this section, we review the application of a perturbation technique, for instance the method of Multiple Scales (MMS) to study the nonlinear dynamic of nonlinear systems.

(a) Perturbation series

Consider the set of ε-dependent ordinary differential equations given by

$$x' = f(x, t, \varepsilon), \tag{5.36}$$

where ε is a small parameter and for $\varepsilon = 0$ these equations can be easily solved. Then, a common approach in finding an approximate solution to Eq. (5.36), for small ε, is to approximate $x(t)$ in the following form of a "perturbation series"

$$x = x_0(t) + \varepsilon x_1(t) + \varepsilon^2 x_2(t) + \cdots \tag{5.37}$$

Then Eq. (5.36) can be expanded in powers of ε

$$\begin{aligned} x_0' + \varepsilon x_1' + \varepsilon^2 x_2' + \cdots &= f(x, t, 0) + \frac{df(x, t, 0)}{d\varepsilon}\varepsilon + \cdots \\ &= f(x_0, t, 0) + \varepsilon\left(\frac{df(x_0, t, 0)}{d\varepsilon} + \frac{df(x_0, t, 0)}{dx}(x_0 + 2\varepsilon x_1 + \cdots) + \cdots\right) + \cdots \end{aligned} \tag{5.38}$$

Thus, an infinite sequence of simpler problems by equating terms with like powers of ε can be obtained. For example, gathering the terms of order 0 (terms multiplying ε^0)

$$x_0' = f(x,\ t,\ 0) \tag{5.39}$$

Once the order 0 problem is solved, the solution for x_0 may be plugged into the order ε^1-equation

$$x_1' = \frac{df}{d\varepsilon}(x_0, t, 0) + \frac{df}{dx}(x_0, t, 0)x_1, \tag{5.40}$$

and, hypothetically, this process can be iterated to solve for x_n at all orders.

In general, proving that a perturbation series converges or that it is asymptotic is hard.

(b) Method of multiple scales

The perturbation series expansion is not guaranteed to work for all dynamical problems. The nonlinear Rayleigh and Van Der Pol oscillators are two classical examples in which such an expansion fails. This is due to the fact that lower-order solutions $x_1, x_2, \ldots$, may grow without bound and eventually become large enough to interact strongly with higher-order terms. When this happens, treating the dynamics at different orders separately no longer makes sense. Generally speaking, this kind of failure occurs when there are two or more important scales in the system. For example, when a sound wave with a short wavelength travels through a slowly spatially varying medium, the resulting wave form looks locally like a sine wave when zooming into the wavelength scale. But on the length scale over which the medium properties vary substantially, the wave may exhibit significant phase drift or amplitude changes.

Solutions of the Rayleigh and Van Der Pol equations also have multiple scales, in the sense that on the order of one wavelength the solutions have frequency $\omega \approx \omega_0$, but over time scales of order ε^{-2} (Van Der Pol and Rayleigh) the solutions will have a significant phase drift.

The idea of this method is to pretend that an ODE is a PDE for a minute, where x depends both on time and on "slow time"; i.e.,

$$x = x_0(t, T) + \varepsilon x_1(t, T) + \varepsilon^2 x_2(t, T) + \dots, \tag{5.41}$$

where $T = \varepsilon t$.
We then derive that

$$\frac{d(.)}{dt} = \frac{\partial(.)}{\partial t} + \frac{dT}{dt}\frac{\partial(.)}{\partial T} = \frac{\partial(.)}{\partial t} + \varepsilon\frac{\partial(.)}{\partial T} \tag{5.42}$$

and

$$\frac{d^2(.)}{dt^2} = \frac{\partial^2(.)}{\partial t^2} + 2\varepsilon\frac{\partial^2(.)}{\partial t \partial T} + \varepsilon^2\frac{\partial^2(.)}{\partial T^2} \tag{5.43}$$

We write the original perturbation expansion in a more reminiscent form

$$x = \tilde{x}_0(t) + \varepsilon\tilde{x}_1(t) + \varepsilon^2\tilde{x}_2(t) + \cdots, \tag{5.44}$$

where

$$\tilde{x}_n(t) = x_n(t, T(t)) \tag{5.45}$$

We then plug these substitutions into our equation, and proceed to solve it iteratively as a perturbation series. However, now the first-order equation will be sufficient only to determine x_0's dependence on time t, and not its dependence on t. This will leaves us free at order ε to tune the T dependence of x_0 to prevent x_1 from growing without bound.

(b) Application of the method of multiple scales

We propose now to approximate the dynamic responses of the Van der Pol and the Rayleigh oscillators using the method of Multiple Scales.

(i) The Van der Pol oscillator

We consider the Van der Pol oscillator in the case of a primary resonance Excitation; $\gamma = O(\varepsilon), \quad \Omega = \omega_0 + O(\varepsilon)$

$$\frac{d^2x}{dt^2} - \varepsilon(1 - x^2)\frac{dx}{dt} + \omega_0^2 x = f(t), \text{ with: } \begin{cases} f(t) = \gamma\varepsilon\cos(\Omega t) \\ \Omega = \omega_0 + \varepsilon\sigma \end{cases} \tag{5.46}$$

We seek an approximate solution of Eq. (5.46) using the method of Multiple Scales. In general, we consider $x(t)$ to be a function of multiple (two in this case) independent time variables or scales. We express x in the form

$$x = x_0(T_0, T_1) + \varepsilon x_1(T_0, T_1), \tag{5.47}$$

where $T_0 = t$ is a fast scale and $T_1 = \varepsilon t$ is a slow scale characterizing the modulation in the amplitude and phase caused by the nonlinearity, damping and resonances. The time derivatives become

$$\frac{d}{dt} = \frac{\partial}{\partial t} + \frac{dT}{dt}\frac{\partial}{\partial T} = D_0 + \varepsilon D_1 \tag{5.48}$$

so that

$$\frac{d^2}{dt^2} = D_0^2 + 2\varepsilon D_0 D_1, \tag{5.49}$$

where $D_n = \partial/\partial T_n$. Substituting Eqs. (5.47)–(5.49) into Eq. (5.46) and equating the coefficient of ε^0 and ε^1 on both sides, we obtain

$$D_0^2 x_0 + \omega_0^2 x_0 = 0, \tag{5.50}$$

$$D_0^2 x_1 + \omega_0^2 x_1 + D_0 x_0 - x_0^2 D_0 x_0 + 2D_0 D_1 x_0 = \gamma \cos(\Omega T_0) \tag{5.51}$$

The solution of Eq. (5.50) can be expressed as

$$x_0 = A(T_1)\exp(i\omega_0 T_0) + \bar{A}(T_1)\exp(-i\omega_0 T_0) \tag{5.52}$$

Therefore, Eq. (5.51) becomes

$$\begin{aligned} D_0^2 x_1 + \omega_0^2 x_1 = & \, i\omega_0[-2A' + A^2\bar{A} - A + \gamma \exp(i\sigma T_1)]\exp(i\omega_0 T_0) + \\ & + i\omega_0 A^3 \exp(3i\omega_0 T_0) + \text{cc}, \end{aligned} \tag{5.53}$$

where "cc" denotes the complex conjugate of the preceding terms. The secular terms can be eliminated from the solution of x_1 if

$$-2A' + A^2\bar{A} - A + \gamma \exp(i\sigma T_1) = 0 \tag{5.54}$$

We let $A = a\exp(i\beta)/2$ in Eq. (5.54), where a and β are real functions of the slow times scale T_1, and separate real and imaginary parts, to obtain

$$a' = \frac{a}{2}(1 - \frac{1}{4}a^2) + \frac{\gamma}{2\omega_0}\sin(\lambda), \tag{5.55}$$

$$a\beta' = -\frac{\gamma}{2\omega_0}\cos(\lambda), \tag{5.56}$$

where a' and β' are derivatives of the slow time scale T_1 and

$$\lambda = \sigma T_1 - \beta, \quad \frac{d\lambda}{dT_1} = \sigma - \frac{d\beta}{dT_1} \tag{5.57}$$

Eliminating β from Eqs. (5.55) and (5.56) gives

$$a\lambda' = a\sigma + \frac{\gamma}{2\omega_0}\cos(\lambda) \tag{5.58}$$

Therefore, to the first-order of approximation, we obtain

$$x = a\cos(\Omega t - \lambda) + O(\varepsilon) \tag{5.59}$$

For steady-state motion, the time variation of the amplitude and phase of the response must vanish $a' = \lambda' = 0$. It follows from Eqs. (5.55) and (5.56) that

$$\frac{a}{2}(1 - \frac{a^2}{4}) = -\frac{\gamma}{2\omega_0}\sin(\lambda), \tag{5.60}$$

$$a\sigma = -\frac{\gamma}{2\omega_0}\cos(\lambda) \tag{5.61}$$

The steady-state solution of Eq. (5.58), obtained by squaring and adding Eqs. (5.60) and (5.61), yields the following frequency-response equation:

$$\rho(\omega_0 - \rho)^2 + 4\sigma^2\rho = \frac{\gamma^2}{4\omega_0}, \quad \text{where} \quad \rho = \frac{a^2}{4} \tag{5.62}$$

The frequency-response curves, in terms of the amplitude $\rho = a^2/4$, appear in Fig. 5.6 for selected values of the forcing amplitude γ. As γ increases, the curves consist of two branches. The first branch runs close to the σ-axis and the second branch is a close curve which can be approximated by an ellipse having its center

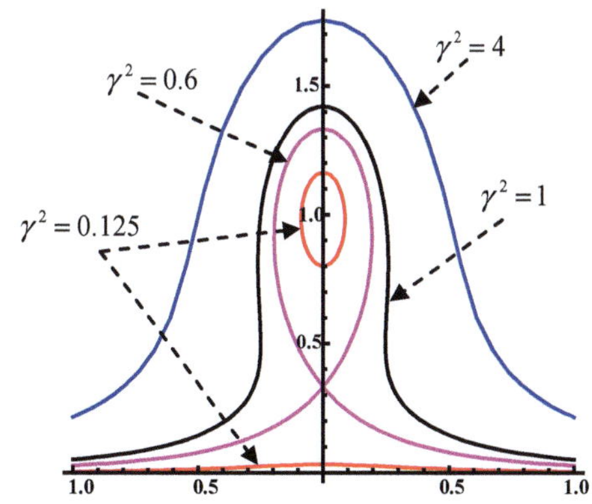

Fig. 5.6 Frequency-response curves for primary resonances of the Van der Pol Oscillator for various forcing amplitudes of $\gamma\,(\omega_0 = 1\ \text{rad/s})$

at the ρ-axis. As γ increases further, the ellipses expand; open and coalesce with the first branch to form a single branch of solutions and finally the response curves are single-valued for all σ. We note that in the $\sigma - \rho$ plane, the frequency-response curves, which are symmetric with respect to the σ axis, have shapes similar to those of the force-response curves.

When solving Eqs. (5.60) and (5.61), we can also obtain the first-order approximate solution in the following form:

$$x(t) = \sqrt{\frac{4\eta}{1+(\frac{4\eta}{a_0^2}-1)e^{-\varepsilon\eta t}}}\cos(\omega_0 t) + \frac{\gamma}{\omega_0^2-\Omega^2}\cos(\Omega t), \text{ where: } \begin{cases} a_0 = x(0) - \dfrac{\gamma}{\omega_0^2-\Omega^2} \\ \eta = 1 - \dfrac{\gamma^2}{2(\omega_0^2-\Omega^2)^2} \end{cases} \tag{5.63}$$

Figure 5.7 shows that the approximate solution, for the Van der Pol oscillator using the perturbation technique, is close to that found using a numerical integration the oscillator's differential equation in Mathematica.

The second-order solution was derived using Mathematica [87] and Fig. 5.8 displays the resulting response.

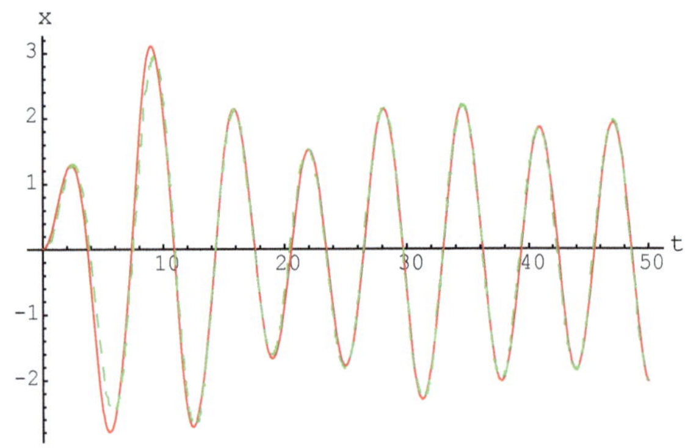

Fig. 5.7 Comparison of the first-order approximate solution with the exact solution for the Van Der Pol oscillator for an initial condition of $x_0 = 0.01$, $\varepsilon = 0.1$. (In the figure: (—) the numerical solution, and the approximate solution using MMS(- - - - -))

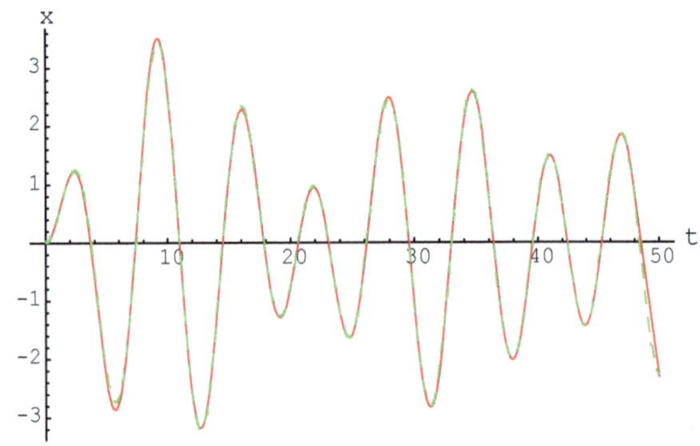

Fig. 5.8 Comparison of the second-order approximate solution with the exact solution for the Van Der Pol oscillator for an initial condition of $x_0 = 0.01$, $\varepsilon = 0.1$. (In the figure: (—) the numerical solution and the approximate solution using MMS(- - - - -))

(ii) The Rayleigh oscillator

Now, let us consider the Rayleigh oscillator in the case of a primary resonance Excitation ($\gamma = O(\varepsilon), \quad \Omega = \omega_0 + O(\varepsilon)$). Its dynamics is described by the following equation:

$$\frac{d^2x}{dt^2} - \varepsilon\left(1 - \left(\frac{dx}{dt}\right)^2\right)\frac{dx}{dt} + \omega_0^2 x = f(t), \text{ with: } \begin{cases} f(t) = \gamma\varepsilon\cos(\Omega t) \\ \Omega = \omega_0 + \varepsilon\sigma \end{cases}, \quad (5.64)$$

We seek an approximate solution of Eq. (5.64) using the method of Multiple Scales. An approach similar to that followed in the van der Pol oscillator is used to determine the following frequency-response equation of this oscillator:

$$\varepsilon^2\rho(2\eta - \frac{3}{4}\rho)^2 + 4\sigma^2\rho = \gamma^2, \text{ where } \begin{cases} \rho = \omega_0^2 a^2 \\ \eta = 1/2 \end{cases} \quad (5.65)$$

Similarly, the approximate solution of Eq. (5.64) is expressed in the following form:

$$x(t) = \sqrt{\frac{8\eta / 3\omega_0^2}{1 + (\frac{8\eta}{3a_0^2\omega_0^2} - 1)e^{-2\varepsilon\eta t}}}\cos(\omega t) + \frac{\gamma}{\omega_0^2\Omega^2}\cos(\Omega t), \quad (5.66)$$

where

$$a_0 = x(0) - \frac{\gamma}{\omega_0^2 - \Omega^2}, \quad \eta = \frac{1}{2} - \frac{3}{4}\frac{\gamma^2\Omega^2}{(\omega_0^2 - \Omega^2)^2} \quad (5.67)$$

Figure 5.9 displays the approximate solution with the one obtained by numerically integrating Eq. (5.64). We can clearly see that the agreement is excellent among both approaches. Frequency-response curves generated from Eq. (5.65) are presented in Fig. 5.10.

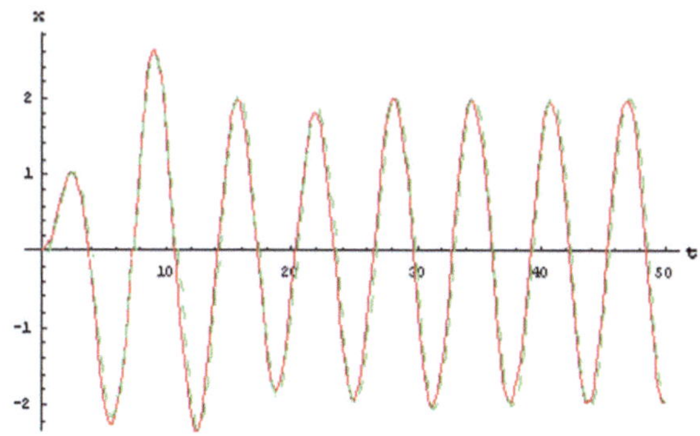

Fig. 5.9 Comparison of the approximate solution with that obtained by integrating the original equation for an initial condition $x_0 = 0.01$, $\varepsilon = 0.1$. (In the figure: (—) the numerical solution and the approximate solution using MMS(-----))

Fig. 5.10 Frequency-response curves for primary resonances of the Rayleigh Oscillator for various forcing amplitudes of $\gamma(\omega_0 = 1\ \mathrm{rad/s}$, and $\varepsilon = 1)$

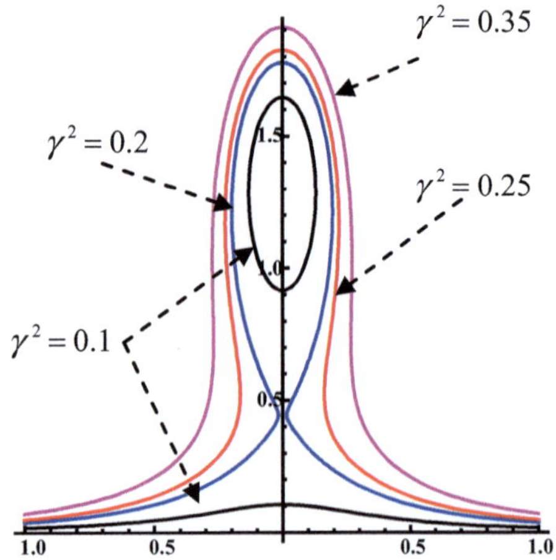

5.2.4 *Carbon Nanotubes*

(a) Structure overview

Single-walled carbon nanotubes (SWCNTs) can be considered to be formed by the rolling of a single layer of graphite (also called a graphene sheet) into a cylinder, Fig. 5.11.

Multi-walled carbon nanotubes (MWCNTs) can similarly be formed by considering coaxial assembly of cylinders of SWCNTs separated by the thickness of each graphene sheet.

Regarding the molecular structure of CNTs, they have a lattice-like structure [88] consisting of bounded carbon atoms, Fig. 5.11. Their geometric properties define their structural topology, which is likely similar to beams for small nanotube radius and to cylindrical shells for large nanotubes radius. This is true under certain specific conditions [34] that will be discussed later in this chapter. Figure 5.12 describes some of the geometric properties of CNTs, which are

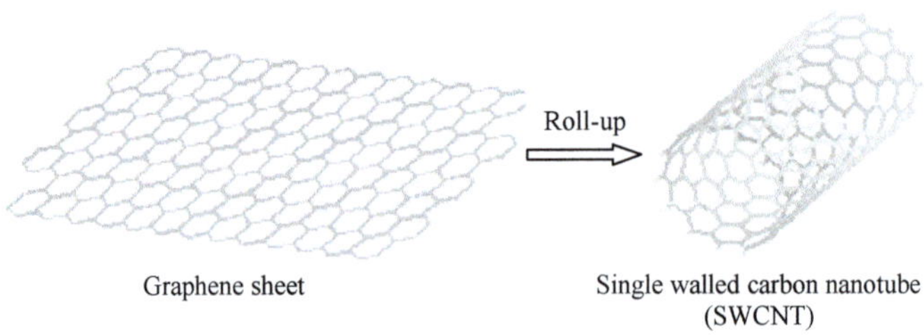

Fig. 5.11 Formation of a single-walled carbon nanotube from a graphene sheet

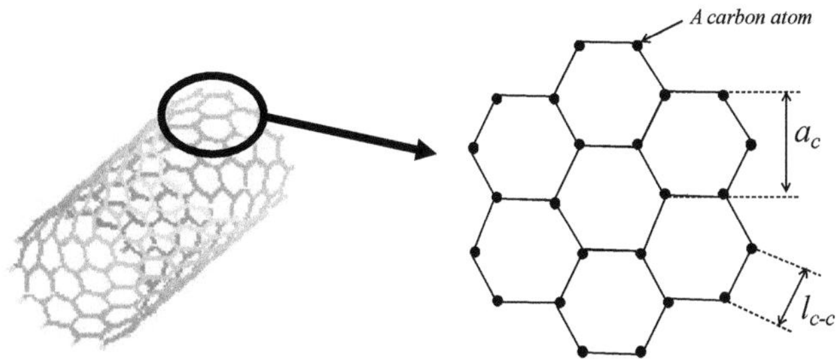

Fig. 5.12 Carbon atoms attachment in a single-walled carbon nanotube

- The distance of the adjacent carbon atoms (the length of the C–C bonds) denoted by l_{c-c} and found typically to be equal to 0.14 nm.
- The width of the hexagonal carbon rings denoted by a_c, which is typically equal to 0.24 nm [89].

Other important properties that define the geometry of a nanotube and that are not represented in Fig. 5.12 are

- The diameter of the nanotube (d_{NT}) which can be ranged from 0.4 nm to 100 nm and even higher [33, 34].
- The interlayer spacing (h) also called the graphene shell thickness found to be equal to 0.34 nm [33, 34]
- The length of the nanotube (L_{NT}) that may also range from 1 nm to 10 micrometer [33].

(b) The applicability of the continuum mechanics in modeling CNTs

In the following, we will present criteria for the applicability of the continuum mechanics in simulating the response of CNTs. Although some simulation results achieved in the literature were based on the molecular dynamics method, as we reviewed in Sect. 4.1, this method is still limited to CNTs with a small number of atoms and is therefore restrained to the study of small-scale modeling such as localized effects on small portions of the CNTs. So, in order to simulate the mechanical behavior of large-sized atomic CNTs, the continuum model was show to be practical in analyzing large-scale and the global mechanical behavior of CNTs [39] but under certain conditions [34].

Based on scaling analysis, Harik [33, 34] proposed following three nondimensional numbers to check the validity of the beam assumption for modeling the mechanics of CNTs:

- The homogenization criteria of the nanotube that ensures the molecular length-scales and described by the following equation:

$$a_c/L_{NT} \prec 1 \tag{5.68}$$

- The aspect ratio criteria described by the following equation:

$$d_{NT}/L_{NT} \prec 1 \tag{5.69}$$

- The linearity of strains criteria described by the following equation:

$$(L_{NT} - (L_{NT})_0)/L_{NT} \prec 1 \tag{5.70}$$

As will be shown later on, we have verified that all the applicability conditions of the continuum theory as stated by [34] are satisfied for the CNTs under consideration in this work.

5.3 Structural Behavior of Straight Carbon Nanotube Resonators

In this section, we investigate the nonlinear static and dynamic of both cantilevered and clamped-clamped straight carbon nanotubes (CNTs). We present numerical approaches and methodologies to predict and simulate the dynamic behavior of CNTs when driven by AC and DC loads ranging from small to large values. We will present numerical approaches to calculate accurately the resonance frequency of clamped-clamped and cantilever CNTs accounting for the effects of their geometric nonlinearities, DC and AC loads. This in turn should lead to accurate calibration and prediction for the mechanical properties of CNTs by relating the measured nonlinear resonance frequencies to the predicted one based on a proper model. Then, we present in-depth investigation for the dynamic response of CNT resonators when driven by large values of AC and DC loads. Instability regimes of the resonators as a function of the frequency and amplitude of the AC load will be shown. The instability regimes can increase the knowledge about the limitations and practical applications of the CNTs when used as resonators.

5.3.1 Problem Formulation

Here, we formulate the problem for the static and dynamic behavior of an electrically actuated CNT resonator. The considered boundary conditions for the CNT are a

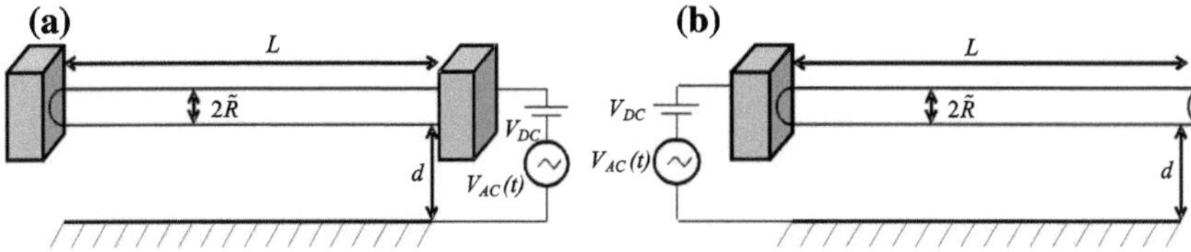

Fig. 5.13 Schematic of the electrically actuated (**a**) clamped-clamped and (**b**) cantilever CNT resonator

clamped-clamped beam, Fig. 5.13a, and a cantilever beam, Fig. 5.13b. The CNT is actuated by an electrode underneath it with a gap width d. It is modeled as an Euler-Bernoulli cylindrical beam of radius $\tilde{R}$, length L, and a quality factor Q. It has a cross-sectional area $A = \pi \tilde{R}^2$, area moment of inertia $I = \pi \tilde{R}^4/4$, natural frequency $\omega_n = \beta^2 \sqrt{EI/\rho A L^4}$ (β is equal to 4.73 for the case of a clamped-clamped beam and equal to 1.875 for the cantilever beam), and damping coefficient $\tilde{c} = \rho A \omega_n / Q$. The carbon nanotube is assumed to have a Young's modulus $E = 1$ *TPa* and a density $\rho = 1.35\,\text{g/cm}^3$ [90].

The equation of motion of a clamped-clamped and cantilever carbon nanotube resonator, Fig. 5.13a, b, can be written as [10, 12]

$$EI\frac{\partial^4 w}{\partial x^4} + \rho A \frac{\partial^2 w}{\partial t^2} + \tilde{c}\frac{\partial w}{\partial t} = F_{non} + F_{elec}, \tag{5.71}$$

where ε_0 is the air permittivity and the geometric nonlinearity term F_{non} is written for clamped-clamped and cantilever CNTs, respectively as

$$F_{non}^{clamped} = \left(\frac{EA}{2L}\int_0^L \left(\frac{\partial w}{\partial x}\right)^2 dx\right)\frac{\partial^2 w}{\partial x^2}, \tag{5.72}$$

$$F_{non}^{can} = EI\frac{\partial}{\partial x}\left[\frac{\partial w}{\partial x}\frac{\partial}{\partial x}\left(\frac{\partial w}{\partial x}\frac{\partial^2 w}{\partial x^2}\right)\right] \tag{5.73}$$

The electrostatic force per unit length expression is given as follows (see background section, Eq. (5.35)

$$F_{elect} = \frac{\pi \varepsilon_0 (V_{DC} + V_{AC}\cos(\tilde{\Omega}))^2}{\sqrt{(d-w)(d-w+2\tilde{R})}\left(\cosh^{-1}\left(1 + \frac{d-w}{\tilde{R}}\right)\right)^2} \tag{5.74}$$

The boundary conditions for the clamped-clamped CNT are

$$w(0,t) = 0, \quad \frac{\partial w}{\partial x}(0,t) = 0, \quad w(L,t) = 0, \quad \frac{\partial w}{\partial x}(L,t) = 0, \tag{5.75}$$

and for the cantilever CNT

$$w(0,t)=0,\quad \frac{\partial w}{\partial x}(0,t)=0,\quad \frac{\partial^2 w}{\partial x^2}(L,t)=0,\quad \frac{\partial^3 w}{\partial x^3}(L,t)=0 \tag{5.76}$$

For convenience, we introduce the following nondimensional variables:

$$\hat{w}=\frac{w}{d},\qquad \hat{x}=\frac{x}{L},\qquad \hat{t}=\frac{t}{T}, \tag{5.77}$$

where T is a time constant defined by $T=\sqrt{\rho AL^4/EI}$. Next, by dropping the hats, the nondimensional equations of motions and associated boundary conditions for the clamped-clamped and cantilever carbon nanotubes are written respectively as

$$\begin{aligned}\frac{\partial^4 w}{\partial x^4}+\frac{\partial^2 w}{\partial t^2}+c\frac{\partial w}{\partial t}&=\alpha_1\left(\int_0^1\left(\frac{\partial w}{\partial x}\right)^2 dx\right)\frac{\partial^2 w}{\partial x^2}+\\&+\frac{\alpha_2(V_{DC}+V_{AC}\cos(\Omega t))^2}{\sqrt{(1-w)(1-w+2R)}\left(\cosh^{-1}\left(1+\frac{1-w}{R}\right)\right)^2},\end{aligned} \tag{5.78}$$

$$w(0,t)=0,\quad \frac{\partial w}{\partial x}(0,t)=0,\quad w(1,t)=0,\quad \frac{\partial w}{\partial x}(1,t)=0, \tag{5.79}$$

$$\begin{aligned}\frac{\partial^4 w}{\partial x^4}+\frac{\partial^2 w}{\partial t^2}+c\frac{\partial w}{\partial t}&=\alpha_3\frac{\partial}{\partial x}\left[\frac{\partial w}{\partial x}\frac{\partial}{\partial w}\left(\frac{\partial w}{\partial x}\frac{\partial^2 w}{\partial x^2}\right)\right]+\\&+\frac{\alpha_2(V_{DC}+V_{AC}\cos(\Omega t))^2}{\sqrt{(1-w)(1-w+2R)}\left(\cos h^{-1}\left(1+\frac{1-w}{R}\right)\right)^2},\end{aligned} \tag{5.80}$$

$$w(0,t)=0,\quad \frac{\partial w}{\partial x}(0,t)=0,\quad \frac{\partial^2 w}{\partial x^2}(1,t)=0,\quad \frac{\partial^3 w}{\partial x^3}(1,t)=0, \tag{5.81}$$

where

$$\alpha_1=2\left(\frac{d}{R}\right)^2,\quad \alpha_2=\frac{\pi\varepsilon_0 L^4}{EId^2},\quad \alpha_3=\left(\frac{d}{L}\right)^2,\quad c=\tilde{c}\frac{L^4}{EI},\quad \Omega=\frac{\tilde{\Omega}}{\omega_n},\quad R=\frac{\tilde{R}}{d} \tag{5.82}$$

The Reduced-Order Model

To simulate the response of the CNT, Eqs. (5.78)–(5.81) are discretized using the Galerkin procedure to yield a ROM. Hence, the deflection of the CNT is approximated as

$$w(x,t)=\sum_{i=1}^{n}u_i(t)\,\phi_i(x), \tag{5.83}$$

where $\phi_i(x)$ are the normalized linear undamped mode shapes of a straight beam and $u_i(t)$ are the nondimensional modal coordinates. To obtain the ROM, we substitute Eq. (5.83) into Eqs. (5.78)–(5.81), multiply by $\phi_i(x)$, use the orthogonality conditions of the mode shapes, and then integrate the outcome from 0 to 1. The results are differential equations in terms of the modal coordinates $u_i(t)$.

Several points need to be clarified about the developed ROM. Unlike the case of MEMS [72], here there is no numerical advantage of multiplying Eq. (5.78) or Eq. (5.80) by the denominator of the electrostatic force term since the mode shape $\phi_i(x)$ will remain embedded inside the square root term and the inverse hyperbolic cosine term even after imposing the orthogonality of the mode shapes. To deal with the complicated integral terms due to the electrostatic force, we evaluated the spatial integrals containing the $\phi_i(x)$ terms numerically, using a trapezoidal method, simultaneously while integrating the differential equations of the modal coordinates $u_i(t)$ with time.

As an example, assuming one mode, the modal equation describing $u_1(t)$ for a clamped-clamped CNT is written as

$$\ddot{u}_1(t) + C\dot{u}_1(t) + Ku_1(t) = \alpha_1 \Gamma u_1^3(t) + \\ + \int_0^1 \frac{\alpha_2 (V_{DC} + V_{AC}\cos(\Omega t))^2 \phi_1(x)}{\sqrt{(1-\phi_1(x)u_1(t))(1-\phi_1(x)u_1(t)+2R)}\left(\cosh^{-1}\left(1+\frac{1-\phi_1(x)u_1(t)}{R}\right)\right)^2}^2 dx, \tag{5.84}$$

where

$$K = \int_0^1 \left(\phi_1(x)\,\phi_1^{iv}(x)\right)dx, \qquad C = c\int_0^1 (\phi_1(x))dx, \tag{5.85}$$

and

$$\Gamma = \left[\int_0^1 \left(\phi_1(x)\,\phi_1''(x)\right)dx\right]\left[\int_0^1 \left(\phi_1'(x)\right)^2 dx\right] \tag{5.86}$$

The Static Analysis

Various case studies of carbon nanotubes are considered for simulations, Table 5.1. We first examine the convergence of the ROM. Figure 5.14 shows the normalized maximum static deflection of a clamped-clamped carbon nanotube $w_{\max} = w_s(x = 0.5)$ for case 2 of Table 5.1 using one, two, and three symmetric modes of the ROM while varying the DC load. It follows from the figure that using one mode yields acceptable converged results. This result shows a clear difference between CNTs and other MEMS beams [72], where at least three symmetric modes need to be used for convergence. One possible justification for this is due to the high stiffness of CNTs making the contribution of the higher-order modes weak compared to the first one.

Next, we verify the obtained results using one mode in the ROM by comparing them with the simulation results of Pugno et al. [15], which are based on a finite-difference method. Following Pugno et al. [15], we first conduct a comparison

Table 5.1 The geometrical properties of the studied carbon nanotube beams

Case #	Boundary conditions	d [nm]	L [nm]	$\tilde{R}$ [nm]	References
1	Clamped-clamped	100	3000	20	[15]
2	Clamped-clamped	100	3000	30	[15]
3	Clamped-clamped	100	2000	30	[15]
4	Cantilever	390	2500	5.45	[91]
5	Cantilever	3000	6800	23.5	[47]

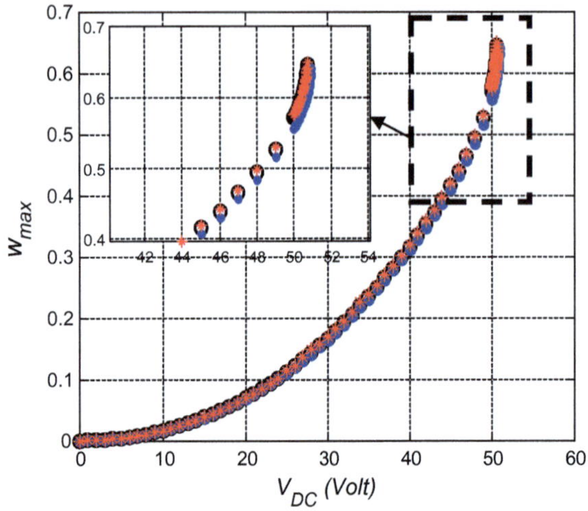

Fig. 5.14 Variation of the normalized static deflection of the carbon nanotube with the DC voltage for case 2 of Table 5.1. In the figure: (∘) one-mode ROM, (•) two modes ROM, and (*) three modes ROM

assuming a linear beam model for the case of clamped-clamped carbon nanotube (without mid-plane stretching). Figure 5.15a shows the results for cases 1 and 2 of Table 5.1 indicating excellent agreement between our results and the results of Pugno et al. [15]. In addition, we show the unstable branches of the equilibrium solutions. We can see that both the unstable and stable branches collide at pull-in. Also, it is worth to note that pull-in occurs here at a normalized deflection approximately equal 0.46, compared to 0.33 in the linear case of electrostatically actuated structures [78]. This can be attributed to the geometrical shape of the carbon nanotube and the difference between the electrostatic force field in this case and the case of parallel-plate rectangular shaped capacitor.

Next, we include the mid-plane stretching term in the simulation and compare with the nonlinear model results of Pugno et al. [15], Fig. 5.15a. The figure also shows excellent agreement; thereby validating the ROM. Comparing Fig. 5.3a, b, one can see the importance of mid-plane stretching in predicting accurately the pull-in voltage of the carbon nanotube. Further, it is noted that pull-in occurs here at $w_{max} = 0.64$, compared to 0.45 in the nonlinear case of an electrostatically actuated microbeam of rectangular cross section [72]. Table 5.2 shows a comparison between

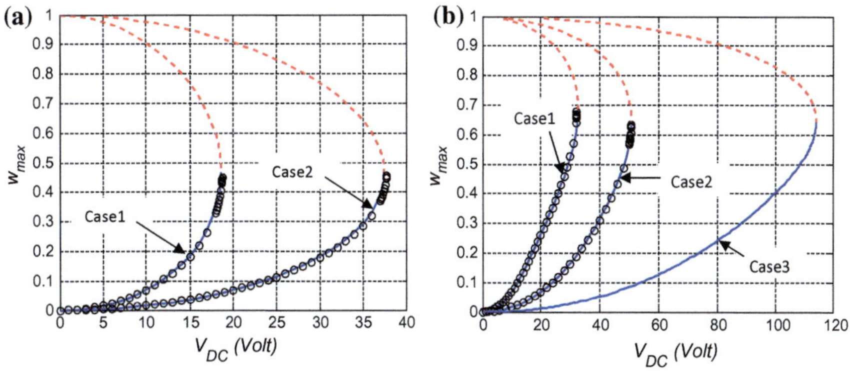

Fig. 5.15 Variation of the normalized static deflection of clamped-clamped carbon nanotubes with the DC voltage, (**a**) without including mid-plane stretching and (**b**) with including mid-plane stretching. In the figure: (—) stable branch, (- - - - -) unstable branch, and (o) results of Pugno et al. [15]

Table 5.2 The calculated pull-in voltages for the cases of Table 5.1

Case #	Pull-in voltage [Volt] (linear theory)	Pull-in voltage [Volt] (nonlinear theory)
1	18.62	32.65
2	37.46	50.72
3	82.47	114.2
4	2.309	2.31
5	50.20	48.26

the calculated pull-in voltage for the cases of Table 5.1 using linear and nonlinear beam theories. As noted from the table, the error in using linear theory to predict the behavior of CNTs can be significant.

Next, we show results for cantilever carbon nanotubes. In Fig. 5.16a, b, we validate the one-mode ROM results for the cantilever CNT case by comparing them with the experimental data reported in Akita et al. [91] and Pugno et al. [15]. In the figures, the normalized maximum static deflection of the cantilever carbon nanotube is $w_{\max} = w_s(x = 1)$. The ROM includes the geometric nonlinearities of the cantilever CNT. In Fig. 5.16a, b, the parameters of cases 4 and 5 of Table 5.1 were used, respectively. The experimentally measured pull-in voltages are 2.33 Volt [91] and 48.26 Volt [15] for cases 4 and 5, respectively, whereas the obtained values using the ROM are 2.31 Volt and 48.26 Volt. It is clear that the experimental data and the ROM results are in good agreement. As noted, the effect of the geometric nonlinearities in the cantilever case is less than it is for the case of clamped-clamped carbon nanotubes.

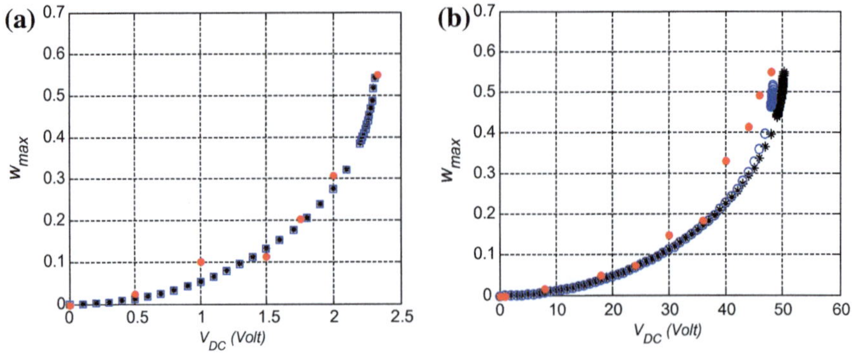

Fig. 5.16 Variation of the normalized static deflection of cantilever carbon nanotubes with the DC voltage for (**a**) Case 4 of Table 5.1 and (**b**) Case 5 of Table 5.1. In the figure: (*) ROM (one mode, linear case), (□) ROM (two modes, linear case), (○) ROM (two modes, nonlinear case), and (•) experimental data of (**a**) Akita et al. [91] and (**b**) Pugno et al. [15]

Free-Vibration Problem Under a DC Voltage

(a) Numerical approaches

In this section, we present two computationally-efficient approaches to solve the free-vibration problem of the CNT under a DC electrostatic load. First, we use the ROM of Sect. 4.3.2. In this approach, the static deflection of the CNT needs to be solved for each DC voltage. For a given voltage, we substitute the stable static solution, found by solving the static equations as we did in Sect. 4.3.3, into the Jacobian matrix of the ROM. Then, the eigenvalues of the calculated Jacobian matrix are solved. By taking the magnitudes of each individual eigenvalue, the natural frequencies of the system are obtained. This is similar to what we have been done in [92, 93].

The second approach that can be used is through solving numerically the original boundary value problem, Eq. (5.71), using numerical software such as Mathematica® [87]. This approach is considered more accurate compared to the ROM since the latter approximates the original problem while the former deals with the problem exactly. This approach can be applied on Eqs. (5.71) and (5.72) for clamped-clamped CNTs combined with a shooting technique to iterate on the integral term of Eq. (5.72) [78]. Also, it can be applied directly on Eqs. (5.71) and (5.73) for cantilever CNTs. However this technique suffers convergence problems for clamped-clamped CNTs because of the mid-plane stretching term. To resolve this problem, one can split the in-plane equation from the out-of-plane equation in Eq. (5.71) and then solve the coupled system. Here, out-of-plane means motion transversal to the beam axis (toward the substrate) and in-plane means across the beam length. Next, we rewrite the beam equation as two coupled partial differential equations governing the transverse, w, and axial, u, deflections of the beam as [94, 95].

$$EI\frac{\partial^4 w}{\partial x^4}+\rho A\frac{\partial^2 w}{\partial t^2}=EA\left(\frac{\partial u}{\partial x}+\frac{1}{2}\left(\frac{\partial w}{\partial x}\right)^2\right)^2\frac{\partial^2 w}{\partial x^2}+ \\ +\frac{\pi\varepsilon_0(V_{DC}+V_{AC}\cos(\tilde{\Omega}t))^2}{\sqrt{(d-w)(d-w+2\tilde{R})}\left(\cosh^{-1}\left(1+\frac{d-w}{\tilde{R}}\right)\right)^2}, \tag{5.87}$$

$$\rho A\frac{\partial^2 u}{\partial t^2}=EA\left(\frac{\partial^2 u}{\partial x^2}+\frac{\partial w}{\partial x}\frac{\partial^2 w}{\partial x^2}\right) \tag{5.88}$$

The boundary conditions governing w and u respectively are

$$w(0,t)=0,\quad \frac{\partial w}{\partial x}(0,t)=0,\quad w(L,t)=0,\quad \frac{\partial w}{\partial x}(L,t)=0, \\ u(0,t)=0,\quad u(L,t)=0 \tag{5.89}$$

Because the natural frequency in the axial direction is much larger than the one in the transverse direction [95], we drop the inertia term of the u component in Eq. (5.88), and get the following equation:

$$\frac{\partial^2 u}{\partial x^2}=-\frac{\partial w}{\partial x}\frac{\partial^2 w}{\partial x^2} \tag{5.90}$$

Equations (5.87) and (5.90) can now be solved simultaneously for u and w. Using the nondimensional variables defined in Eq. (5.77) and dropping the hats yield

$$\frac{\partial^4 w}{\partial x^4}+\frac{\partial^2 w}{\partial t^2}=\alpha_1\left(\frac{\partial w}{\partial x}\right)^2\frac{\partial^2 w}{\partial x^2}+\alpha_4\frac{\partial u}{\partial x}\frac{\partial^2 w}{\partial x^2}+ \\ +\frac{\partial_2(V_{DC}+V_{AC}\cos(\Omega t))^2}{\sqrt{(1-w)(1-w+2R)}\left(\cosh^{-1}\left(1+\frac{1-w}{R}\right)\right)^2}, \tag{5.91}$$

$$\frac{\partial^2 u}{\partial x^2}=-\alpha_3\frac{\partial w}{\partial x}\frac{\partial^2 w}{\partial x^2}, \tag{5.92}$$

where $\alpha_1, \alpha_2, \alpha_3$ are given by Eq. (5.82) and $\alpha_4 = 4(L/\tilde{R})^2$.

We split the transverse and axial deflections into static components due to the DC voltage, denoted by $w_s(x)$ and $u_s(x)$ respectively, and dynamic components denoted by $w_d(x,t)$ and $u_d(x,t)$ respectively, that is,

$$w(x,t)=w_s(x)+w_d(x,t), \tag{5.93}$$

$$u(x,t)=u_s(x)+u_d(x,t) \tag{5.94}$$

Substituting Eqs. (5.93) and (5.94) into Eqs. (5.91) and (5.92), dropping the AC forcing term, and retaining only the linear terms in w_d and u_d yields the following equations describing the small free vibration of the clamped-clamped carbon nanotube:

$$w_d{}''{}'' + \ddot{w}_d = \alpha_1 (w_s'^2 w_d{}'' + 2 w_s{}' w_s{}'' w_d^1) + \alpha_4 (u_s' w_d{}'' + w_s{}'' u_d^1) +$$

$$+ \left(\frac{\alpha_2 V_{DC}^2 \left(2\sqrt{(1-w_s)(1-w_s+2R)} + (1-w_s+R)\cosh^{-1}\left(1+\frac{1-w_s}{R}\right)\right)}{((1-w_s(1-w_s)(1-w_s+2R))^{\frac{3}{2}} \left(\cosh^{-1}\left(1+\frac{1-w_s}{R}\right)\right)^3} \right) w_d, \quad (5.95)$$

$$u_d{}'' = -\alpha_3 w_s{}' w_d{}'' - \alpha_3 w_s{}'' w_d' \quad (5.96)$$

To derive the eigenvalue problem, we let

$$w_d(x,t) = \Phi(x)\, e^{i\omega t}, \quad (5.97)$$

$$u_d(x,t) = \psi(x)\, e^{i\omega t}, \quad (5.98)$$

where $\Phi(x)$ and $\psi(x)$ are the associated transverse and axial eigenfunctions respectively and ω is the nondimensional natural frequency. Substituting Eqs. (5.97) and (5.98) into Eqs. (5.95) and (5.96) and solving the resulting equations yield the mode shapes Φ and ψ and their associated natural frequency ω.

(b) Results

To start, we use a one mode approximation in the ROM of Sect. 4.3.2 to calculate the natural frequencies and then compare the results to those obtained by solving the boundary value problem directly. In Fig. 5.17, we compare the variation of the first natural frequency of a clamped-clamped CNT (case 2 of Table 5.1) using the two methods. We can conclude from the figure that the one-mode ROM converges and follow the CNT behavior even near the pull-in instability.

Figure 5.18a depicts the fundamental natural frequency calculated using the ROM (o) and the results of Dequesnes et al. [12] for a clamped-clamped carbon nanotube of length = 20.7 nm, radius = 0.9 nm, and gap width = 3 nm. The figure shows that the natural frequency decreases slightly for small ranges of the DC voltage, then increases again as the effect of mid-plane stretching of the CNT increases, and then drops suddenly to zero near pull-in. As seen in the figure, the model shows high robustness in tracking the increase and then the drop of the natural frequency to zero near pull-in. Also, the figure indicates good agreement with the results of Dequesnes et al. [12], which were obtained by a molecular dynamics model.

Because the gap width of this case study is too small (below 3 nm), van der Walls forces can have significant effect on the results [12]. To investigate this effect, we add the van der Walls forces term to the beam equation, which becomes [10, 12]

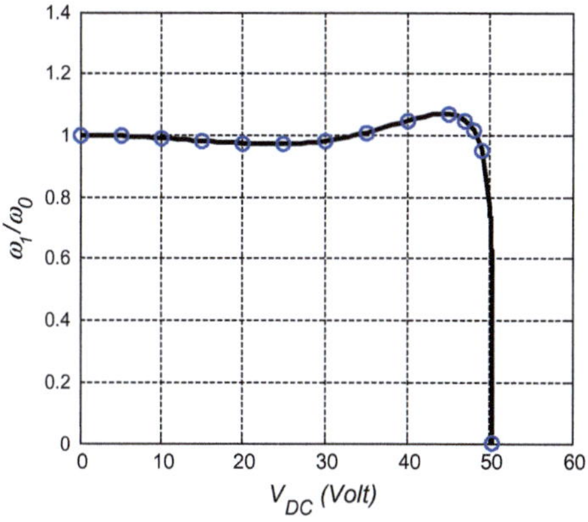

Fig. 5.17 Variation of the fundamental natural frequency ω_0 normalized with that at zero voltage ω_0 for various values of DC voltage. Results are shown for the carbon nanotube of case 2 of Table 5.1. In the figure: (o) are the results obtained by solving the boundary value problem directly and (—) are those obtained using a one-mode ROM

$$EI\frac{\partial^4 w}{\partial x^4}+\partial A\frac{\partial^2 w}{\partial t^2}+\tilde{c}\frac{\partial w}{\partial t}=\left(\frac{EA}{2L}\int_0^L\left(\frac{\partial w}{\partial x}\right)^2 dx\right)\frac{\partial^2 w}{\partial x^2}+F_{elec}+F_{vdw}, \tag{5.99}$$

where

$$F_{elect}=\frac{\pi\varepsilon_0(V_{DC}+V_{AC}\cos(\tilde{\Omega}t))^2}{\sqrt{r(r+2\tilde{R})}\left(\cosh^{-1}\left(1+\frac{r}{\tilde{R}}\right)\right)^2},$$
$$F_{vdw}=-\frac{C_6\pi^2\sigma^2\tilde{R}\sqrt{r(r+2\tilde{R})}(8r^4+32r^3\tilde{R}+72r^2\tilde{R}^2+80r\tilde{R}^3+35\tilde{R}^4)}{2r^5(r+2\tilde{R})^5}, \tag{5.100}$$

where $r=d-w$ is the distance between the CNT and the substrate, $C_6=15.2eV\overset{o}{A}$ is a constant characterizing the interaction between two carbon atoms and $\sigma=38\,\text{nm}^{-2}$ is the substrate layer (graphite) surface density. By applying the procedure of Sects. 4.3.2 and 4.3.4 on Eqs. (5.99) and (5.100), we obtain the natural frequencies under the effect of both the electrostatic and van der Walls forces.

Figure 5.18a compares the obtained results with and without the van der Walls forces. As seen, in this case the van der Walls forces have negligible effect. Figure 5.18b shows another case of smaller gap width, $d=1$ nm, for a clamped-clamped carbon nanotube of length $=20.7$ nm and radius $=0.9$ nm. The figure shows comparisons among the results of the ROM and Dequesnes et al. [12] with and without van der Walls forces. There are good agreements among all results. In this case, as shown from Fig. 5.18b, the van der Walls forces have significant impact on changing the natural frequencies and the pull-in voltage. It is worth to mention that

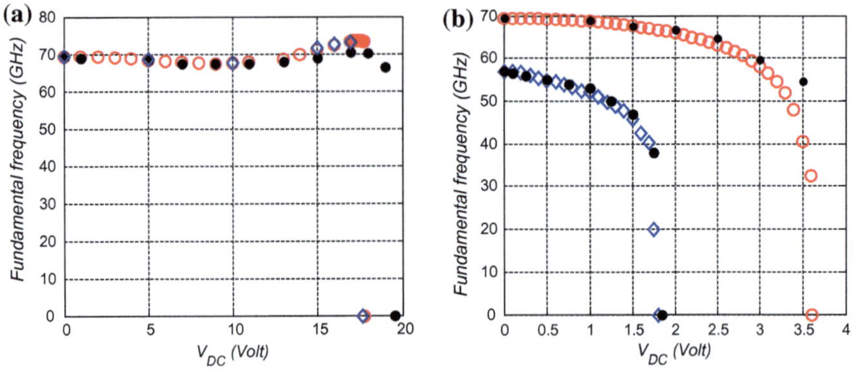

Fig. 5.18 Variation of the fundamental natural frequency with the DC voltage for two clamped-clamped carbon nanotubes. Figure 5.18a is for a CNT of length 20.7 nm and gap width 3 nm and Fig. 5.18b is for a CNT of length 20.7 nm and gap width 1 nm. In the figure: (∘) ROM without including van der Wall forces, (◇) ROM with including van der Wall forces, and (•) results of Desquesnes et al. [12]

Fig. 5.19 Variation of the normalized fundamental natural frequency ω_1 with the DC voltage for clamped-clamped carbon nanotubes for case 1 (∘), case 2 (*), and case 3 (•) of Table 5.1

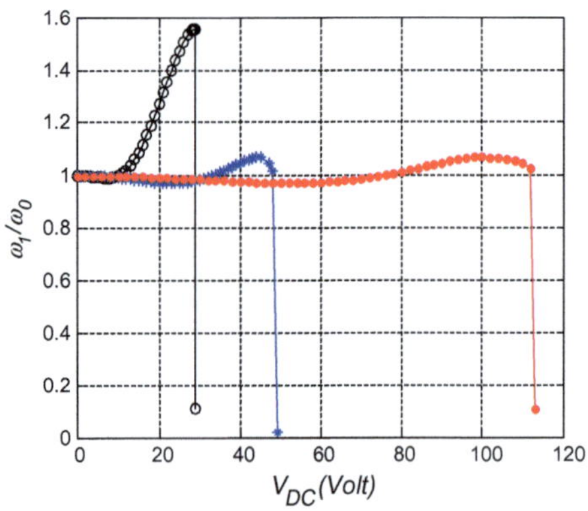

from a computational point of view, the ROM is very cheap compared the molecular dynamics model of Dequesnes et al. [12].

Figure 5.19 shows the variation of the fundamental natural frequency of the clamped-clamped CNTs for cases 1, 2, and 3 of Table 5.1. As seen in the figure, the fundamental frequency increases to higher values as the electrostatic force increases for all the considered cases and then decreases to zero when pull-in occurs. This is due to the fact that the effect of mid-plane stretching dominates that of the electrostatic force. This agrees with the reported experimental data of [4]. One can see from Fig. 5.19 that both parameters (the length and the radius) of the CNT can be used to tune the fundamental natural frequency to be almost unchanged over an extended range of DC voltages (see the curve of case 3). This attractive feature can be promising for the implementation of CNTs as resonant sensors.

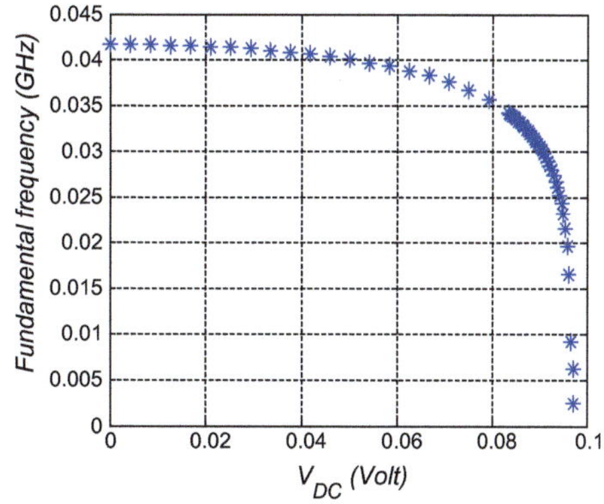

Fig. 5.20 Variation of the fundamental natural frequency with the DC voltage for a cantilever carbon nanotube (case 4 of Table 5.1)

We end this section by showing the variation of the natural frequency of a cantilever CNT with the DC voltage. Because of the fact that the geometric nonlinearity is weak in this case, the electrostatic force is always dominant and hence the CNT experiences a softening behavior. Figure 5.20 shows simulation results for the cantilever beam of case 4 of Table 5.1. We notice in the figure a monotonic decrease in the natural frequency until it reaches zero at pull-in.

Dynamic Response to DC and AC Loads

(a) Primary resonance of the fundamental mode

Here, we simulate the frequency-response of the carbon nanotube when excited by a DC load superimposed to an AC harmonic load of frequency near its fundamental natural frequency (primary-resonance excitation, $\Omega \approx \omega_1$). In the dynamic analysis, long-time integration for the reduced-order model equations of motion can be used. However, this method suffers convergence problems near bifurcations and instabilities and in general is not considered a robust method for studying nonlinear vibrations. Hence, a second method will be used, which is called the shooting technique [96]. The shooting method is a numerical technique to find periodic solutions, analyze their stability, and also locate and identify bifurcation points. This method will be used in conjunction with the Floquet theory [96] to study the stability of the captured periodic orbits.

We first investigate the response of carbon nanotubes to small AC and DC load (Figs. 5.21 and 5.22). In all subsequent figures, dashed lines refer to unstable solutions. Figure 5.21 shows a frequency-response curve of a clamped-clamped CNT of case 2 of Table 5.1. The figure compares the results of the shooting technique, based on one mode approximation, to the long-time integration technique using one and two modes in the ROM. The figure verifies the convergence of the ROM for the dynamic simulations and the fact that one mode yields adequate accuracy. Figure 5.9 shows a hardening behavior of the clamped-clamped CNT even for this small value of electric load, which is away from the pull-in voltage (50.7 Volt). This represents another sign for the dominant effect of mid-plane stretching of clamped-clamped

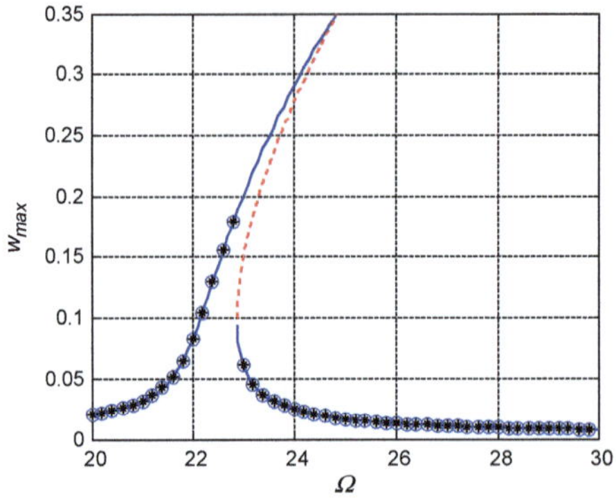

Fig. 5.21 Frequency-response curve of the CNT for case 2 of Table 5.1 shows the shift in the nonlinear resonance frequency in the hardening-behavior case. Results are shown for $V_{DC} = 5$ Volt, $V_{AC} = 2$ Volt, and Q = 100. In the figure, $w_{\max} = w(0.5, t)$ is the mid-point/maximum deflection of a clamped-clamped CNT, (—) shooting method (stable branch), (- - - - -) shooting method (unstable branch), (*) long-time integration (one-mode ROM), and (o) long-time integration (two modes ROM)

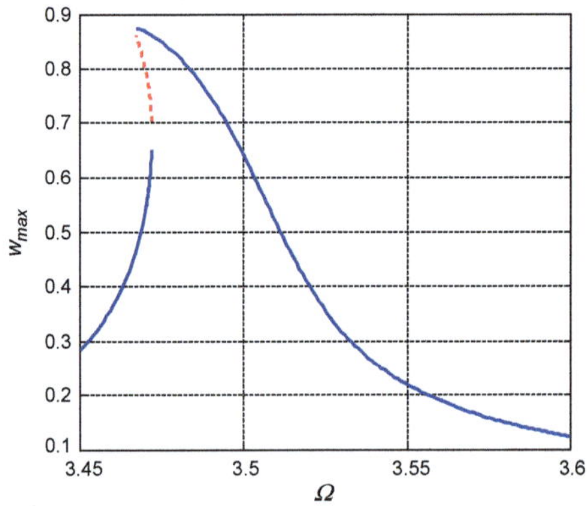

Fig. 5.22 Frequency-response curve of the CNT for case 4 of Table 5.1 showing the shift in the nonlinear resonance frequency in the softening-behavior case. Results are shown for $V_{DC} = 0.5$ Volt, $V_{AC} = 0.13$ Volt, and $Q = 100$. In the figure, $w_{\max} = w(1, t)$ is the tip/maximum deflection of a cantilever CNT, (—) stable branch, and (- - - - -) unstable branch

CNTs. Figure 5.22 shows a frequency-response curve of a cantilever CNT of case 4 of Table 5.1. The figure shows a weak softening behavior even for large values of V_{AC}.

Next, we show by simulation and using the shooting technique the shift in the resonance frequency Ω_r [79], which is the quantity being measured experimentally and is influenced by the AC amplitude, compared to the linear natural frequency ω_1, which depends on the DC voltage only. Figure 5.23a, b show the variation of the normalized nonlinear resonance frequency Ω_r/ω_1 for clamped-clamped and cantilever

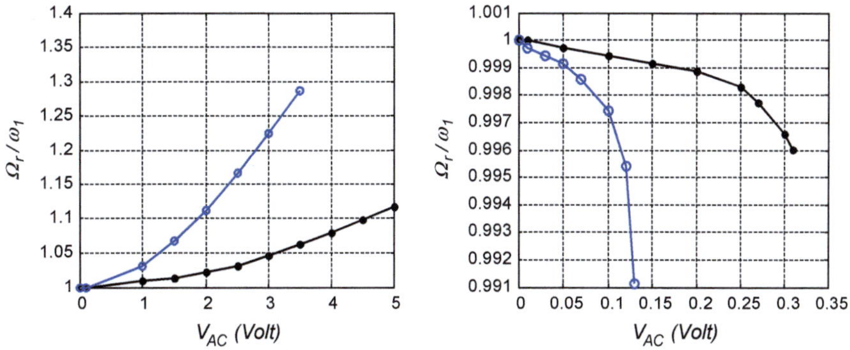

(a) Case 2 of Table 5.1 for (•) V_{DC}=2 Volt, and (•) V_{DC}=5 Volt.

(b) Case 4 of Table 5.1 for (•) V_{DC}=0.2 Volt, and (•) V_{DC}=0.5 Volt.

Fig. 5.23 The normalized nonlinear resonance frequency for two cases of CNTs calculated using the shooting technique for $Q = 100$

CNTs, respectively, for various values of AC and DC loads. We can see that the resonance frequency increases considerably with the AC load in the case of hardening behavior for the clamped-clamped CNT but decreases slightly in the case of softening behavior for the cantilever CNT. It is concluded that the strong hardening behavior of the clamped-clamped CNTs makes them almost useless in the linear regime, in agreement with the observation of Postma et al. [3].

Next, we examine the case of the CNT of Fig. 5.23 when V_{DC} is increased to 25 Volt, Fig. 5.24. The figure shows also a hardening-type behavior. In addition, the figure shows that the upper branch of the frequency-response curve opens up and loses stability through a saddle-node bifurcation at an excitation frequency near 28, where its slope approaches infinity and one Floquet multiplier approaches unity. This is an indication of a dynamic pull-in, as has been found for the case of clamped-clamped microbeams [84].

Fig. 5.24 Frequency-response curve of the clamped-clamped carbon nanotube of case 2 of Table 5.1 and for $V_{DC} = 25$ Volt, $V_{AC} = 5$ Volt, and $Q = 100$. In the figure: (——) stable branch, (- - - - -) unstable branch

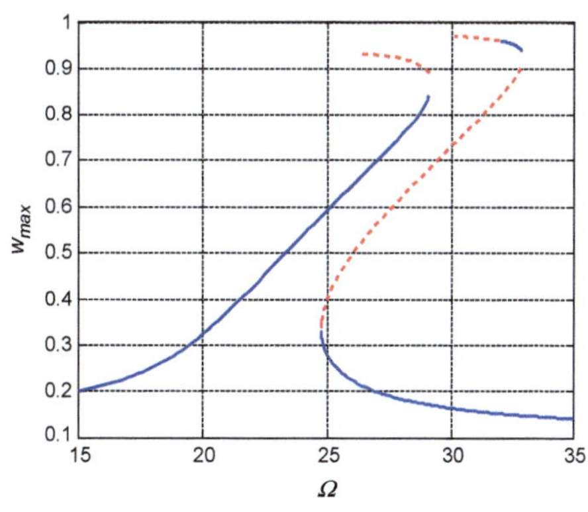

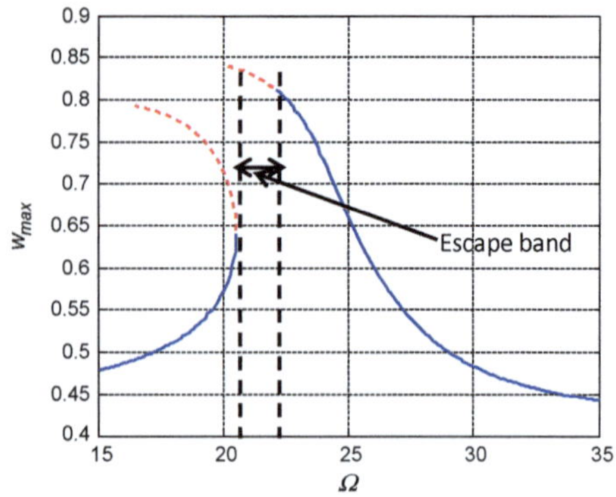

Fig. 5.25 Frequency-response curve of the clamped-clamped carbon nanotube of case 3 of Table 5.1 and for $V_{DC} = 100$ Volt, $V_{AC} = 5$ Volt, and $Q = 100$. In the figure: (—–) stable branch, (- - - - -) unstable branch

As the value of the DC voltage is increased further, the frequency-response of the carbon nanotube in this case remains of hardening type until reaching the static pull-in voltage (it does not reverse to softening behavior as in MEMS microbeams [84]). This is because in this considered case, the effect of mid-plane stretching is always dominant over the electrostatic force, except at pull-in. This seems to be a unique feature for CNTs over other structures.

To demonstrate a softening-type behavior, we choose another case study, case 3 of Table 5.1, where mid-plane stretching has a slightly weaker effect. Hence, at some DC load before static pull-in value, the electrostatic force nonlinearity dominates that of mid-plane stretching. Figure 5.25 shows the results. The figure shows a dynamic pull-in through period doubling bifurcations, where we found a Floquet multiplier reaching negative one [84, 96]. Figure 5.25 also shows a band of frequencies where there is no stable solution for the system. This band is called an inevitable escape band [85], where the oscillator is forced to escape its potential well. In this case the carbon nanotube is forced to escape to pull-in (see end of Sect. 4.2.2).

Increasing the AC load further in this case results in a larger escape band of frequencies. We use the shooting technique to calculate the carbon nanotube escape bands for the primary resonance case while varying the AC voltage and frequency (instability tongues). Figure 5.26a, b show the results for case 2 of Table 5.1 with $V_{DC} = 25$ Volt (hardening behavior) and for case 3 of Table 5.1 with $V_{DC} = 100$ Volt (softening behavior), respectively. Operating the resonator within those instability limits leads to definite pull-in. It is clear also from the figures that if the AC amplitude increases, the escape band limits increase.

We end this section by investigating the response of cantilever carbon nanotubes. In this case, the frequency-response curve is always of softening-type because of the weak effect of the geometric nonlinearity and the strong effect of the electrostatic nonlinearity. Figure 5.27 shows the frequency-response curve of the carbon nanotube resonator for case 4 of Table 5.1. The figure shows also a dynamic pull-in through period doubling bifurcations, where we found Floquet multipliers exceeding the unit circle through negative one.

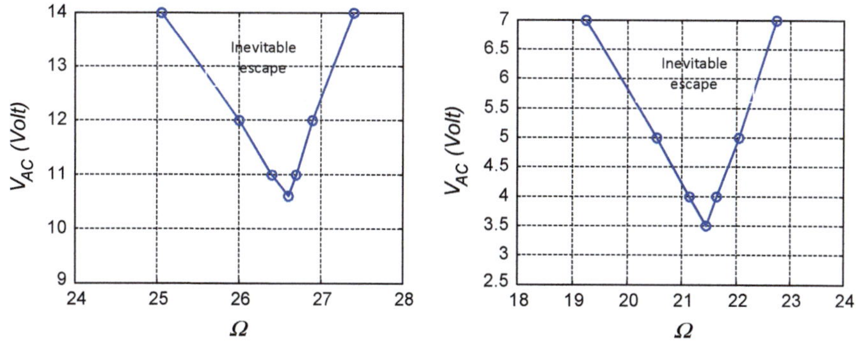

(a) Hardening. (b) Softening.

Fig. 5.26 The calculated instability tongues (inevitable escape bands) in the case of primary resonance of a clamped-clamped carbon nanotube (**a**) for the case 2 of Table 5.1 with $V_{DC} = 25$ Volt, (**b**) for the case 3 of Table 5.1 with $V_{DC} = 100$ Volt and $Q = 100$

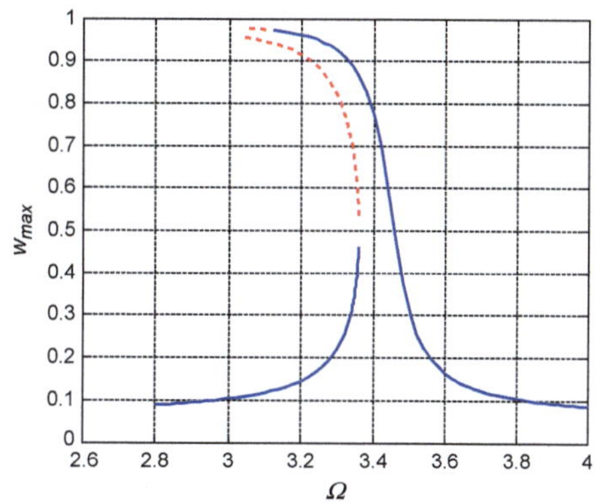

Fig. 5.27 Frequency-response curve of a cantilever carbon nanotube for case 4 of Table 5.1 and for $V_{DC} = 1$ *Volt*, $V_{AC} = 0.1$ Volt, and $Q = 100$. In the figure, $w_{\max} = w(1, t)$ is the tip/maximum deflection of the carbon nanotube, (—) stable branch, and (-----) unstable branch

(b) Secondary resonances of the fundamental mode

This section examines the carbon nanotube dynamics near superharmonic resonance of order two (excitation near half the fundamental natural frequency) and subharmonic resonances of order one half and one third (excitation near twice and three times the fundamental natural frequency).

Figure 5.28 shows the response of the clamped-clamped carbon nanotube of case 2 of Table 5.1 when excited near superharmonic resonance of order two of the fundamental mode. The figure shows a hardening-type behavior and a qualitatively similar behavior to that near primary resonance of Fig. 5.24. Also, the figure indicates the occurrence of dynamic pull-in through a saddle-node bifurcation in the upper branch of the curve.

We now examine the response of the same carbon nanotube near subharmonic resonances. It is worth to note that the activation of subharmonic resonance requires exceeding specific thresholds of AC load and quality factor. We excited the carbon nanotube by $V_{DC} = 20$ Volt and $V_{DC} = 1.2$ Volt near twice its natural frequency

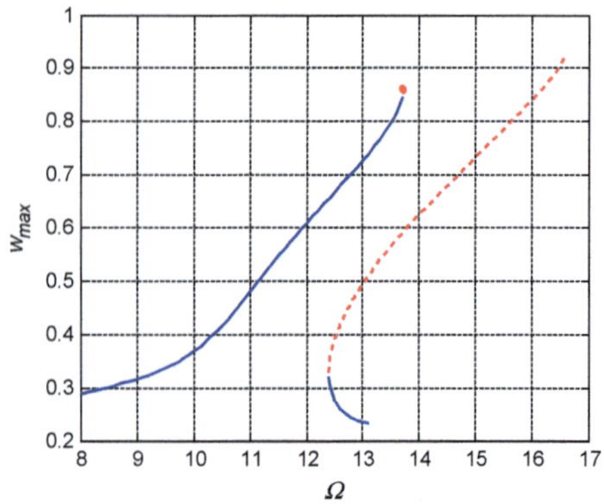

Fig. 5.28 Frequency-response curve of the clamped-clamped carbon nanotube for the superharmonic resonance of order two for case 2 of Table 5.1 for $V_{DC} = 20$ Volt, $V_{AC} = 16$ Volt and $Q = 150$. In the figure: (—) stable branch, (- - - - -) unstable branch

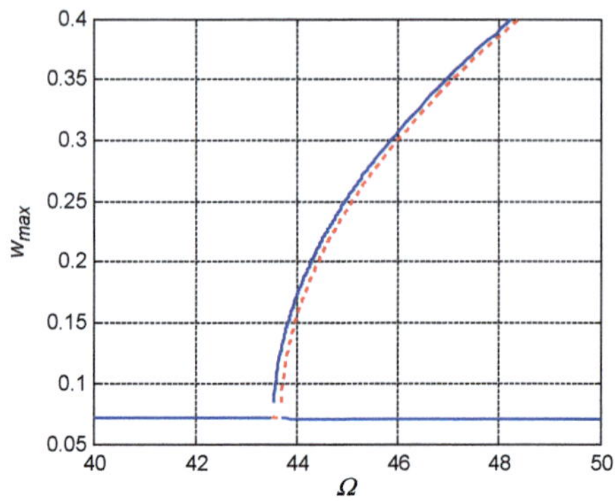

Fig. 5.29 Frequency-response curve for the subharmonic resonance of order one half of the carbon nanotube of case 2 of Table 5.1 for $V_{DC} = 20$ Volt, $V_{AC} = 1.2$ Volt and $Q = 150$. In the figure: (—) stable branch, (- - - - -) unstable branch

and found that a subharmonic resonance of order one half is activated, Fig. 5.17. Increasing the AC load further to $V_{AC} = 16$ Volt leads to a dynamic pull-in, Fig. 5.30.

Figures 5.28, 5.29, and 5.30a demonstrate secondary resonances of order two and one half due to the influence of the electrostatic force, which is quadratic in nature. However, it is possible also to activate secondary resonances of order one third and three due to the effect of mid-plane stretching, which is cubic in nature. An example of this case is depicted in Fig. 5.30b. Also, in this figure, the curve undergoes dynamic pull-in characterized by a Floquet multiplier approaching unity. One can note from Figs. 5.29 and 5.30 the fact that subharmonic resonance remains activated over a wide range of frequency. This is another significant difference between the dynamics of CNTs and MEMS beams, where subharmonic resonance is activated over a very narrow range of frequency [80].

Next, we use the shooting technique to calculate the inevitable escape bands of frequencies for the subharmonic resonance of order one half while varying the AC voltage and frequency, as we did in the primary resonance case. Figure 5.31a, b show the results for cases 2 and 3 of Table 5.1, respectively. To illustrate the increase in the escape band as the AC voltage increases, the frequency-response curves for the upper and the lower limits of the instability tongue of Fig. 5.31a are depicted in Fig. 5.32a, b, respectively.

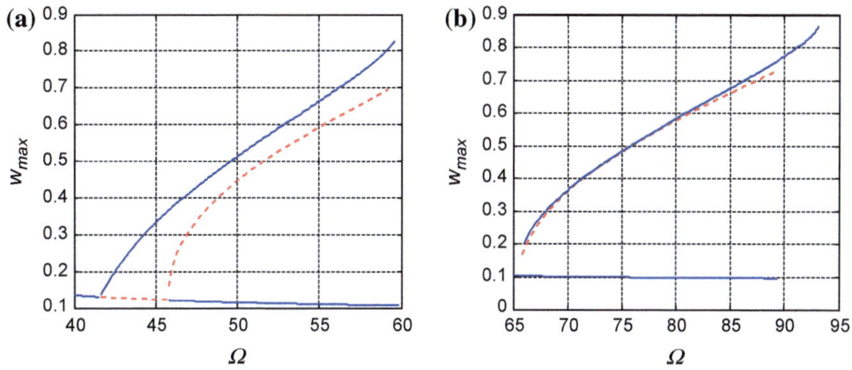

Fig. 5.30 Frequency-response curve for the subharmonic resonance of order (**a**) one half and (**b**) one third of the carbon nanotube of case 2 of Table 5.1 for $V_{DC} = 20$ Volt, $V_{AC} = 16$ Volt, and $Q = 150$. In the figure: (——) stable branch, (- - - - -) unstable branch

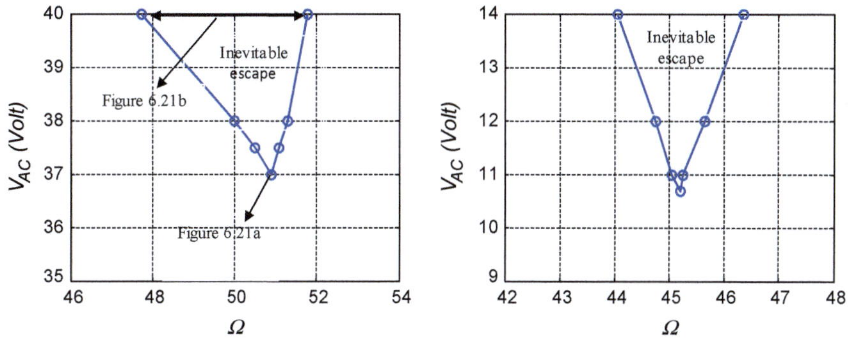

(a) Hardening. (b) Softening.

Fig. 5.31 The calculated instability tongues in the case of subharmonic resonance of order one half (**a**) for case 2 of Table 5.1 with $V_{DC} = 20$ Volt, (**b**) for case 3 of Table 5.1 with $V_{DC} = 100$ Volt and $Q = 150$

Next we show the subharmonic response of a cantilever carbon nanotube. Figure 5.33 depicts a softening-type behavior of the subharmonic resonance of order one half of the fundamental mode. The curve shows dynamic pull-in characterized by period doubling bifurcation.

5.4 Dynamics of Slacked Carbon Nanotube Resonators

In the previous section, we presented a model utilizing a nonlinear beam equation to simulate the static and dynamic behaviors of electrically actuated straight CNT resonators. It is worth to mention that the adopted model does not account for the effect of slack (curvature), buckling, or initial deformation of carbon nanotubes.

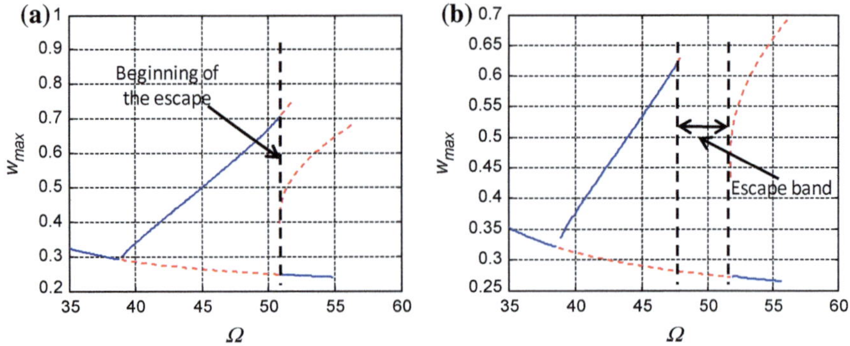

Fig. 5.32 Frequency-response curves for the subharmonic resonance of order one half of the carbon nanotube for case 2 of Table 5.1 when $V_{DC} = 20$ Volt and $Q = 150$ and for (**a**) $V_{AC} = 37$ Volt and (**b**) $V_{AC} = 40$ Volt. In the figure: (——) stable branch and (- - - - -) unstable branch

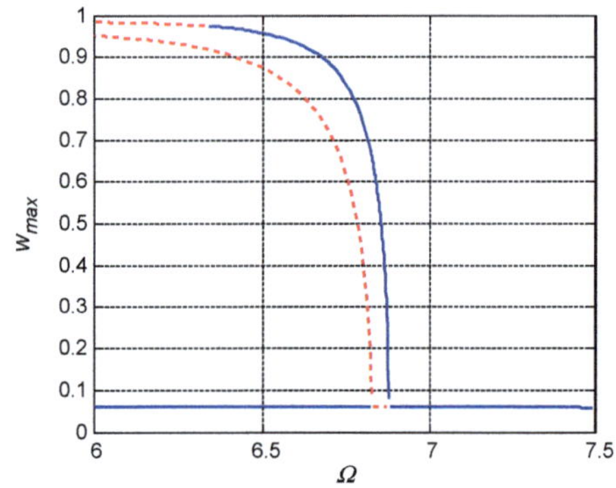

Fig. 5.33 Frequency-response curve for the subharmonic resonance of order one half of the cantilever carbon nanotube for case 4 of Table 5.1 when $V_{DC} = 1$ Volt, $V_{AC} = 0.2$ Volt, and $Q = 150$. In the figure: (——) stable branch, (- - - - -) unstable branch

This can have significant influence on the stability, natural frequencies, and pull-in calculations presented so far in the previous chapter. In this section, a 2-D nonlinear curved beam model (arch) is utilized to simulate the motion of a slacked CNT. The variation of the natural frequencies, mode shapes, and effective nonlinearity of a CNT with various levels of slack and DC electrostatic loads is investigated. Various scenarios are shown for mode crossing and mode veering as the levels of slack and DC load are varied. In addition, the forced vibration of the slacked CNT when actuated by small DC and AC loads is analyzed to show the transfer of energy among the vibration modes involved in the veering phenomenon.

5.4.1 Problem Formulation

We start by formulating the 2-D problem governing the static and dynamic behavior of an electrically actuated slacked carbon nanotube resonator, Fig. 5.34. The nanotube

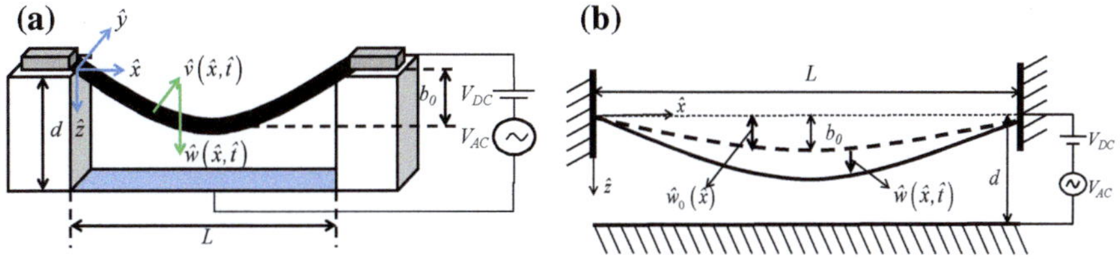

Fig. 5.34 (**a**) 3-D schematic of the electrically actuated slacked carbon nanotube resonator, (**b**) in-plane view of the nanotube

is actuated by an electrode underneath it with a gap width d. It is modeled as a hollow cylinder Euler-Bernoulli beam of radius $\tilde{R}$, shell thickness h, and length L. It has a cross-sectional area A and an area moment of inertia I. The nanotube is assumed to have a Young's modulus $E = 1.2\,TPa$ and a mass density $\rho = 1.3\,\text{g/cm}^3$. The CNT is considered here to be initially curved in the direction of the lower electrode with an initial shape [65].

The 2-D equations describing the in-plane deflection $\hat{w}\left(\hat{x},\hat{t}\right)$ and out-of-plane deflection $\hat{v}\left(\hat{x},\hat{t}\right)$ of the clamped-clamped CNT resonator can be written as [69]

$$EI\frac{\partial^4\hat{w}}{\partial\hat{x}^4} + \partial A\frac{\partial^2\hat{w}}{\partial\hat{t}^2} + \tilde{c}\frac{\partial\hat{w}}{\partial\hat{t}} = \hat{F}_{elec} + \\ +\frac{EA}{2L}\left[\int_0^L\left\{\left(\frac{\partial\hat{w}}{\partial\hat{x}}\right)^2 + \left(\frac{\partial\hat{v}}{\partial\hat{x}}\right)^2 - 2\left(\frac{\partial\hat{w}}{\partial\hat{x}}\frac{d\hat{w}_0}{d\hat{x}}\right)\right\}dx\right]\left[\frac{\partial^2\hat{w}}{\partial\hat{x}^2} - \frac{d^2\hat{w}_0}{d\hat{x}^2}\right], \tag{5.101}$$

$$EI\frac{\partial^4\hat{v}}{\partial\hat{x}^4} + \partial A\frac{\partial^2\hat{v}}{\partial\hat{t}^2} + \tilde{c}\frac{\partial\hat{v}}{\partial\hat{t}} = \frac{EA}{2L}\left[\int_0^L\left\{\left(\frac{\partial\hat{w}}{\partial\hat{x}}\right)^2 + \left(\frac{\partial\hat{v}}{\partial\hat{x}}\right)^2 - 2\left(\frac{\partial\hat{w}}{\partial\hat{x}}\frac{d\hat{w}_0}{d\hat{x}}\right)\right\}dx\right]\frac{\partial^2\hat{v}}{\partial\hat{x}^2}, \tag{5.102}$$

where

$$\hat{F}_{elec} = \frac{\pi\varepsilon_0(V_{DC} + V_{AC}\cos(\tilde{\Omega}t))^2}{\sqrt{(d-\hat{w}-\hat{w}_0)(d-\hat{w}-\hat{w}_0+2\tilde{R})}\left(\cosh^{-1}\left(1+\dfrac{d-\hat{w}-\hat{w}_0}{\hat{R}}\right)\right)^2}, \tag{5.103}$$

and where $\tilde{c}$ is the viscous damping coefficient and ε_0 is the air permittivity.

The respective boundary conditions are

$$\hat{w}\left(0,\hat{t}\right) = 0,\quad \frac{\partial\hat{w}}{\partial\hat{x}}\left(0,\hat{t}\right) = 0,\quad \hat{w}\left(L,\hat{t}\right) = 0,\quad \frac{\partial\hat{w}}{\partial\hat{x}}\left(L,\hat{t}\right) = 0, \\ \hat{v}\left(0,\hat{t}\right) = 0,\quad \frac{\partial\hat{v}}{\partial\hat{x}}\left(0,\hat{t}\right) = 0,\quad \hat{v}\left(L,\hat{t}\right) = 0,\quad \frac{\partial\hat{v}}{\partial\hat{x}}\left(L,\hat{t}\right) = 0 \tag{5.104}$$

For convenience, we introduce the following nondimensional variables:

$$w=\frac{\hat{w}}{d},\qquad v=\frac{\hat{v}}{d},\qquad x=\frac{\hat{x}}{L},\qquad t=\frac{\hat{t}}{T},\tag{5.105}$$

where T is a time constant defined by $T=\sqrt{(\rho AL^4)/(EI)}$. By substituting Eq. (5.105) into Eqs. (5.101)–(5.104), the nondimensional equations of motions and associated boundary conditions of the considered clamped-clamped CNT are written as

$$\frac{\partial^4 w}{\partial x^4}+\frac{\partial^2 w}{\partial t^2}+\tilde{c}\frac{\partial w}{\partial t}=\alpha_2 F_{elec}+$$
$$+\alpha_1\left[\int_0^L\left\{\left(\frac{\partial w}{\partial x}\right)^2+\left(\frac{\partial v}{\partial x}\right)^2-2\left(\frac{\partial w}{\partial x}\frac{dw_0}{dx}\right)\right\}dx\right]\left[\frac{\partial^2 w}{\partial x^2}-\frac{d^2 w_0}{dx^2}\right],\tag{5.106}$$

$$\frac{\partial^4}{\partial x^4}+\frac{\partial^2 v}{\partial t^2}+\tilde{c}\frac{\partial v}{\partial t}=\alpha_1\left[\int_0^L\left\{\left(\frac{\partial w}{\partial x}\right)^2+\left(\frac{\partial v}{\partial x}\right)^2-2\left(\frac{\partial w}{\partial x}\frac{dw_0}{dx}\right)\right\}dx\right]\frac{\partial^2\hat{v}}{\partial x^2},\tag{5.107}$$

$$\begin{aligned}&w(0,t)=0,\quad \frac{\partial w}{\partial x}(0,t)=0,\quad w(1,t)=0,\quad \frac{\partial w}{\partial x}(1,t)=0,\\ &v(0,t)=0,\quad \frac{\partial v}{\partial x}(0,t)=0,\quad v(1,t)=0,\quad \frac{\partial v}{\partial x}(1,t)=0,\end{aligned}\tag{5.108}$$

where

$$F_{else}=\frac{(V_{DC}+V_{AC}\cos(\Omega t))^2}{\sqrt{(1-w-w_0)(1-w-w_0+2R)}\left(\cosh^{-1}\left(1+\frac{1-w-w_0}{R}\right)\right)^2},\tag{5.109}$$

$$w_0(x)=\frac{b_o}{d}\sin(\pi x),$$

$$\alpha_1=\frac{Ad^2}{2I},\ \alpha_2=\frac{\pi\varepsilon_0 L^4}{EId^2},\ c=\frac{\tilde{c}L^4}{EIT},\ \Omega=\frac{\tilde{\Omega}}{\sqrt{(EI)/(\rho AL^4)}},\ R=\frac{\tilde{R}}{d}\tag{5.110}$$

5.4.2 The Reduced-Order Model

To solve the obtained nondimensional equation of motions of the slacked CNT, Eqs. (5.106)–(5.109) are discretized using the Galerkin procedure to yield a ROM. Hence, the in-plane and out-of-plane responses of the CNT are approximated, respectively, as

$$w(x,t) = \sum_{i=1}^{n} u_i(t)\,\phi_i(x), \;\; v(x,t) = \sum_{i=1}^{m} \xi_i(t)\,\phi_i(x), \tag{5.111}$$

where $\phi_i(x)$ are the normalized linear undamped mode shapes of a straight beam and $u_i(t)$ and $\xi_i(t)$ are the nondimensional modal coordinates of the in-plane and out-of-plane motions, respectively.

To obtain the ROM, we substitute Eq. (5.111) into Eqs. (5.106)–(5.108), multiply by $\phi_i(x)$, use the orthogonality conditions of the mode shapes, and then integrate the outcome from 0 to 1. The results are differential equations in terms of the modal coordinates $u_i(t)$ and $\xi_i(t)$. We should mention here that, as in Sect. 4.3.2, the mode shapes $\phi_i(x)$ will remain embedded inside the denominator of the electrostatic force term, Eq. (5.109), in the ROM. To deal with the complicated integral terms due to that electrostatic force, we evaluate the spatial integrals containing the mode shapes $\phi_i(x)$ numerically simultaneously while integrating the differential equations of the modal coordinates $u_i(t)$ and $\xi_i(t)$.

The Static Response

As a case study, a CNT of $L = 3000$ nm, h $= 0.34$ nm, $\tilde{R} = 1$ nm, $d = 500$ nm, and initial rise $b_o = 100$ nm is considered. Such dimensions are typical of CNTs with slack [4–6]. We have shown in Sect. 4.3 that using one mode only in the ROM is enough to capture the static response of a CNT. Next a one mode is used in the ROM to compare the variation of the maximum in-plane static deflection of the CNT with and without slack, Fig. 5.35.

The figure indicates that for small DC load, the CNT with slack is relatively stiffer due to the linear stiffness term added from the initial curvature. It can be also seen from the figure that the CNT with slack undergoes the pull-in instability at a lower DC load than the one without slack. This is expected since in the slacked configuration, the CNT is nearer to the electrode. Note here that the static response of the out-of-plane motion is zero since the in-plane deflection appears as a homogenous term in the out-of-plane equation (Eq. (5.107)). The stability of the obtained solutions in Fig. 5.35 is studied by calculating the eigenvalues of the Jacobian matrix of the ROM

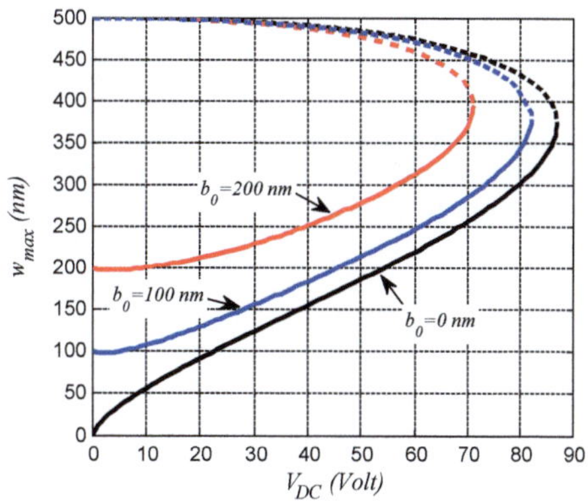

Fig. 5.35 Variation of the maximum in-plane static deflection of the CNT with the DC voltage for various levels of initial curvature b_0. Solid (——) and dashed (-----) lines denote the stable and unstable branches, respectively

evaluated at these solutions [72]. The results show that one of the eigenvalues of the Jacobian matrix corresponding to the upper branches is always positive indicating unstable solution (dashed line in Fig. 5.35). Also, all of the eigenvalues of the lower branches are pure imaginary indicating stable solutions (continuous line in Fig. 5.35). At pull-in, both stable and unstable branches collide and destroy each other with one eigenvalue tending to zero corresponding to a saddle-node bifurcation.

The Eigenvalue Problem

Next, we investigate the eigenvalue problem of the slacked CNT by calculating the variation of the in-plane and out-of-plane natural frequencies and mode shapes with and without slack and under the actuation of the DC voltage. Toward this, we consider the same procedure of Sect. 4.3.4, where:

$$X = [u_1, u_2, \ldots, u_n, \xi_1, \xi_2, \ldots, \xi_m] \tag{5.112}$$

is the considered modal amplitudes vector.

First, we consider the case study of Üstünel et al. [6], which is a non-hollow CNT of $L = 1750$ nm, $\tilde{R} = 1$ nm, and $d = 500$ nm. For the following results, we will denote the in-plane natural frequencies by ω_i and the out-of-plane frequencies byλ_i. In Fig. 5.36, the variation of the in-plane and out-of-plane natural frequencies is calculated at zero DC load for various values of initial rise of the CNT or slack percentages. The slack percentage is defined as $\%slack = (L - \tilde{L})/L$, where $\tilde{L}$ is the length of the CNT in the deformed (curved) position. Figure 5.36 compares the results with those reported in Üstünel et al. [6], which indicates good qualitative agreement. However, there is a quantitative discrepancy due to the fact that the model used in Üstünel et al. [6] is applicable only when the strain is small enough that the nonlinear elastic effects may be ignored. Unlike their model, our approach

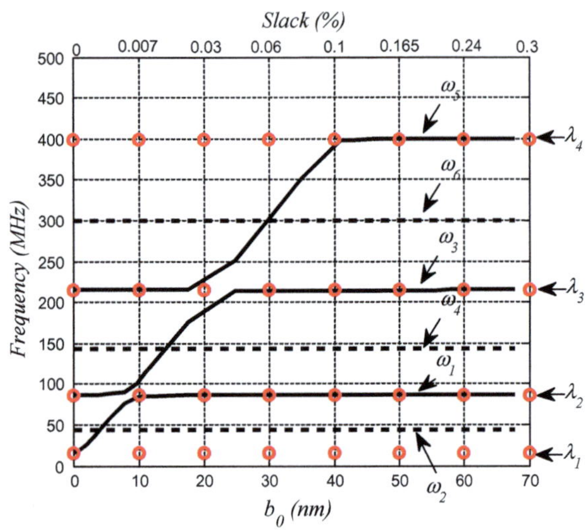

Fig. 5.36 Variation of the first few in-plane and out-of-plane natural frequencies of a CNT with slack for the case study of Üstünel et al. [6]. *Solid line* (——), *dashed line* (-----), and *circles* (o) denote, respectively, the odd in-plane, even in-plane and out-of-plane, and the odd out-of-plane frequencies

is a continuous model that can describe the nonlinear oscillations of the CNT under its various parameters without dividing the CNT behavior into different regimes.

In Fig. 5.37, we investigate in more depth the effect of varying the initial rise of the CNT of Fig. 5.35, assuming zero DC load, on the in-plane and out-of-plane natural frequencies. We can see clearly that the even in-plane (dashed lines in Fig. 5.37a) and

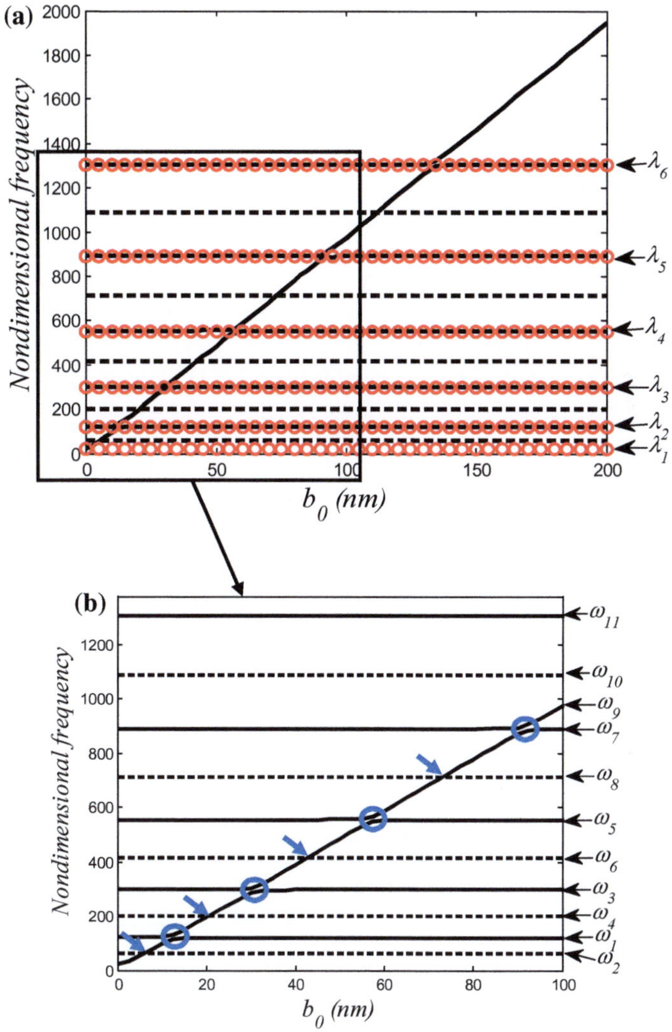

Fig. 5.37 (**a**) Variation of the in-plane and out-of-plane natural frequencies of a CNT with various levels of slack at zero DC load. (**b**) A zoomed view of Fig. 5.4a showing the crossings and veering of the in-plane frequencies (the odd out-of-plane are not shown for clarity). *Solid line* (——), *dashed line* (- - - - -), and circles (o) denote, respectively, the odd in-plane, the even in-plane and the out-of-plane, and the odd out-of-plane frequencies

out-of-plane frequencies (circles in Fig. 5.37a) are insensitive to the variation of slack whereas the odd in-plane frequencies vary with slack (continuous lines in Fig. 5.37a). These frequencies appear in the large scale to intersect (Fig. 5.37a). When enlarging the apparent intersection zones, one can see that they do not intersect (blue circles in Fig. 5.37b); they diverge in a manner called curve veering [97, 98]. This phenomenon is common and has been cited for the natural frequencies of a rectangular membrane when varying the ratio of its lengths' sides [98] and also for the natural frequencies of cables when varying their sagging levels [99, 100].

A frequency veering occurs when the loci of two eigenvalues, in an eigenvalue problem, approach each other when a parameter is varied and then veer away when being too close like two repulsive charges [101]. In a frequency veering, the eigenfunctions associated with the eigenvalues on each locus before veering is interchanged during the veering [98]. To further clarify this, we plot the corresponding eigenfunctions of the odd in-plane natural frequencies in Table 5.3. We can see that the modes are exchanging shapes and (bold squares in Table 5.3) when varying the slack level. This table describes all the veering scenarios depicted in Fig. 5.37b, which appear following a straight line. In this line, the shape of the first mode is transferred into the different odd modes, from the lower to the higher modes, depending on the slack level. A final note to be mentioned here is that the odd in-plane frequencies intersects the even ones for certain levels of slack offering many possibilities of internal resonances and exchange of energy among higher- and lower-order modes.

Table 5.3 The simulated eigenfunctions of the first five odd in-plane modes for various slack levels

	ω_1	ω_3	ω_5	ω_7	ω_9
$b_0=0$ μm					
$b_0=10$ μm					
$b_0=20$ μm					
$b_0=40$ μm					
$b_0=70$ μm					
$b_0=100$ μm					

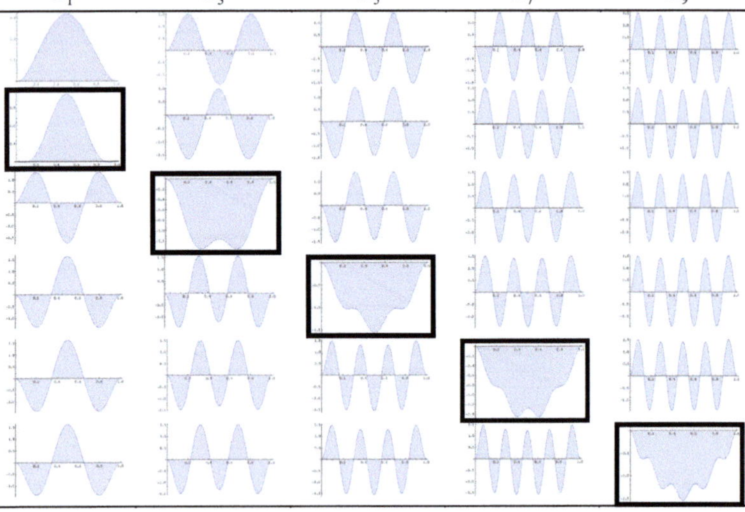

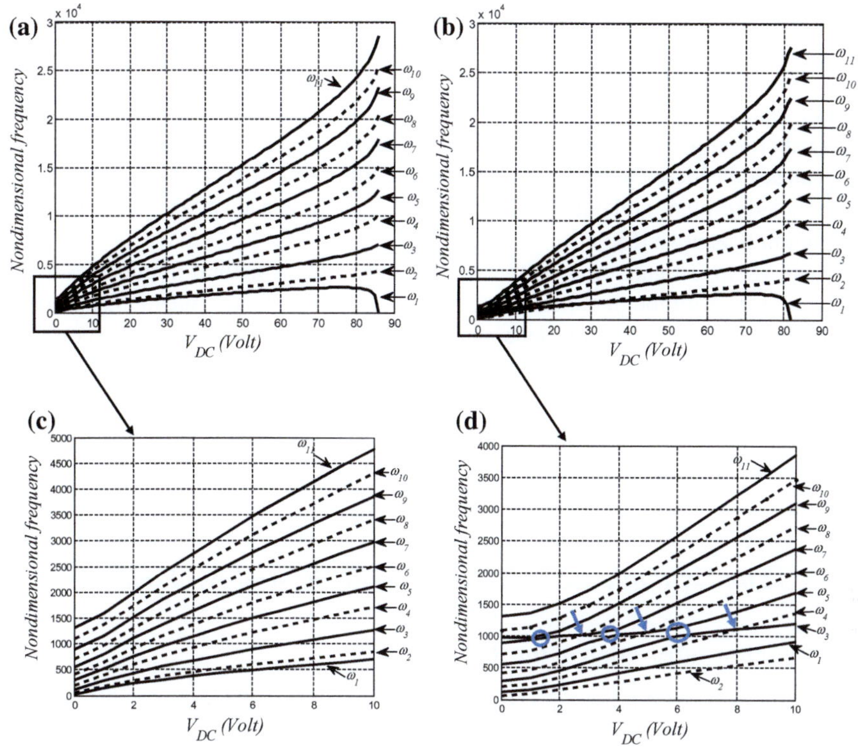

Fig. 5.38 (**a**) Variation of the in-plane natural frequencies with the DC load of a CNT (**a**) without and (**b**) with slack. In both cases, a zoomed view near small values of voltages are shown (**c**) and (**d**). *Solid line* (——) and *dashed line* (-----) denote the odd in-plane and the even in-plane frequencies, respectively

Next we investigate the variation of the in-plane natural frequencies of the CNT with the DC load. First, we consider the unslacked case. We can see from Fig. 5.38a that all the natural frequencies are sensitive to the variation of the DC voltage. All of them are increasing dramatically, with a sublinear variation behavior, except for the first frequency near pull-in where it drops to zero. This increase in the frequencies suggests a tunable resonator over a wide range of frequencies. Practically, when designing a resonator made with such CNT, one can easily tune the frequencies from MHz range to GHz and even THz range. In addition, one can see from Fig. 5.38c that there is no possibility of modes veering or modes crossing even for small range of voltages. These scenarios however appear when slack is added, Fig. 5.38b. The figure shows the variation of the first few in-plane frequencies of the slacked CNT ($b_0 = 100\,\text{nm}$) with the DC load. We can see that the odd modes exhibit the frequency veering phenomenon (blue circles in Fig. 5.38d), but in this case from the higher to the lower modes where the energy is transferred to the lowest fundamental frequency that eventually drops to zero at pull-in. There are also possibilities of odd and even in-plane modes crossings (blue arrows in Fig. 5.38d).

It is worth to mention here that our demonstrated results can justify and predict some of observed experimentally phenomena in Sazanova [5]. Other models of [5, 6] could not explain the previously mentioned phenomenon. For example, Fig. 5.38 shows what was cited as sublinear dispersion of the frequency in Sazanova [5]. The avoided crossings for some frequency range in the case of slack cited in Sazanova [5], can be considered to be the same as veering, Fig. 5.39. Also, our results, based on the coupled in-plane and out-of-plane motions, have shown abundance of resonances as cited in Sazanova [5].

Next, we investigate the sensitivity of the out-of-plane natural frequencies to the DC load, which although not directly actuated, they are affected due to the nonlinear coupling between the out-of-plane and in-plane deflections. Figure 5.39 shows the first five in-plane and out-of plane natural frequencies as they vary with the DC load. We can note that, except for the first mode, the first few out-of-plane frequencies are larger than those of the in-plane frequencies especially at high DC loads. For the higher-order modes, both in-plane and out-of-plane natural frequencies are equal.

5.4.3 *The Dynamic Response for Small DC and AC Loads*

(a) Long-time integration of the ROM

Next, we use the ROM developed in Sect. 4.4.2 to integrate the differential equations of motion in time to obtain the dynamic response of the slacked CNT under a very small DC and AC harmonic load. The choice of the very small voltage loads is to guarantee linear forced vibration response. The response to small electric loads is important to enable precise prediction of the resonance frequency in the linear regime. The resonance frequency is the frequency that is commonly measured experimentally for CNTs when driven by AC and DC loads. Here we use a nondimentional damping coefficient $c = 1.196$. Figure 5.40a, b show the response of the 100 nm and the 200 nm slacked CNTs, respectively. We can see from the figures that the linear dynamic response is significant in the neighborhood of the first natural frequency, ω_1, and the frequencies that are located on the veering straight line, as predicted from Fig. 5.37. Those frequencies are ω_9 in the case of the 100 nm slack and ω_{11} in the case of the 200 nm slack. It is clear from Fig. 5.40a, b that the higher-order modes located on the veering line are sharing the energy of vibration with the fundamental mode. This conclusion might explain one of the reasons behind the low quality factor reported experimentally for CNTs [5] when driven harmonically at resonance near their fundamental modes.

To further clarify this point, we calculate in Fig. 5.41 the participation of each individual odd in-plane mode shape in the dynamic response of the 100 nm slacked CNT. It is clear from the figure that the participations of the first mode (the lowest mode) and the ninth mode (the one located on the veering line) are the most important ones. This indicates that even when exciting the CNT near its first natural frequency, significant participation is anticipated of the mode located on the veering line.

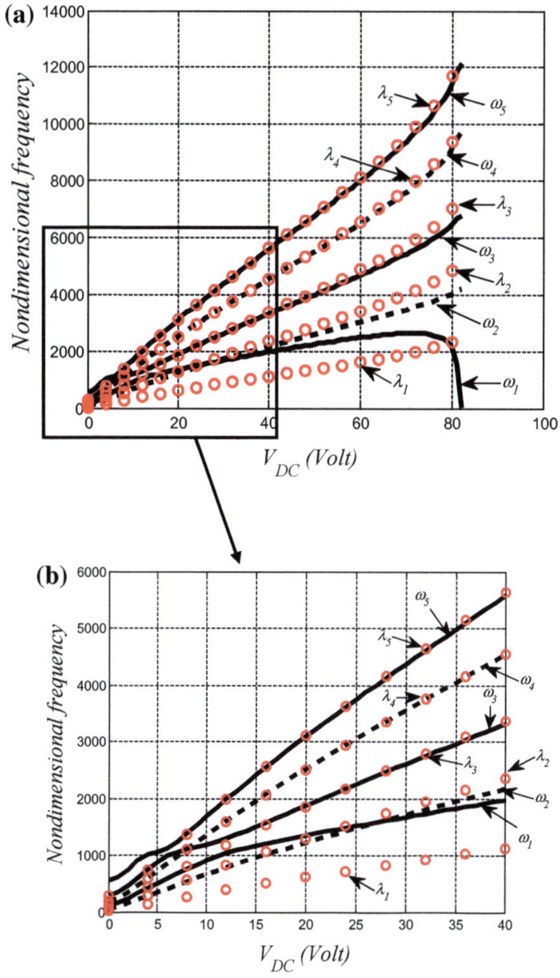

Fig. 5.39 **a** Variation of the first five in-plane and out-of-plane natural frequencies with the DC load of a CNT with a 100 nm slack. **b** A zoomed view of Fig. 5.39a near small values of voltages. *Solid line* (——), *dashed line* (- - - - -), and circles (○) denote, respectively, the odd in-plane, the even in-plane and the out-of-plane, and the odd out-of-plane frequencies

(b) Perturbation analysis

(i) Derivations

In this section, perturbation analysis is carried out using the method of multiple scales and a direct attack of the in-plane equations of motion same as we did in Sect. 4.5.1. To this end, we define the variables for the time scale (T_i), their derivatives (D_i), the influence of the damping coefficient and the forcing amplitude, as in Eqs. (4.117)

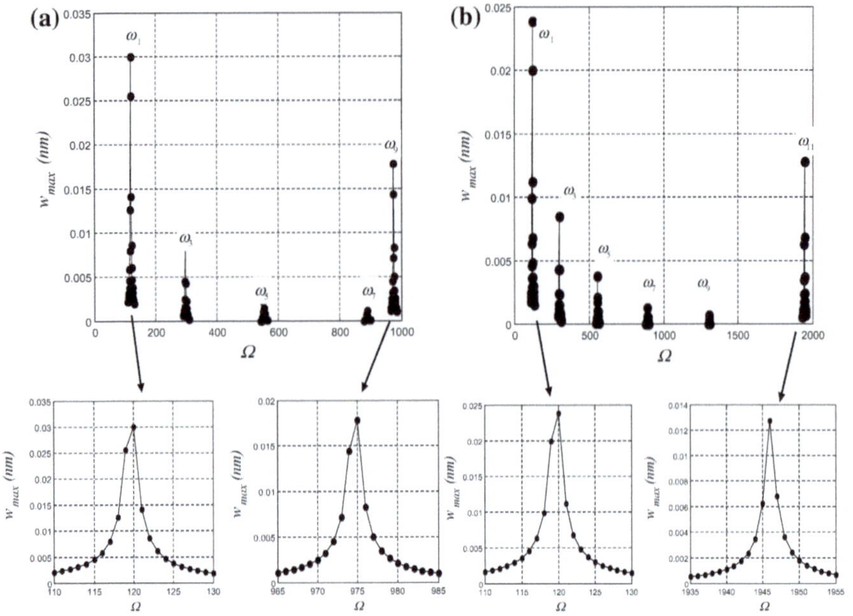

Fig. 5.40 Frequency-response curves of (**a**) 100 nm slacked CNT and (**b**) 200 nm slacked CNT at the odd in-plane natural frequencies. Results are shown for $V_{DC} = 0.01$ Volt, $V_{AC} = 0.01$ Volt, and 100 quality factor

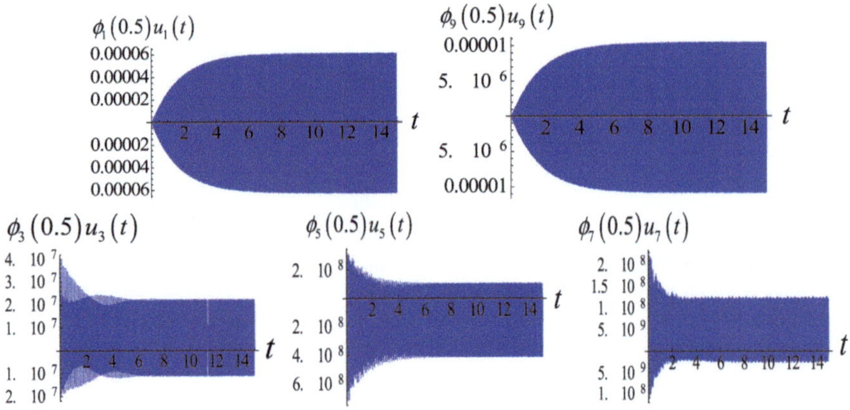

Fig. 5.41 Time-response curves of the 100 nm slacked CNT at $\Omega \approx \omega_1$ showing the participation of the odd in-plane mode shapes. Results are shown for $V_{DC} = 0.01$ Volt, $V_{AC} = 0.01$ Volt, and 100 quality factor

and (4.118) respectively, and we seek a solution in the form of Eq. (4.119). Next, we expand the electrostatic force term, Eq. (5.109), into Taylor series up to the third order as

$$\frac{1}{\sqrt{(1-u-w_s-w_0)(1-u-w_s-w_0+2\tilde{R})}\left(\cosh^{-1}\left(1+\frac{1-u-w_s-w_0}{\tilde{R}}\right)\right)^2} \approx \tag{5.113}$$
$$\approx F_s + F_1 u + F_2 u^2 + F_3 u^3 + \cdots ,$$

Substituting Eqs. (4.117)–(4.120) and Eq. (5.113) into Eqs. (5.106) and (5.108) and then equating like powers of ε, we obtain

- **Order ε^0: (the static equation)**

$$\begin{gathered} w_s^{iv} = \alpha_1[w_s'' - w_0''][\Gamma(w_s, w_s) - 2\Gamma(w_s, w_0)] + F_s, \\ w_s(0) = w_s(1) = 0, \qquad w_s'(0) = w_s'(1) = 0, \end{gathered} \tag{5.114}$$

- **Order ε^1:**

$$\begin{gathered} L\,(u_1) = D_0^2 u_1 + u_1^{iv} - \alpha_1 \left[\Gamma\,(w_s, w_s) - 2\Gamma\,(w_s, w_0)\right] u_1'' \\ -2\alpha_1 \left[w_s'' - w_0''\right] \left[\Gamma\,(w_s, u_1) - \Gamma\,(w_0, u_1)\right] - \alpha_2 F_1 u_1 = 0, \end{gathered} \tag{5.115}$$

- **Orderε^2:**

$$L\,(u_2) = \alpha_1 \Gamma\,(u_1, u_1) \left(w_s'' - w_0''\right) + 2\alpha_1 \left[\Gamma\,(w_s, u_1) - \Gamma\,(w_0, u_1)\right] u_1'' + F_2 u_1^2, \tag{5.116}$$

- **Order ε^3:**

$$\begin{aligned} L\,(u_3) = & -2D_0 D_2 u_1 - c D_0 u_1 + 2\alpha_1 \Gamma\,(u_1, u_2) \left(w''_s - w''_0\right) \\ & +2\alpha_1 \left[\Gamma\,(w_s, u_1) - \Gamma\,(w_0, u_1)\right] u''_2 \\ & +2\alpha_1 \left[\Gamma\,(w_s, u_2) - \Gamma\,(w_0, u_2)\right] u''_1 + \alpha_1 \Gamma\,(u_1, u_1)\, u''_1 \\ & +2\alpha_2 V_{\mathrm{DC}} V_{\mathrm{AC}} F_s \cos\,(\Omega t) + \alpha_2 F_2 u_1 u_2 + \alpha_2 F_1 u_1^3, \end{aligned} \tag{5.117}$$

The solution of Eq. (5.115) is assumed to consist of only the directly excited mode, $\Phi_i\,(x)$, because in the absence of internal resonances, all the other modes die out with the damping [102]. Accordingly, we express the first dynamic component u_1 as

$$u_1(x, T_0, T_2) = \left[A(T_2) e^{i\omega_i T_0} + \bar{A}(T_2) e^{-i\omega_i T_0}\right] \Phi_i\,(x)\,, \tag{5.118}$$

where $A(T_2)$ is a complex-valued function, the over bar denotes the complex conjugate, and ω_i and $\Phi_i\,(x)$ are the natural frequency and corresponding eigenfunction of the directly excited mode, respectively. Substituting Eq. (5.118) into Eq. (5.16), we obtain

$$L(u_2) = \left(2A\bar{A} + A^2 e^{2i\omega_i T_0} + \bar{A}^2 e^{-2i\omega_i T_0}\right) h\,(x)\,, \tag{5.119}$$

where

$$\begin{gathered} h\,(x) = 2\alpha_1 \left(\Gamma(w_s, \Phi_i) - \Gamma(w_0, \Phi_i)\right) \Phi_i'' + \\ +\alpha_1 \Gamma(\Phi_i, \Phi_i) \left(w''_s - w''_0\right) + \alpha_2 F_2 \Phi_i^2, \end{gathered} \tag{5.120}$$

The solution of Eq. (5.119) can be expressed as follows:

$$\begin{aligned} u_2(x, T_0, T_2) = \psi_1(x)\, A^2(T_2) e^{2i\omega_i T_0} + \\ +2\psi_2(x)\, A(T_2)\bar{A}(T_2) + \psi_1(x)\, \bar{A}^2(T_2) e^{-2i\omega_i T_0}, \end{aligned} \tag{5.121}$$

where ψ_1 and ψ_2 are the solutions of the following boundary value problems:

$$\begin{aligned} H\left(\psi_j, 2\omega_i\delta_{1j}\right) = h(x), \quad j = 1, 2, \\ \psi_j = 0 \text{ and } \psi_j' = 0 \text{ at } x = 0 \text{ and } x = 1, \quad j = 1, 2, \end{aligned} \tag{5.122}$$

where δ_{ij} is the Kronecker delta operator and the linear differential operator H is defined as

$$\begin{aligned} H(f(x), \omega) = f^{iv}(x) - \omega^2 f(x) \\ -2\alpha_1\left(w_s'' - w_0''\right)\left[\Gamma(f(x), w_s) - \Gamma(f(x), w_0)\right] \\ -\alpha_1\left[\Gamma(w_s, w_s) - 2\Gamma(w_0, w_s)\right] f''(x) - \alpha_2 F_1 f(x) \end{aligned} \tag{5.123}$$

Note here that the eigenfunction $\Phi_i(x)$ is solution of

$$H(\Phi_i, \omega) = 0, \tag{5.124}$$

In order to describe the nearness of the excitation frequency Ω to the fundamental natural frequency ω_i, we introduce a detuning parameter σ defined by

$$\Omega = \omega_i + \varepsilon^2\sigma, \tag{5.125}$$

Substituting Eqs. (5.118), (5.121), and (5.125) into Eq. (5.117) we obtain

$$L(u_3) = \left[-i\omega_i\left(2A' + cA\right)\Phi_i(x) + \chi(x)\, A^2\bar{A} + \bar{F}(x)\, e^{i\sigma T_2}\right] e^{i\omega_i T_0} + cc + NST, \tag{5.126}$$

where

$$\bar{F}(x) = 2\alpha_2 V_{DC} V_{AC} F_s \tag{5.127}$$

In Eq. (5.126), A' denotes the derivative of A with respect to T_2, "*cc*" denotes the complex conjugate of the preceding terms, *NST* stands for the terms that do not produce secular terms, and $\chi(x)$ is defined by

$$\begin{aligned} \chi(x) = \alpha_1\Phi_i''\left[3\Gamma(\Phi_i, \Phi_i) + 2\Gamma(w_s, \psi_1) - 2\Gamma(w_0, \psi_1) + 4\Gamma(w_s, \psi_2) - 4\Gamma(w_0, \psi_2)\right] \\ +\alpha_1\left[w_s'' - w_0''\right]\left[2\Gamma(\Phi_i, \psi_1) + 4\Gamma(\Phi_i, \psi_2)\right] \\ +2\alpha_1\left[\psi_1'' + 2\psi_2''\right]\left[\Gamma(\Phi_i, w_s) - \Gamma(\Phi_i, w_0)\right] \\ +3\alpha_2 F_3\Phi_i^3 + 2\alpha_2 F_2\Phi_i\psi_1 + 4\alpha_2 F_2\Phi_i\psi_2, \end{aligned} \tag{5.128}$$

Note that the function $\chi(x)$ can be divided into two coefficients that describe the two sources of nonlinearity presented in the considered electrically actuated CNT slacked resonator as follows:

$$\chi(x) = \chi_c(x) + \chi_q(x), \tag{5.129}$$

where

$$\begin{aligned}
\chi_c(x) &= 3\alpha_1 \Phi_i'' \Gamma(\Phi_i, \Phi_i) + 3\alpha_2 F_3 \Phi_i^3, \\
\chi_q(x) &= 3\alpha_1 \Phi_i'' \left[2\Gamma(w_s, \psi_1) - 2\Gamma(w_0, \psi_1) + 4\Gamma(w_s, \psi_2) - 4\Gamma(w_0, \psi_2)\right] + \\
&\quad + \alpha_1 \left[w_s'' - w_0''\right] \left[2\Gamma(\Phi_i, \psi_1) + 4\Gamma(\Phi_i, \psi_2)\right] + \\
&\quad + 2\alpha_1 \left[\psi_1'' + 2\psi_2''\right] \left[\Gamma(\Phi_i, w_s) - \Gamma(\Phi_i, w_0)\right] + \\
&\quad + 2\alpha_2 F_2 \Phi_i \psi_1 + 4\alpha_2 F_2 \Phi_i \psi_2.
\end{aligned} \tag{5.130}$$

The subscripts c and q denote the cubic nonlinear term and the quadratic nonlinear term, respectively.

Next, multiplying the right-hand side of Eq. (5.126) by $\Phi_i(x) e^{-i\omega_i T_0}$, where Φ_i is normalized such that $\int_0^1 \Phi_i^2 dx = 1$, integrating the result from $x = 0$ to $x = 1$ and equating the secular terms to zero, yields the following solvability condition:

$$-i\omega\left(2A' + cA\right) + SA^2\bar{A} + Fe^{i\sigma T_2} = 0, \tag{5.131}$$

where

$$F = \int_0^1 \Phi_i \bar{F} dx, \text{ and } S = S_c + S_q, \tag{5.132}$$

where the nonlinear coefficients are expressed as

$$S_c = \int_0^1 \Phi_i \chi_c dx, \; S_q = \int_0^1 \Phi_i \chi_q dx \tag{5.133}$$

Next, we express A in the polar form $A = ae^{i\beta}/2$, where $a = a(T_2)$ and $\beta = \beta(T_2)$ are real-valued functions, representing, respectively, the amplitude and phase of the response. Substituting the expression of A into Equation (5.31) and letting $\gamma = \sigma T_2 - \beta$, we obtain

$$\left[\left(-ia' + a\beta' - \frac{1}{2}ica\right)\omega_i + \left(\frac{a^3}{8}\right)S\right]e^{i\beta} + Fe^{i\sigma T_2} = 0 \tag{5.134}$$

Separating the real and imaginary parts in Eq. (5.134), we obtain the following modulation equations:

$$a' = -\frac{1}{2}ca + \frac{\sin\gamma}{\omega_i}F, \tag{5.135}$$

$$a\gamma' = a\sigma + \frac{a^3}{8\omega_i}S + \frac{\cos\gamma}{\omega_i}F \tag{5.136}$$

Substituting Eqs. (5.118) and (5.121) into the Method of Multiple Scales approximated solution and setting $\varepsilon = 1$, we obtain, to the second-order approximation, the following CNT response to the external excitation as follows:

$$\begin{aligned} w(x,t) = w_s(x) + a\cos(\Omega t - \gamma)\,\Phi_i(x) + \\ + \tfrac{1}{2}a^2\left[\psi_1(x)\cos 2(\Omega t - \gamma) + \psi_2(x)\right] + \cdots, \end{aligned} \tag{5.137}$$

It follows from Eq. (5.137) that periodic solutions correspond to constant a and γ ; that is, the fixed points (a_0, γ_0) of Eqs. (5.135) and (5.136). Thus, letting $\gamma' = 0$ and $a' = 0$ in Eqs. (5.135) and (5.136), and eliminating γ_0 yield the following frequency-response equation:

$$\frac{F^2}{\omega_i^2} = a_0^2\left[\left(\sigma + \frac{a_0^2}{8\omega_i}S\right)^2 + \frac{1}{4}c^2\right], \tag{5.138}$$

where $c = 2\zeta\omega_i$.

(ii) Results

As seen before when analyzing the dynamic behavior of MEMS arches, the important advantage of the perturbation analysis described above is that it enables studying the variation of the effective nonlinearity of the system and its quadratic and cubic components analytically. To calculate the variation of the effective nonlinearity of the slacked CNT along with its two cubic and quadratic nonlinear components, we evaluate numerically the parameters ω_i, Φ_i, ψ_1, ψ_2, and w_s associated with Eq. (5.132) using a ROM similar to what has been done in Sect. 4.3.3.

First, motivated by studying the effect of the DC load on the CNT nonlinearity for various unslacked CNT radiuses, we calculate its effective nonlinearity variation with the electric load for three different radiuses (10, 20, and 30 nm) of a non-hallow CNT of $L = 1000$ nm, $d = 100$ nm, Young modulus $E = 1\,TPa$, and mass density $\rho = 1.3$ g/cm^3, Fig. 5.42. The figure shows clearly that the percentage of the DC load needed to switch from a hardening behavior to a softening one over the pull-in value increases with the increase of the CNT radius. This percentage is calculated to be 16 % for the CNT of 10 nm radius, 43 % for the CNT of 20 nm radius, and 65 % for the CNT of 30 nm radius. One clear possibility of this behavior is that with the increase of the CNT radius, the ratio of the nondimentional cubic nonlinearity parameter over the quadratic one of Eq. (5.106), i.e. α_1/α_2, increases and hence letting the hardening behavior becoming more dominant for CNT with large radius. Add to all of this, we can see clearly that with the increase of the CNT radius, the increase of the cubic nonlinear coefficient (S_c), which turns out to be more dominant over a wide range of DC load. For those cases, the effective nonlinearity switches from positive to negative only when the quadratic nonlinear terms (S_q) starts to vary with the DC load. This coefficient starts to vary for small DC load compared to the DC pull-in value for

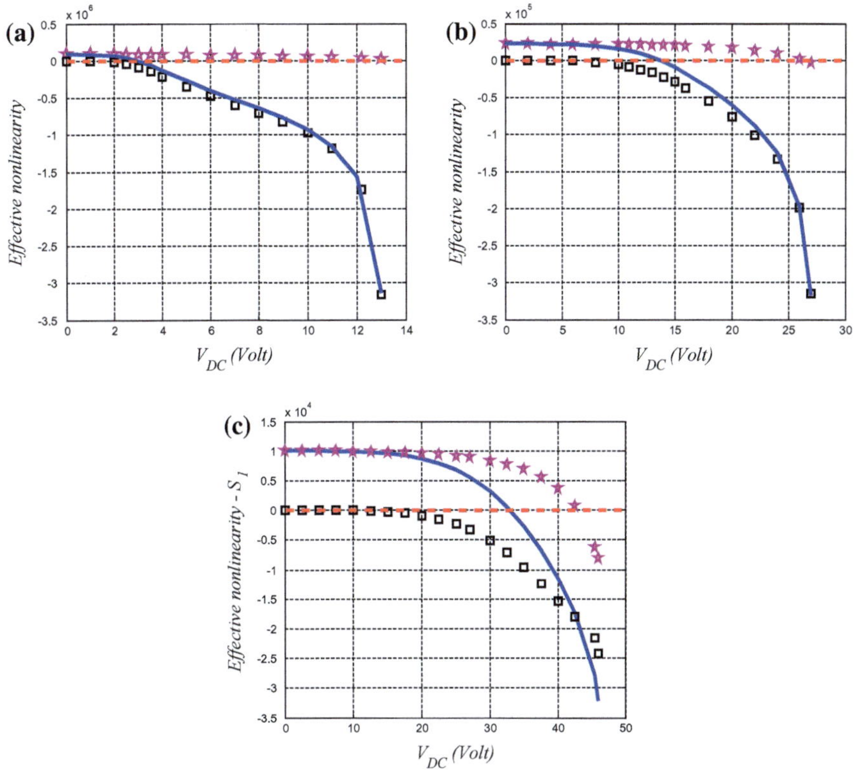

Fig. 5.42 Variation of the effective nonlinearity of the fundamental mode (——), the cubic nonlinearity coefficient (☆), and the quadratic nonlinearity coefficient (□) for an unslacked CNT with the DC voltage: **a** $\tilde{R} = 10\,\text{nm}$, **b** $\tilde{R} = 20\,\text{nm}$, and (**c**) $\tilde{R} = 30\,\text{nm}$. The *dashed line* (-----) separates the positive and the negative regimes of the effective nonlinearity describing, respectively, the hardening and the softening behaviors of the CNT

smaller CNT radius and vice versa. Those trends explain clearly the difference in the dynamical behavior between CNT with small radius (high aspect ratio) and those with high radius (small aspect ratio).

Next, as a case study for slacked CNT, a non-hallow CNT of $L = 1000\,\text{nm}$, $\tilde{R} = 5\,\text{nm}$, $d = 200\,\text{nm}$, $E = 1\,\text{TPa}$, and $\rho = 1.3\,\text{g/cm}^3$ is considered. In Fig. 5.43, we show the variation of the effective nonlinearity coefficient of the fundamental mode (S) of the CNT with the slack. For the case of small slack ($b_o < 4\,\text{nm}$), S is positive indicating a hardening behavior and then it switches to negative sign ($b_o > 4\,\text{nm}$) indicating softening-type behavior. Dynamically point of view, this indicates that for tiny slack level, the CNT is locally dominated by a softening type behavior which means that the quadratic nonlinearity coming from the initial curvature dominates the dynamic behavior of the slacked CNT (S_q is dominant for high values of b_o).

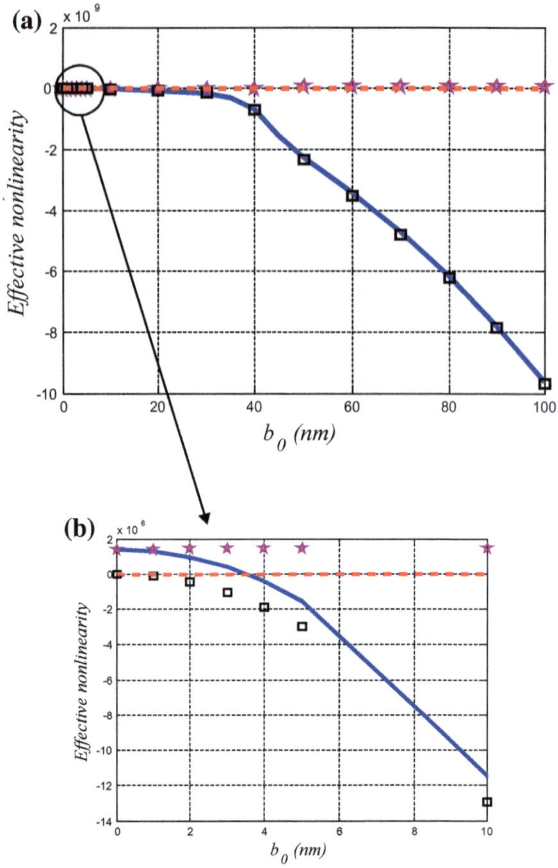

Fig. 5.43 **a** Variation of the effective nonlinearity of the fundamental mode (—), the cubic nonlinearity coefficient (☆), and the quadratic nonlinearity coefficient (□) of the CNT with the slack level. **b** A zoomed view of Fig. 5.43a showing the switch of the effective nonlinearity from positive to negative values

To verify the results of Fig. 5.43, we use the ROM developed in Sect. 6.2 and integrate with time the differential equations of motion to obtain the forced response of the slacked CNT under small DC voltage and an AC harmonic load. Figure 5.44a, b show the responses of a 2 nm and 5 nm slacked CNT, respectively. We can see from the figures that the dynamic response of the CNT switched from being hardening for the first case to softening for the second case confirming what was found in Fig. 5.43.

Next, we show the variation of the effective nonlinearity coefficient (S) of the CNT as well as the nonlinear coefficients with DC load for the case of no initial curvature, Fig. 5.45. For small DC load ($V_{DC} < 10$ nm), S is positive indicating a hardening behavior and then it switches to negative sign ($V_{DC} > 10$ nm) indicating softening type behavior, Fig. 5.45. This shows that the considered CNT is dominated by a softening once the DC load exceeds a small voltage compared to the pull-in value. Here we also generate frequency-response curves based on the ROM, Fig. 5.46, to verify the results of Fig. 5.45.

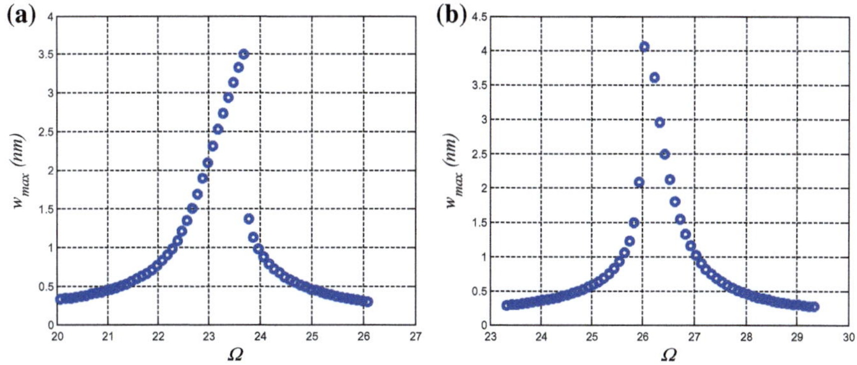

Fig. 5.44 Frequency-response curves of (**a**) 2 nm and (**b**) 5 nm slacked CNT showing hardening and softening behaviors, respectively. Results are shown for $V_{DC} = V_{AC} = 1$ Volt, and 100 quality factor

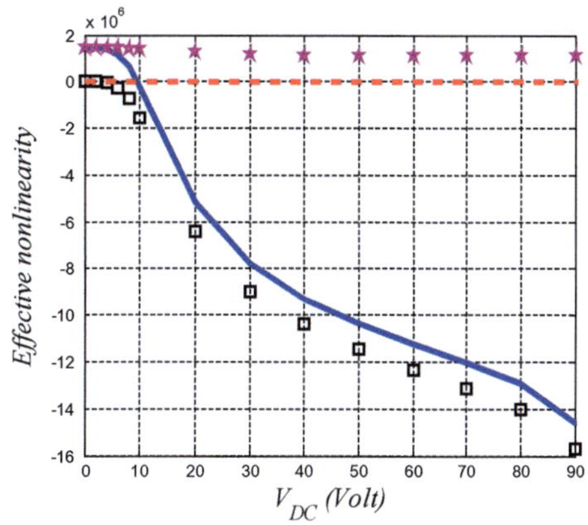

Fig. 5.45 Variation of the effective nonlinearity of the fundamental mode (—), the cubic nonlinearity coefficient (☆), and the quadratic nonlinearity coefficient (□) of the unslacked CNT with the DC voltage

Figure 5.47 shows that for nonzero slack (80 nm), S is always negative, with an increase except near the pull-in zone, indicating a softening type behavior of the slacked CNT. This result indicates the fundamental mode of a slacked CNT is locally dominated by a softening type behavior demonstrating that the quadratic nonlinearities coming from the initial curvature and the electrostatic force dominate the dynamic behavior of the slacked CNT.

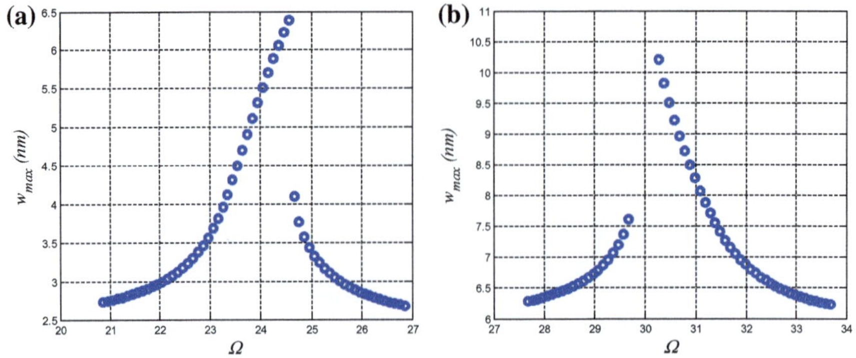

Fig. 5.46 Frequency-response curves of unslacked CNT for (**a**) $V_{DC} = 8$ Volt, and (b) $V_{DC} = 14$ Volt showing hardening and softening behaviors respectively. Results are shown for $V_{AC} = 0.2$ Volt, and 100 quality factor

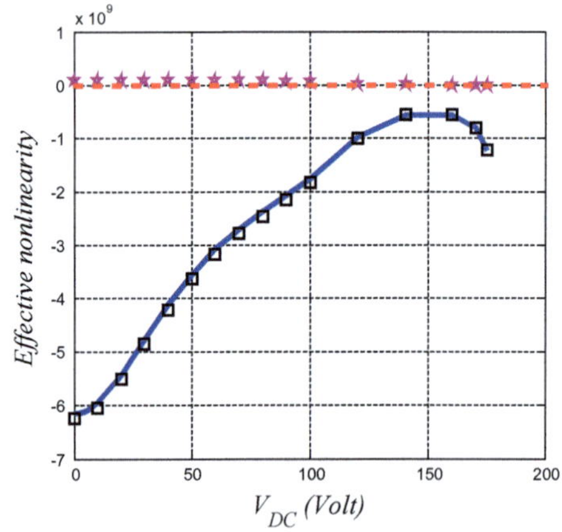

Fig. 5.47 Variation of the effective nonlinearity of the fundamental mode (——), the cubic nonlinearity coefficient (☆), and the quadratic nonlinearity coefficient (□) for the 80 nm slacked CNT with the DC voltage

5.4.4 *The String Model*

(a) Motivation and formulation

For some specific CNT geometry, especially those with small radius (i.e., with big aspect ratio), the beam model experiences some numerical problems especially when trying to solve the boundary value problems of Eqs. (5.126)–(5.128). The algorithm that we used to calculate the effective nonlinearity coefficient based on the beam model of Sect. 4.4.1 works only for small DC load (i.e., the bending dominated regime). This is expected since, for the case of CNT with small radius, the bending term is dominant only for small DC load [4, 6]. Hence, we propose to use a string model, in which we eliminate the bending term in Eq. (5.106). This yields the following nondimentional equation of motion and associated boundary conditions:

$$\frac{\partial^2 w}{\partial t^2}+c\frac{\partial w}{\partial t}=\alpha_2 F_e+\alpha_1\left[\int_0^1\left\{\left(\frac{\partial w}{\partial x}\right)^2-2\left(\frac{\partial w}{\partial x}\frac{dw_0}{dx}\right)\right\}dx\right]\left[\frac{\partial^2 w}{\partial x^2}-\frac{d^2 w_0}{dx^2}\right], \quad (5.139)$$

$$w(0,t)=0,\quad w(1,t)=0, \quad (5.140)$$

where F_e and w_0 are given by Eq. (5.109), and the nondimensional parameters are defined as

$$\alpha_1=\frac{EAd^2}{2L^2},\quad \alpha_2=\frac{\pi\varepsilon_0 L^2}{d^2},\quad c=\frac{\tilde{c}}{\sqrt{\rho AL^2}},\ \Omega=\tilde{\Omega}\sqrt{\rho AL^2},\ R=\frac{\tilde{R}}{d} \quad (5.141)$$

(b) Results

As a case study for a CNT with small radius and high aspect ratio, a hallow CNT of $L = 3000$ nm, $h = 0.34$ nm, $\tilde{R} = 1$ nm, $d = 500$ nm, $E = 1.2\,TPa$, and $\rho = 1.35\,\mathrm{g/cm^3}$ is considered.

We first compare between the dimensional fundamental natural frequency of the considered CNT with small radius while using both models (the beam and the string models). One can see from Fig. 5.48 that for a wide range of DC load (5 Volt $<$ $V_{DC} < 60$ Volt), between the bending dominated regime and the pull-in zone, the string and the beam model are in a good agreement. Hence, we will use now the perturbation analysis derived in Sect. 4.4.5b while considering the beam model for small DC load, and use the string model for higher values of the DC load.

Next, we simulate the variation of the effective nonlinearity with the curvature level. In this case, the beam model is used since the nonlinearity coming from the electrostatic load is small. We can see from Fig. 5.49 that S is positive in this case for very small range of initial curvature level ($b_o < 1$ nm) indicating a hardening

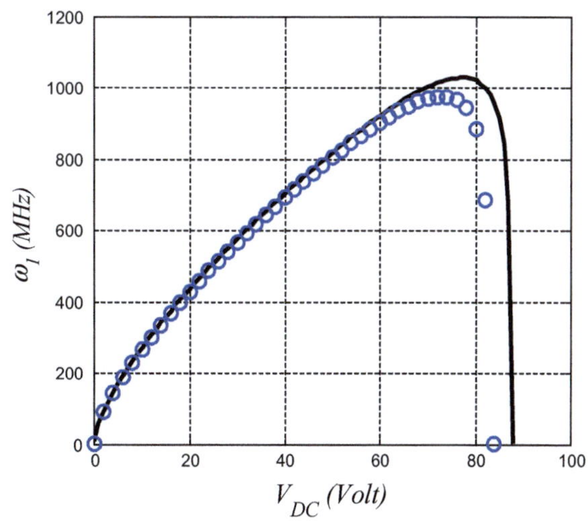

Fig. 5.48 Comparison between the variations of the dimensional fundamental natural frequency of the unslacked CNT with the DC voltage using the beam model (——) and the string model (o)

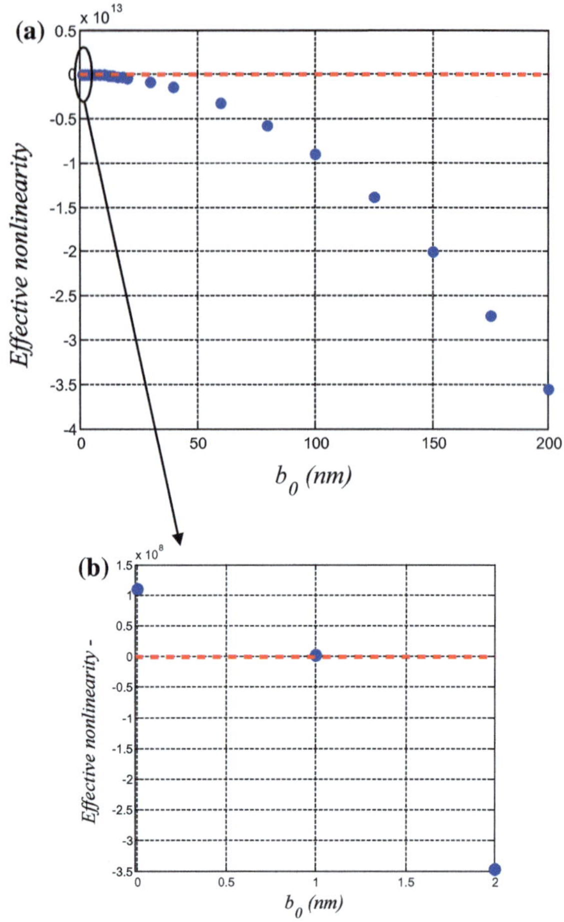

Fig. 5.49 Variation of the effective nonlinearity of the CNT with slack

behavior. This means that for CNT with large aspect ratio, a tiny slack level will soften the fundamental mode. To verify the results of Fig. 5.49, we integrate with time the differential equations of motion to obtain the dynamic response of the slacked CNT under small DC and AC harmonic load. Figures 5.50a, b show the response of a 0.5 nm and 1.5 nm slacked CNT, respectively, which verify the perturbation results.

The variation of the effective nonlinearity with the DC load for the unslaked CNT case is shown in Fig. 5.51. In this case, the beam model applies up to $V_{DC} < 1$ Volt, after which the string model is used. We can see from Fig. 5.51a, b that S is only positive in this case in a very small range of DC load indicating a hardening behavior. The results of the beam model were compared by those of the string model, in which we can see that the CNT behavior is of softening type. Also the results are verified using time integration of the equation of motion using the ROM, Fig. 5.51c, d.

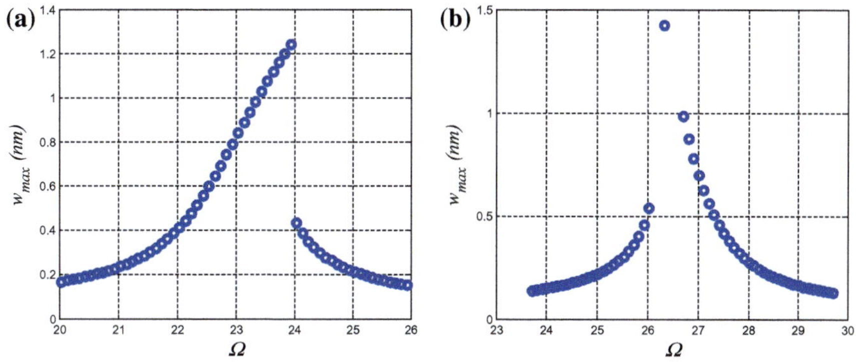

Fig. 5.50 Frequency-response curves of (**a**) 0.5 nm and (**b**) 1.5 nm slacked CNT showing hardening and softening behaviors respectively. Results are shown for $V_{DC} = V_{AC} = 0.01$ Volt, and 100 quality factor

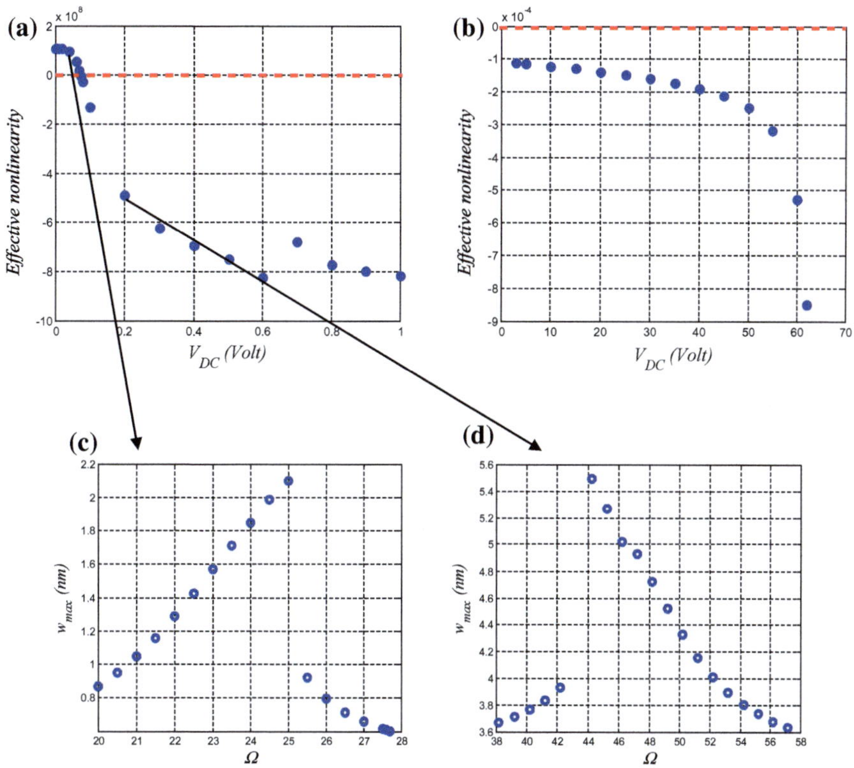

Fig. 5.51 Variation of the effective nonlinearity coefficient S of the unslacked CNT with the DC voltage while using (**a**) a beam model, and (**b**) a string model. Frequency-response curves of unslacked CNT for (**c**) $V_{DC} = 0.04$ Volt, and (d) $V_{DC} = 0.2$ Volt showing hardening and softening behaviors respectively. Results of Fig. 6.19c, d are shown for $V_{AC} = 0.01$ Volt, and 100 quality factor

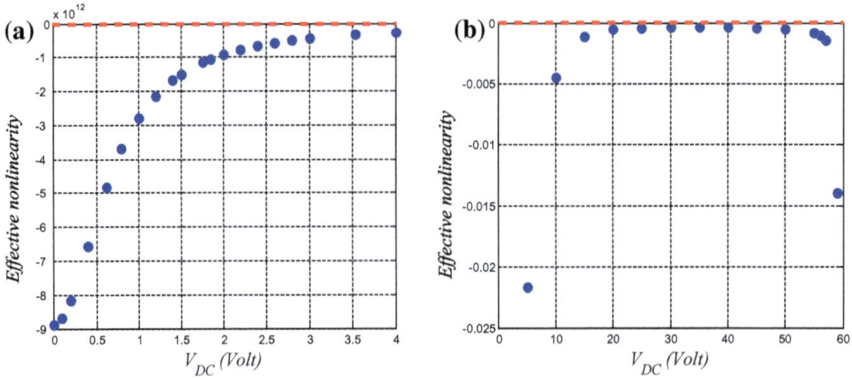

Fig. 5.52 Variation of the effective nonlinearity coefficient S of the 100 nm slacked CNT with the DC voltage while using (**a**) a beam model, and (**b**) a string model

Finally, the results of Fig. 5.51a, b are simulated for a 100 nm slacked CNT in Fig. 5.52a, b. Here also we notice that the beam model applies in the bending regime, i.e., $V_{DC} < 4$ Volt. Above this value, the string model is used. We can see also that we get the same behavior as the case of CNT with higher radius, Fig. 5.47. In this case, the CNT is locally dominated by a softening behavior due to the presence of the huge quadratic nonlinearity from the slack level. We notice also an increase of the effective nonlinearity, which is due to the presence of a source of cubic nonlinearity coming from the mid-plane stretching of the CNT. Finally, the effective nonlinearity starts to decrease near the pull-in zone where the electrostatic force starts to dominate the behavior of the slacked CNT.

References

1. Iijima, S. (1991). Helical microtubules of graphitic carbon. *Nature, 354*, 56–58.
2. Craighead, H. G. (2000). Nanoelectromechanical systems. *Science, 290*, 1532–1535.
3. Postma, H., Kozinsky, I., Husain, A., & Roukes, M. (2005). Dynamic range of nanotube- and nanowire-based electromechanical systems. *Applied Physics Letters, 86*, 223105(1–3).
4. Sazonova, V., Yaish, Y., Üstünel, H., Roundy, D., Arias, T. A., & McEuen, P. L. (2004). A tunable carbon nanotubes electromechanical oscillator. *Nature, 431*, 284–287.
5. Sazonova, V. A. (2006). A tunable carbon nanotube resonator, Ph.D. Thesis, Department of Physics, Cornell University.
6. Üstünel, H., Roundy, D., & Arias, T. A. (2005). Modeling a suspended nanotube oscillator. *Nano Letter, 5*, 523–526.
7. Gibson, R. F., Ayorinde, E. O., & Wen, Y. F. (2007). Vibrations of carbon nanotubes and their composites: A review. *Composites Science and Technology, 67*, 1–28.
8. Gao, R. P., Wang, Z. L., Bai, Z. G., de Heer, W. A., Dai, L. M., & Gao, M. (2000). Nanomechanics of individual carbon nanotubes from pyrolytically grown arrays. *Physical Review Letters, 85*, 622–625.

9. Wang, Z. L., Gao, R. P., Poncharal, P., de Heer, W. A., Dai, Z. R., & Pan, Z. W. (2001). Mechanical and electrostatic properties of carbon nanotubes and nanowires. *Materials Science and Engineering C*, *16*, 3–10.
10. Dequesnes, M., Rotkin, S. V., & Aluru, N. R. (2002). Calculation of pull-in voltages for carbon-nanotube-based nanoelectromechanical switches. *Nanotechnology*, *13*, 120–131.
11. Sapmaz, S., Blanter, Y. M., Gurevich, L., & van der Zant, H. S. J. (2003). Carbon nanotubes as nanoelectromechanical systems. *Physical Review B*, *67*, 14–2354.
12. Dequesnes, M., Tang, S., & Aluru, N. R. (2004). Static and dynamic analysis of carbon nanotube-based switches. *Journal of Engineering Materials and Technology*, *126*, 230–237.
13. Lefèvre, R., Goffman, M. F., Derycke, V., Miko, C., Forró, L., Bourgoin, J. P., et al. (2005). Scaling law in carbon nanotube electromechanical devices. *Physical Review Letters*, *95*, 55–59.
14. Ke, C. H., & Espinosa, H. D. (2005). Numerical analysis of nanotube-based NEMS devices—Part I: Electrostatic charge distribution on multiwalled nanotubes. *Journal of Applied Mechanics*, *72*, 721–725.
15. Pugno, N., Ke, C. H., & Espinosa, H. D. (2005). Analysis of doubly clamped nanotube devices in the finite deformation regime. *Journal of Applied Mechanics*, *72*, 445–449.
16. Witkamp, B., Poot, M., & van der Zant, H. S. J. (2006). Bending-mode vibration of a suspended nanotube resonator. *Nano Letter*, *6*, 2904–2908.
17. Poot, M., Witkamp, B., Otte, M. A., & van der Zant, H. S. J. (2007). Modeling suspended carbon nanotube resonators. *Physica Status Solidus (b)*, *244*, 4252–4256.
18. Peng, H. B., Chang, C. W., Aloni, S., Yuzvinsky, T. D., & Zettl, A. (2007). Microwave electromechanical resonator consisting of clamped carbon nanotubes in an abacus arrangement. *Physical Review B*, *76*, 354–359.
19. Srivastava, D., & Barnard, S. T. (1997). Molecular dynamics simulation of large-scale carbon nanotubes on a shared-memory architecture. In *Proceedings of the 1997 ACM/IEEE Conference on Supercomputing*. San Jose, CA.
20. Sears, A. T. (2006). Carbon nanotube mechanics: continuum model development from molecular mechanics virtual experiments. Ph.D. Thesis, Virginia Polytechnic Institute and State University, Blacksburg, VA.
21. Hwang, H. J., & Lee, J. H. (2006). Molecular dynamics modeling of eectromechanical nanotube memory. *Journal of the Korean Physical Society*, *49*(3), 1136–1142.
22. Brodka, A., Kołoczek, J., Burian, A., Dore, J. C., Hannon, A. C., & Fonseca, A. (2006). Molecular dynamics simulation of carbon nanotube structure. *Journal of Molecular Structure*, *792–793*, 78–81.
23. Kang, J. W., Hwang, H. J., & Jiang, Q. (2006). A molecular dynamics study on oscillation of a carbon nanotube inside an encapsulating Boron-Nitride nanotube. *Journal of Computational and Theoretical Nanoscience*, *3*(6), 880–884(1–5).
24. Kang, J. W., Kang, D. Y., Choi, Y. G., Lee, S., & Hwang, H. J. (2009). Molecular dynamics study of tunable double-walled carbon nanotube oscillator. *Journal of Computational and Theoretical Nanoscience*, *6*(7), 1580–1584(1–5).
25. Kang, J. W., Won, C. S., Ryu, G. H., & Choi, Y. G. (2009). Molecular dynamics study on resonance characteristics of gigahertz carbon nanotube motor. *Journal of Computational and Theoretical Nanoscience*, *6*(1), 178–186(1–9).
26. Greaney, P. A., & Grossman, J. C. (2007). Nanomechanical energy transfer and resonance effects in single-walled carbon nanotubes. *Physical Review Letters*, *98*, 125503–125507.
27. Lu, J.-M., Wang, Y.-C., Chang, J.-G., Su, M.-H., & Hwang, C.-C. (2008). Molecular-dynamic investigation of buckling of double-walled carbon nanotubes under uniaxial compression. *Journal of the Physical Society of Japan*, *77*(4), 044603(1–7).
28. Shayan-Amin, S., Dalir, H., & Farshidianfar, A. (2009). Molecular dynamics simulation of double-walled carbon nanotube vibrations: comparison with continuum elastic theories. *Journal of Mechanics*, *25*(4).
29. Wang, X. Y., & Wang, X. (2004). Numerical simulation for bending modulus of carbon nanotubes and some explanations for experiment. *Composites Part B*, *35*, 79–86.

30. Lau, K. T., Chipara, M., Ling, H., & Hui, D. (2004). On the effective elastic moduli of carbon nanotubes for nanocomposite structures. *Composites Part B*, *35*, 95–101.
31. Garg, M. (2005). Mechanics of deformation of carbon nanotubes. MS thesis, Department of Mechanical Engineering, Massachuset Institute of Technology.
32. Yakobson, B., Brabec, C., & Bernholc, J. (1996). Nanomechanics of carbon tubes: instabilities beyond linear response. *Physical Review Letters*, *76*, 2511–2514.
33. Harik, V. M. (2002). Mechanics of carbon nanotubes: applicability of the continuum-beam models. *Computational Materials Science*, *24*, 328–342.
34. Harik, V. M. (2001). Ranges of applicability of the continuum beam model in the mechanics of carbon nanotubes and nanorods. *Solid State Communication*, *120*, 331–335.
35. Liu, J. Z., Zheng, Q., & Jiang, Q. (2001). Effect of a rippling mode on resonances of carbon nanotubes. *Physical Review Letters*, *86*, 43–46.
36. Pantano, A., Parks, D. M., & Boyce, M. C. (2004). Mechanics of deformation of single- and multi-wall carbon nanotubes. *Journal of the Mechanics and Physics of Solids*, *52*, 789–821.
37. Arroyo, M. (2004). Continuum mechanics and carbon nanotubes. In *Proceedings of the XXI ICTAM*. Warsaw, Poland.
38. Arroyo, M., & Belytschko, T. (2005). Continuum mechanics modeling and simulation of carbon nanotubes. *Meccanica*, *40*(4–6), 455–469.
39. Wang, L., Hu, H., & Guo, W. (2010). Thermal vibration of carbon nanotubes predicted by beam models and molecular dynamics. In *Proceedings of the Royal Society A, Mathematical, Physical, and Engineering Sciences*, rspa.2009.0609v1-rspa20090609.
40. Sears, A., & Batra, R. C. (2010). Carbon nanotube mechanics: molecular simulations and continuum models for carbon nanotubes. In *Virginia Space Grant Consortium 2010 Student Research Conference, Department of Engineering Science and Mechanics: Virginia Polytechnic Institute and State University Blacksburg*.
41. Conley, W. G., Raman, A., Krousgrill, C. M., & Mohammadi, S. (2008). Nonlinear and nonplanar dynamics of suspended nanotube and nanowire resonators. *Nano Letters*, *8*, 1590–1595.
42. Elishakoff, I., & Pentaras, D. (2009). Fundamental natural frequencies of double-walled carbon nanotubes. *Journal of Sound and Vibration*, *322*, 652–664.
43. Elishakoff, I., & Pentaras, D. (2009). Natural frequencies of carbon nanotubes based on simplified Bresse-Timoshenko theory. *Journal of Computational and Theoritical Nanoscience*, *6*, 1527–1531.
44. Georgantzinos, S. K., Giannopoulos, G. I., & Anifantis, N. K. (2009). An efficient numerical model for vibration analysis of single-walled carbon nanotubes. *Journal of Computational Mechanics*, *43*, 731–741.
45. Hawwa, M. A., & Al-Qahtani, H. M. (2010). Nonlinear oscillations of a double-walled carbon nanotube. *Computational Material Science*, *48*, 140–143.
46. Ke, C. H., & Espinosa, H. D. (2006). In situ electron microscopy electromechanical characterization of a bistable NEMS device. *Small*, *2*(12), 1484–1489.
47. Ke, C. H., Espinosa, H. D., & Pugno, N. (2005). Numerical analysis of nanotube-based NEMS devices—Part II: Role of finite kinematics, stretching and charge concentrations. *Journal of Applied Mechanics*, *72*, 726–731.
48. Ke, C. H., Pugno, N., Peng, B., & Espinosa, H. D. (2005). Experiments and modeling of carbon nanotube-based NEMS devices. *Journal of the Mechanics and Physics of solids*, *53*, 1314–1333.
49. Isacsson, A., Kinaret, J. M., & Kaunisto, R. (2007) Nonlinear resonance in a three-terminal carbon nanotube resonator. *Nanotechnology*, *18*, 95203(1–8).
50. Isacsson, A., & Kinaret, J. M. (2009) Parametric resonances in electrostatically interacting carbon nanotube arrays. *Physical Review B*, *79*, 165418(1–11).
51. Krishnan, A., Dujardin, E., Ebbesen, T. W., Yianilos, P. N., & Treacy, M. M. J. (1998). Young's modulus of single-walled nanotubes. *Physical Review B*, *58*, 14013–14019.
52. Kim, P., & Lieber, C. M. (1999). Nanotube nanotweezers. *Science Magazine*, *286*(5447), 2148–2150.

53. Poncharal, P., Wang, Z. L., Ugarte, D., & de Heer, W. A. (1999). Electrostatic deflections and electromechanical resonances of carbon nanotubes. *Science*, *283*, 1513–1516.
54. Babic, B., Furer, J., Sahoo, S., Farhangfar, S., & Schonenberger, C. (2003). Intrinsic thermal vibrations of suspended doubly clamped single-wall carbon nanotubes. *Nano Letters*, *3*, 1577–1580.
55. Dujardin, E., Derycke, V., Goffman, M. F., Lefèvre, R., & Bourgoin, J. P. (2005). Self-assembled switches based on electroactuated multiwalled nanotubes. *Applied Physics Letters*, *87*, 1931–1938.
56. Rabieirad, L., Kim, S., Shim, M., & Mohammadi, S. (2005). Doubly clamped single-walled carbon nanotube resonators operating in MHz frequencies. In *Proceedings of 2005 5th IEEE Conference on Nanotechnology*. Nagoya, Japan.
57. Bak, J. H., Kim, Y. D., Hong, S. S., Lee, B. Y., Lee, S. R., Jang, J. H., et al. (2008). High-frequency micromechanical resonators from aluminium-carbon nanotube nanolaminates. *Nature Materials*, *7*, 459–463.
58. San Paulo, A., Black, J, García-Sanchez, D., Esplandiu, M. J., Aguasca, A., Bokor, J., F. Perez-Murano, F., & Bachtold, A. (2008). Mechanical detection and mode shape imaging of vibrational modes of micro and nanomechanical resonators by dynamic force microscopy. *Journal of Physics: Conference Series*, *100*, 052009(1–5).
59. Amlani, I., Lee, K. F., Deng, J., & Wong, H. S. P. (2009). Measuring frequency response of a single-walled carbon nanotube common-sourcea. *IEEE Transactions on Nanotechnology*, *8*, 226–233.
60. Ren, L., Pint, C. L., Booshehri, L. G., Rice, W. D., Wang, X., Hilton, D. J., et al. (2009). Carbon nanotube terahertz polarizer. *Nano Letters*, *9*, 2610–2613.
61. Kienle, D., & Léonard, F. (2000). Terahertz response of carbon nanotube transistors. *Physical Review Letters*, *103*, 026601(1–4).
62. Lu, R. F., Lu, Y. P., Lee, S. Y., Ha, K. L., & Deng, W. Q. (2009). Terahertz response in single-walled carbon nanotube transistor: A real-time quantum dynamics simulation. *Nanotechnology*, *20*, 505401(1–4).
63. Kang, J. W., Lee, J. H., Lee, H. J., & Hwang, H. J. (2005). A study on carbon nanotube bridge as a electromechanical memory device. *Physica E*, *27*, 332–340.
64. Garcia-Sanchez, D., San Paulo, A., Esplandiu, M. J., Perez-Murano, F., Forrò, L., Aguasca, A., & Bachtold, A. (2007). Mechanical detection of carbon nanotube resonator vibrations. *Physical Review Letters*, *99*, 085501(1–4).
65. Mayoof, F. N., & Hawwa, M. A. (2009). Chaotic behavior of a curved carbon nanotube under harmonic excitation. *Journal of Chaos, Solitons & Fractals*, *42*, 1860–1867.
66. Meirovitch, L. (2001). *Fundamentals of vibrations*. New York: McGraw Hill.
67. Rao, S. S. (2004). *Mechanical vibrations* (4th ed.). New Jersey: Prentice Hall.
68. Dawe, D. J. (1971). The Ttransverse vibartion of shallow arches using the displacement method. *International Journal of Mechanical Sciences*, Pergamon Press, *13*, 713–720.
69. Nayfeh, A. H. (2000). *Nonlinear interactions*. New-York, United-States: Wiley Interscience.
70. Itô, K. (Ed.). (1993). Methods other than difference methods. *§303I in encyclopedic dictionary of mathematics* (2nd ed.) (vol. 2, pp. 1139–1980). Cambridge, MA: MIT Press.
71. Ouakad, H. M. (2013). The response of a micro-electro-mechanical system (MEMS) cantilever-paddle gas sensor to mechanical shock loads. *Journal of Vibration and Control*, in press. doi:10.1177/1077546313514763.
72. Younis, M. I., Abdel-Rahman, E. M., & Nayfeh, A. H. (2003). A Reduced-order model for electrically actuated microbeam-based MEMS. *Journal of Microelectromechanical Systems*, *12*, 672–680.
73. Reddy, J. N. (2002). *Energy principles and variational methods in applied mechanics*. New York: Wiley and Sons.
74. Hayt, W. H., & Buck, J. A. (2001). *Engineering electromagnetics*. New York, United-States: McGraw-Hill.
75. Nathanson, H. C., & Wickstrom, R. A. (1965). A resonant gate silicon surface transistor with high Q bandpass properties. *IEEE Applied Physics Letters*, *7*, 84–86.

76. Nathanson, H. C., Newell, W. E., Wickstrom, R. A., & Davis, J. R. (1967). The Resonant gate transistor. *IEEE Transactions on Electron Devices, 14*, 117–133.
77. Newell, W. (1968). Miniaturization of tuning forks. *Science, 161*(3848), 1320–1326.
78. Abdel-Rahman, E. M., Younis, M. I., & Nayfeh, A. H. (2002). Characterization of the mechanical behavior of an electrically actuated microbeam. *Journal of Micromechanics and Microengineering, 12*, 759–766.
79. Younis, M. I., & Nayfeh, A. H. (2003). A study of the nonlinear response of a resonant microbeam to an electric actuation. *Nonlinear Dynamics, 31*, 91–117.
80. Nayfeh, A. H., & Younis, M. I. (2005). Dynamics of MEMS resonators under superharmonic and subharmonic excitations. *Journal of Micromechanics and Microengineering, 15*, 1840–1847.
81. Nayfeh, A. H., Younis, M. I., & Abdel-Rahman, E. M. (2005). Reduced-order models for MEMS applications. *Nonlinear Dynamics, 41*, 211–236.
82. Krylov, S., & Maimon, R. (2004). Pull-in dynamics of an elastic beam actuated by continuously distributed electrostatic force. *Journal of Vibration and Acoustics, 126*, 332–342.
83. Elata, D., & Bamberger, H. (2006). On the dynamic pull-in of electrostatic actuators with multiple degrees of freedom and multiple voltage sources. *Journal of Microelectromechanical Systems, 15*, 131–140.
84. Nayfeh, A. H., Younis, M. I., & Abdel-Rahman, E. M. (2007). Dynamic pull-in phenomenon in MEMS resonators. *Nonlinear Dynamics, 48*, 153–163.
85. Thompson, J. M. T., & Stewart, H. B. (2001). *Nonlinear dynamics and chaos*. New York, United-States: Wiley.
86. Alsaleem, F. M., Younis, M. I., & Ouakad, H. M. (2009). On the nonlinear resonances and dynamic pull-in of electrostatically actuated resonators. *Journal of Micromechanics and Microengineering, 19*(4), 045013.
87. Wolfram, S. (2000). *The Mathematica Book* (Vol. 100, pp. 7237–61820). New York, NY, USA: Cambridge University Press and Wolfram Research Inc.
88. Noor, A. K., & Nemeth, M. P. (1980). Micropolar beam models for lattice grids with rigid joints. *Computer Methods in Applied Mechanics and Engineering, 21*(2), 249–263.
89. Harris, P. J. F. (1999). *Carbon nanotubes and related structures*. Cambridge, MA, United-States: Cambridge University Press.
90. Yu, M. F. (2004). Fundamental mechanical properties of carbon nanotubes: current understanding and the related experimental studies. *Journal of Engineering Materials and Technology, 126*, 271–278.
91. Akita, S., Nakayama, Y., Mizooka, S., Takano, Y., Okawa, T., Miyatake, Y., et al. (2001). Nanotweezers consisting of carbon nanotubes operating in an atomic force microscope. *Applied Physics Letters, 79*(11), 1691–1694.
92. Ouakad, H. M., & Younis, M. I. (2010). The dynamic behavior of MEMS arch resonators actuated electrically. *International Journal of Non-Linear Mechanics, 45*(7), 704–713.
93. Ouakad, H. M. (2014). Static response and natural frequencies of microbeams actuated by out-of-plane electrostatic fringing-fields. *International Journal of Non-Linear Mechanics, 63*, 39–48.
94. Abdel-Rahman, E. M., Emam, S. A., & Nayfeh, A. H. (2003). A generalized model of electrically actuated microbeam-based MEMS devices. In *Proceedings of the DETC.03 ASME 2003 Design Engineering Technical Conference and Computers and Information in Engineering Conference*. Chicago, Illinois, USA.
95. Nayfeh, A. H., & Pai, P. F. (2004). *Linear and nonlinear structural mechanics*. New York, United-States: Wiley.
96. Nayfeh, A. H., & Balachandran, B. (1995). *Applied nonlinear dynamics*. New York: Wiley.
97. Kuttler, J. R., & Sigillito, V. G. (1981). On curve veering. *Journal of Sound and Vibration, 75*, 585–588.
98. Perkins, N. C., & Mote, C. D, Jr. (1986). Comments on curve veering in eigenvalue problems. *Journal of Sound and Vibration, 106*, 451–463.

99. Arafat, H. N., & Nayfeh, A. H. (2003). Non-linear responses of suspended cables to primary resonance excitations. *Journal of Sound and Vibration*, *266*, 325–354.
100. Rega, G. (2004). Nonlinear vibrations of suspended cables-Part I: Modeling and analysis. *Journal of Applied Mechanics Review*, *57*, 443–478.
101. Lin, J., & Parker, R. G. (2001). Natural frequency veering in planetary gears. *Mechanics of Structures and Machines*, *29*, 411–429.
102. Nayfeh, A. H. (1981). *Introduction to perturbation techniques*. New York, United-States: Wiley Interscience.

Chapter 6
Nonlinear Dynamics of Ambient Noise-Driven Graphene Nanostructured Devices for Energy Harvesting

A. El Aroudi, M. López-Suárez, E. Alarcón, R. Rurali and G. Abadal

6.1 Introduction

Recently, there has been an increasing of interest in self-powered devices in remote environment applications where energy becomes an important system requirement and the use of rechargeable batteries becomes problematic such as in hard-to-access locations or, because the lifetime of the batteries is much shorter than of that of the system to be powered. Many solutions have been suggested in the literature to remedy this problem by using energy harvesting techniques [1]. Energy harvesting is considered a key factor in the development of autonomous sensors and micro- and nanogenerators with extended lifetimes. There are different energy harvesting technologies that can be used depending on the kind of the available energy. Among the energy sources, vibrational kinetic energy is the most used for applications such as development of microgenerators [1] and noise harvesters for nanosensors [2]. In this approach, mechanical energy is converted to electricity by using piezoelectric, capacitive, or inductive transducers. Vibration-based energy harvesting research has

A. El Aroudi (✉)
University Rovira i Virgili, Tarragona, Spain
e-mail: abdelali.elaroudi@urv.cat

M. López-Suárez · G. Abadal
Universitat Autónoma de Barcelona, Barcelona, Spain
e-mail: miquel.lopez@nipslab.org

G. Abadal
e-mail: gabriel.abadal@uab.cat

E. Alarcón
Universitat Politècnica de Catalunya, Barcelona, Spain
e-mail: eduard.alarcon@upc.edu

R. Rurali
Institut de Ciència de Materials de Barcelona (ICMAB-CSIC),
Campus de Bellaterra, Barcelona, Spain
e-mail: rrurali@icmab.es

E. Blokhina et al. (eds.), *Nonlinearity in Energy Harvesting Systems*,
DOI 10.1007/978-3-319-20355-3_6

largely focused on linear electromechanical devices excited at resonance. To overcome problems related to linear energy harvester working as resonators, nonlinear energy harvesters have been proposed in certain applications [3]. In the nonlinear energy harvesting approach, rather than resonance frequency tuning [4, 5], the nonlinearity of the system is exploited to improve the performances of the energy harvester within a wide frequency range outperforming classical resonant energy harvesters [3, 6]. Energy harvesting can be performed both at the macro-[3], micro-[7] and even at nanoscales [2]. Unfortunately, most of the available power sources have very low frequency oscillations while the resonant frequency of any nano-mechanical device is paradoxically high ($\propto$ GHz) which makes many issues related to their performances as energy harvesters yet to be solved, particularly that of extracting energy at low frequencies. Nonlinearities in energy harvesting nanodevices may play a vital role to solve these issues. Recent works have used nanodevices such as piezoelectric zinc oxide (ZnO) nanowires [8] where mechanical energy coming from light wind and body movement was converted to electrical energy by a coupled piezoelectric-semiconductor process. Due to their extremely low power consumption, nanosystems that harvest their energy from their environment is attractive for applications such as wireless sensing, personal electronics for battery charging, implantable biological devices, biomedical science, and environmental monitoring such as animal migration tracking and gas and chemical species sensing [9–11]. The nanogenerators could also be used to harvest energy created by tire pressure change and mechanical vibration due to moving vehicles [9].

6.2 Graphene-Based Nanomaterials for Energy Harvesting

Graphene is a newly discovered nanomaterial that has interesting peculiar electronic and mechanical properties making it a promising candidate for future electronic applications [12]. It consists of a flat monolayer of carbon atoms tightly packed into a two-dimensional honeycomb-like lattice. It can be considered as a building block for graphite materials of all other dimensionality [13]. This material exhibits high conductivity, large surface area, and electrochemical window. All these properties make it an advantageous material for energy storage devices. A literature review concerning the electrochemical application of graphene in energy storage/generation devices can be found in [14]. Some applications of graphene in solar cells, batteries, and fuel cells can be found in [15]. In [16] biosensors and biosensing systems employing graphene have been described.

A review providing a scientific progress of different type of graphene materials is [17]. Their structural, thermal, optical, and electrical properties along with their potential applications and the impact of graphene and related materials on the environment, its toxicological effects and its future prospects have been also discussed.

In [18] it has been shown that a strained nanostructured graphene posses an intrinsic mechanical nonlinearity and it can be used for nonlinear energy harvesting at the nanoscale. There, it has been also shown that when weakly compressed, the graphene sheet behaves like a double-well system and that for a critical value of the compression ratio, the harvested power is optimal. In [19] some discussions on some recent studies on graphene-based NGs have been provided. In [20], tests involving water flow with various molarities of hydrochloric acid over few-layered graphene are performed and the authors report higher induced voltages for graphene as compared to nanotubes. In [21] graphene oxide films have been fabricated as a low-cost and flexible nanogenerator to convert acoustic energy into electricity with a about of 12 % conversion efficiency.

6.3 Chapter Outline

The aim of this chapter is to study the nonlinear dynamic behavior of a strained nanostructured graphene membrane first considered in [18] for possible use in energy harvesting applications. Based on the initial study by the authors that was presented in [22] and [23], in this chapter we thoroughly study and expand the previous analysis and we fully explain the reported phenomena. To accomplish this aim, the rest of the paper is organized as follows. In Sect. 6.2, a brief description of the mathematical model of the graphene harvester is provided. Section 6.3 presents the dynamical behavior of the device in the absence of any external excitation. Equilibrium points and their stability are studied in the same section. In Sect. 6.4, deterministic excitation is considered and the dynamics of the system is studied when the external force intensity and frequency are varied. In Sect. 6.5, the dynamics of the system is studied under a random excitation in the form of a white noise with a Gaussian distribution and a limited bandwidth. In Sect. 6.6, the performances of the system under random excitation are studied in terms of RMS levels. The study is extended in Sect. 6.7 to an array of three graphene membrane-based harvesters. Finally, conclusions are provided in the last section.

6.4 Mathematical Model of the Graphene Vibrating Membrane for Energy Harvesting Applications

6.4.1 Nonlinear Mathematical Model

Figure 6.1 shows the system under study in this chapter. It consists of a flat suspended graphene membrane with a compression ratio defined as $\varepsilon = (L_0 - L)/L_0$, where L is the length of the graphene membrane and L_0 its equilibrium value in the absence of any compression. Let us consider different values of the compression ratio ε between 0 % (uncompressed graphene) and 10 %.

Let us suppose that the graphene membrane is submerged in a noisy environment. Let $\xi(t) = \sigma^2\eta(t)$, where $\eta(t)$ is a white noise excitation with a Gaussian distribution

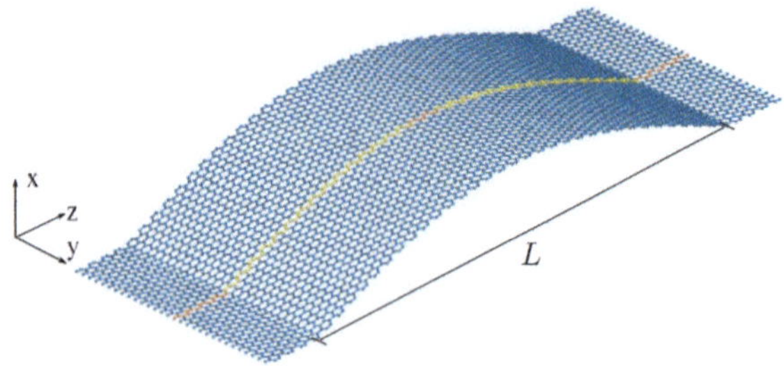

Fig. 6.1 Uncompressed or weakly compressed graphene membrane [18]

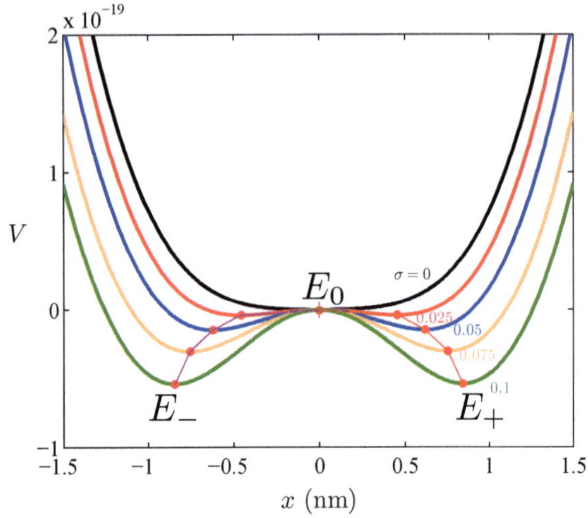

Fig. 6.2 Potential $V(x)$ for different values of the compression factor ε. For $\varepsilon \neq 0$, equilibrium points are also plotted. E_+ and E_- are attractors while E_0 is a saddle point

noise whose mean value is zero, its variance is σ^2. By developing the Euler–Bernoulli equation for clamped beams, the following mass-spring dynamic model is obtained for the considered graphene membrane [18]

$$m\ddot{x} = -\frac{\partial V}{\partial x} - b\dot{x} + \sigma^2\eta(t) \tag{6.1}$$

where x is the displacement, m represents the effective mass of the layer, b stands for the damping factor, and $V(x)$ is the potential energy which is given by [18]

$$V(x) = \frac{1}{56}v_8x^8 + \frac{1}{30}v_6x^6 + \frac{1}{12}v_4x^4 + \frac{1}{2}v_2x^2 \tag{6.2}$$

The potential energy is depicted in Fig. 6.2 for different values of the compression factor ε. Equilibrium points represented by extremum points of $V(x)$ are also plotted in the same figure. For $\varepsilon = 0$, the potential energy corresponds to a harmonic linear oscillator presenting a single equilibrium point E_0 at the origin. For $\varepsilon \neq 0$, the points E_+ and E_- are minima and therefore they are attractors while E_0 is a maximum and corresponds to a saddle point.

Table 6.1 Coefficients v_i $(i = 2, 4, 6, 8)$ for the potential energy for different values of the compression ratio ε

ε	$v_8/56$	$v_6/30$	$v_4/12$	$v_2/2$
0	6.57×10^{50}	-5.73×10^{33}	6.74×10^{16}	0.0096
0.025	1.30×10^{51}	-1.36×10^{34}	9.26×10^{16}	−0.036
0.05	2.56×10^{51}	-2.29×10^{34}	1.14×10^{17}	−0.078
0.075	2.76×10^{51}	-2.34×10^{34}	1.16×10^{17}	−0.112
0.1	3.89×10^{51}	-3.43×10^{34}	1.49×10^{17}	−0.165

The coefficients v_i $(i = 2, 4, 6, 8)$ are given in Table 6.1 for different values of compression coefficient ε. These values have been obtained by means of ab initio density-functional calculations with the SIESTA code [18, 24], where the graphene sheet is modeled at a fully atomistic level and its electronic structure solved quantum mechanically. Equation (6.1) can be rearranged as follows

$$\ddot{x} = -\frac{b}{m}\dot{x} - \frac{1}{m}\frac{\partial V}{\partial x} + \frac{\sigma^2}{m}\eta(t) \tag{6.3}$$

If we let $v = \dot{x}$. Hence, the previous model can also be written as a set of two first-order differential equation in the state space form

$$\dot{x} = v \tag{6.4}$$

$$\dot{v} = -\frac{b}{m}v - \frac{1}{m}\frac{\partial V}{\partial x} + \frac{\sigma^2}{m}\eta(t) \tag{6.5}$$

whose divergence is

$$\nabla = \frac{\partial \dot{x}}{\partial x} + \frac{\partial \dot{v}}{\partial v} + \frac{\partial \dot{t}}{\partial t} = -\frac{b}{m} \tag{6.6}$$

which is unconditionally negative and therefore the system is dissipative for every set of parameter values.

6.4.2 *Dynamics of the Unforced System*

Before embarking on the study of the noise-driven system, let us consider first, the situation where $\xi(t) = 0$ in such a way that the system is free from any excitation. Under this circumstance the dynamics of the system will depend upon the parameters of the nanodevice. Namely, mass, damping factor, and compression ratio. Let us consider a compressed graphene membrane with compression coefficient $\varepsilon \neq 0$ in such a way that the origin is a saddle point (non-stable).

$$\dot{x} = v \tag{6.7}$$

$$\dot{v} = -\frac{b}{m}v - \frac{1}{m}\frac{\partial V}{\partial x} \tag{6.8}$$

Therefore the model (6.7) and (6.8) has three equilibrium points which are

$$E_0 = (0,0), \quad E_- = (X_-, 0), \quad E_+ = (X_+, 0) \tag{6.9}$$

where X_- and X_+ are the non-null displacement equilibria that can be obtained in closed form. Moreover, one can observe that the potential is symmetric and that if X_q is an equilibrium position $-X_q$ will also be. Therefore one has $X_- = -X_+$. Only real equilibrium points are of practical interest. These equilibrium points can also be obtained by forcing the derivative of $V(x)$ to be zero and solving for x. Let us define μ as follows

$$\begin{aligned} \mu &= 14700v_4v_6v_8 - 94500v_2v_8^2 - 2744v_6^3 \\ &\quad + 140v_8\sqrt{5}\sqrt{(3500v_4^3v_8 - 735v_4^2v_6^2 - 28350v_4v_6v_8v_2 + 91125v_2^2v_8^2 + 5292v_2v_6^3} \end{aligned} \tag{6.10}$$

Therefore the expressions of the non-null equilibrium points are given by the following expression

$$X_- = -\frac{1}{30v_8\sqrt[3]{\mu}}\sqrt{-30v_8\sqrt[3]{\mu}(14v_6\sqrt[3]{\mu} - \sqrt[3]{\mu} + 700v_4v_8 - 196v_6^2)} \tag{6.11}$$

$$X_+ = \frac{1}{30v_8\sqrt[3]{\mu}}\sqrt{-30v_8\sqrt[3]{\mu}(14v_6\sqrt[3]{\mu} - \sqrt[3]{\mu} + 700v_4v_8 - 196v_6^2)} \tag{6.12}$$

The Jacobian matrix of (6.7) and (6.8) is

$$\mathbf{J} = \begin{pmatrix} 0 & 1 \\ -\frac{1}{m}\frac{\partial^2 V}{\partial x^2} & -\frac{b}{m} \end{pmatrix} \tag{6.13}$$

and the corresponding characteristic polynomial can be expressed as follows

$$p(\lambda, x) = \lambda^2 + \frac{b}{m}\lambda + \frac{1}{m}\frac{\partial^2 V}{\partial x^2} \tag{6.14}$$

The two equilibrium points E_- and E_+ with a displacement different from zero will be both attracting since their corresponding Jacobian matrices do not have eigenvalues on the right-half side of the complex plane. For the equilibrium point E_0 with null displacement $x = 0$, the characteristic polynomial will have one eigenvalue in the right-half plane and another one in the left side which corresponds to a saddle point. This is a typical situation for double-well systems where two attracting points can be reached depending on the initial conditions while the separatrix of the saddle point will establish the boundary of the basins of attraction. In the bistable case, the two equilibrium points E_- and E_+ are separated by a barrier whose maximum is at $x = 0$ and whose height is $\Delta V = -V(X_+) = -V(X_-)$. The dynamics of the system is mainly characterized by the oscillation around the two minima X_+ and X_-, and crossing over the potential barrier ΔV. This fact will be exploited by making the displacement to swing between the two minima and accordingly to make the RMS value to increase proportionally to the distance between the equilibrium point which under sufficient excitation intensity, this in turn will imply a maximization of the harvested energy when the system is noise driven [25].

6.5 Dynamical Behavior of the Noise-Driven System from Numerical Simulations

Consider now that the system is subject to a random excitation with a normal distribution and a maximum bandwidth ω_{bw}. When such noise is applied to the system, the probability to swing between the two equilibrium points X_- and X_+ increases. Consequently noise can be used to increase the RMS value of the displacement and accordingly to raise the harvested energy. Let us also consider a compressed graphene sheet with compression coefficient $\varepsilon \neq 0$ in such a way that the origin is a saddle point (non stable). The nonlinear model of the system is numerically solved. A MATLAB Simulink model was built for the system equation modeling the graphene nano-harvester device according to the mathematical model given in (6.1). The MATLAB Simulink block diagram used for simulations is shown in Fig. 6.3.

For some specific values of noise level, the system was simulated during 500 ns and the time series, phase space, probability density, and amplitude spectra are plotted. The parameter values used are as follows, damping coefficient $b = 9.9869 \times 10^{-15}$

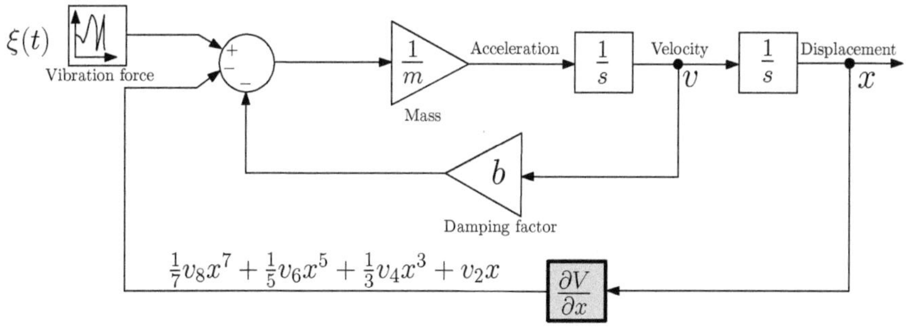

Fig. 6.3 Simulink block diagram of the graphene membrane

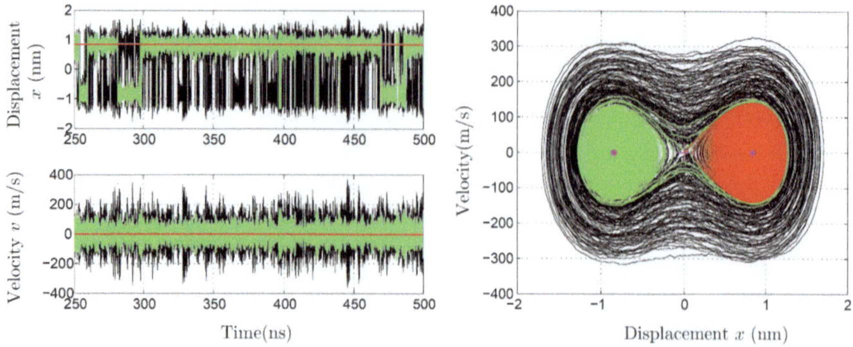

(a) Time series. (b) Phase plane (x, v).

Fig. 6.4 Dynamics of the graphene membrane under a random excitation $\sigma^2 = 1$ pN (*red*), $\sigma^2 = 50.5$ pN (*green*) and $\sigma^2 = 100$ pN (*black*)

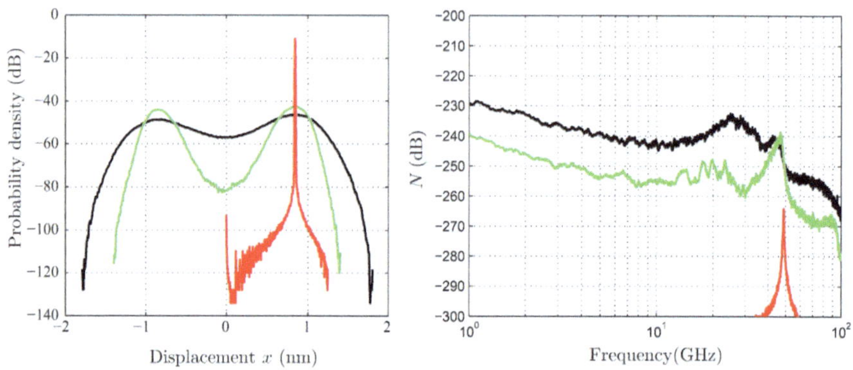

(a) Probability density function of x. (b) The FFT of x and $\sigma^2\eta(t)$.

Fig. 6.5 Spectrum and probability density function corresponding to Fig. 6.4. Note that the spectrum is spread down to low frequency regions when the noise intensity is sufficiently large so that the potential barrier can be crossed

and $m = 5.2982 \times 10^{-24}$. Figures 6.4 and 6.5 show the dynamics of the system for different values of noise strengths. The following particular cases are plotted in these figures

- $\sigma^2 = 1$ pN, the noise intensity is weak and the system evolves to the vicinity of one of the stable equilibrium points X_- or X_+ depending on the initial conditions. The probability to swing between the two equilibria is almost zero. The vicinities of X_- and X_+ correspond to the unique set with nonzero values of the probability density function. The energy of the system is concentrated in a limited range of frequency near the resonant frequency ($\approx$50 GHz).
- $\sigma^2 = 50.5$ pN, the noise intensity is sufficient to make the barrier to be beat and make the system to leave the vicinity of the stable equilibrium points X_- or X_+ swinging between them. The range within which the displacement can be

obtained includes the interval (X_-, X_+). The equilibria X_- and X_+ correspond to the maximum of the probability density function.

- $\sigma^2 = 100$ pN, the noise intensity is in such a way that the probability of swinging between the stable equilibrium points X_- or X_+ is larger and the system presents larger velocities and displacements and therefore the harvested energy will also be larger.

6.6 Performance of the System in Terms of Design Parameters

The RMS value of the displacement and noise amplitude spectra for varying noise level and compression ratio are computed and the amount of power that can be harvested versus noise level and compression ratio are plotted. This power is divided by the corresponding noise amplitude. The results are shown in Fig. 6.6. One can observe in Fig. 6.6a that as the noise level and the compression ratio are increased, the RMS value of the displacement also increases.

Regarding the power that can be harvested, shown in Fig. 6.6b, it can be observed that there is an optimal zone where this ratio is maximal. In particular, it can be noticed that for the linear case ($\varepsilon \approx 0$), the power that can be harvested is low even for relatively high values of the noise level. Nonlinear effects in the device ($\varepsilon \neq 0$), improve the potentials of the device for energy harvesting even at low noise intensity levels. From Fig. 6.6, one can observe that with $\varepsilon \neq 0$, the RMS value of x and therefore the power that can be harvested is always larger than the corresponding linear case ($\varepsilon = 0$)

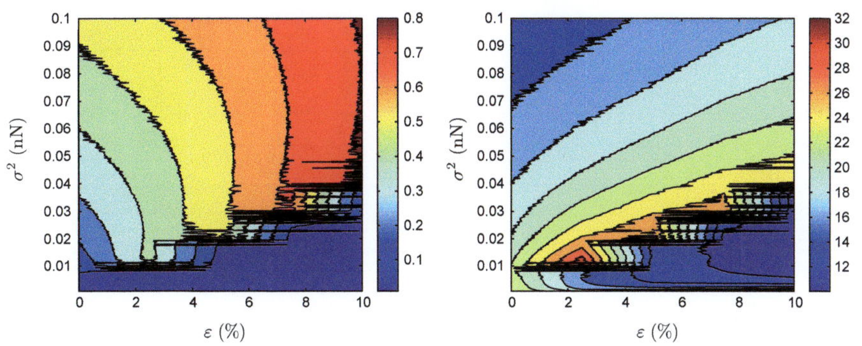

(a) The RMS value of x (in nm) (b) Power/σ^2 (in dB)

Fig. 6.6 Performances of the system in terms of σ^2 and ε

6.7 Coupled Graphene Vibrating Membrane for Energy Harvesting

6.7.1 Nonlinear Mathematical Model

We consider four coupled flat suspended graphene sheets with a compression ratio ε_i, effective mass m_i with an equivalent damping factor b_i and a potential energy $V_i, i = 1 \ldots 4$. The equivalent mechanical scheme is shown in Fig. 6.7. The coupled system can be described by the following set of coupled differential equations

$$m_1\ddot{x}_1 = -\left.\frac{\partial V(x)}{\partial x}\right|_{x=x_1} - \left.\frac{\partial V_1(x)}{\partial x}\right|_{x=x_2-x_1} - b\dot{x}_1 + \sigma^2\eta(t) \tag{6.15}$$

$$m_2\ddot{x}_2 = -\left.\frac{\partial V(x)}{\partial x}\right|_{x=x_2-x_3} - b\dot{x}_2 + \left.\frac{\partial V_2(x)}{\partial x}\right|_{x=x_1-x_2} - b\dot{x}_2 \tag{6.16}$$

$$m_3\ddot{x}_3 = -\left.\frac{\partial V(x)}{\partial x}\right|_{x=x_3-x_4} - b\dot{x}_3 + \left.\frac{\partial V_3(x)}{\partial x}\right|_{x=x_2-x_3} - b\dot{x}_2 \tag{6.17}$$

$$m_4\ddot{x}_4 = -\left.\frac{\partial V(x)}{\partial x}\right|_{x=x_4} - b\dot{x}_4 + \left.\frac{\partial V_4(x)}{\partial x}\right|_{x=x_3-x_4} - b\dot{x}_2 \tag{6.18}$$

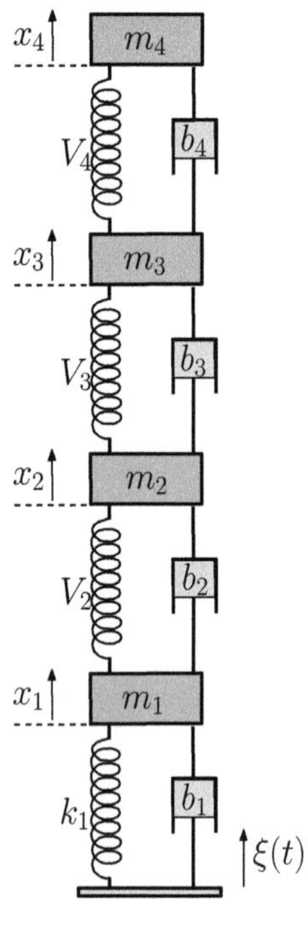

Fig. 6.7 Mechanical equivalent scheme of three coupled graphene membranes

where x_i, $i = 1 \ldots 4$ are the displacements for the coupled graphene membranes, m_i for $i = 1 \ldots 4$ represents the effective mass of each of the layer i, k_i its equivalent stiffness and b_i stands for their damping factors. For simplicity it is assumed that all the membranes are identic, i.e., they have the same mass, stiffness, the same damping coefficient and compression ratio and therefore the same potential energy V which is the same one given (6.2).

6.8 Dynamics of the Coupled Membranes from Numerical Simulations

In this section, only two values of noise intensity are considered, namely $\sigma^2 = 1$ and 50 pN, and different diagrams in the form of times series, state planes, FFT spectra, and probability density function of the displacement are obtained as for the single membrane case.

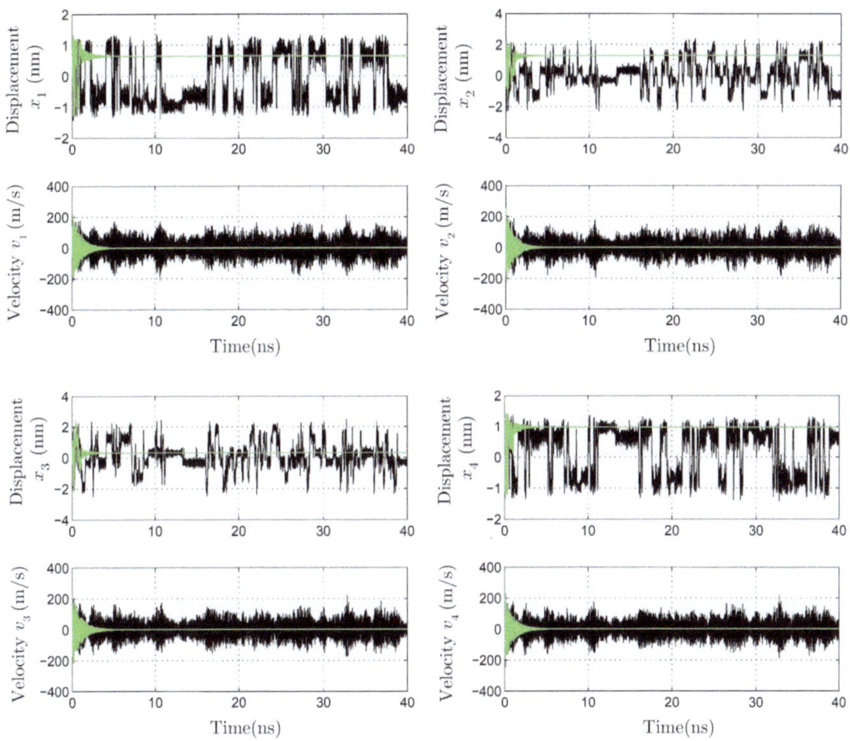

Fig. 6.8 Time domain waveforms of the displacements x_i and velocities v_i, ($i = 1 \ldots 4$) corresponding to the four graphene membranes for two values of noise intensities

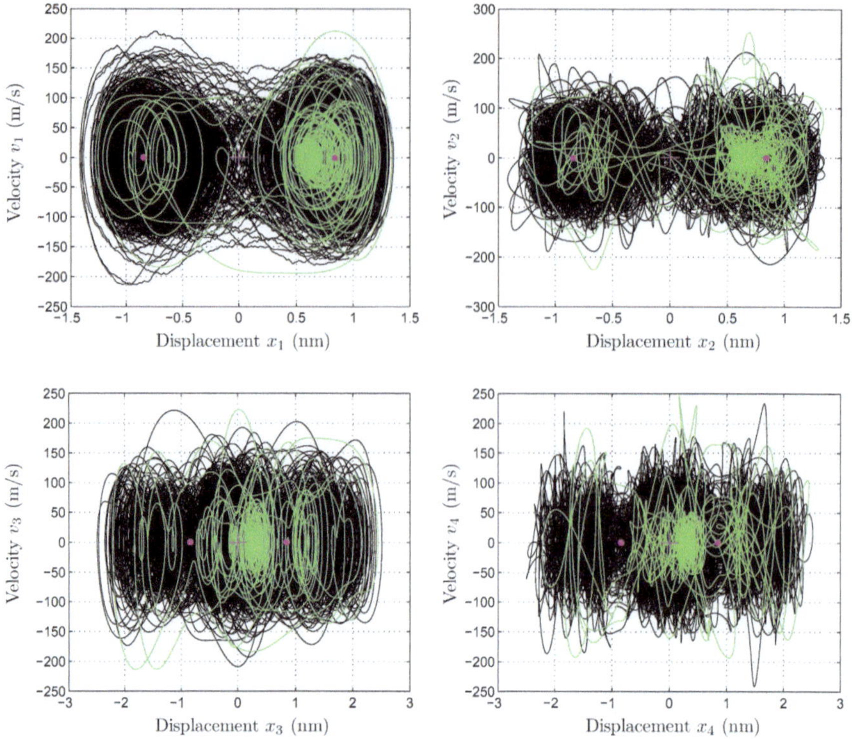

Fig. 6.9 The state space in the planes (x_i, v_i), $(i = 1 \ldots 4)$ corresponding to the four graphene membranes for two values of noise intensities

Figure 6.8 shows the time domain waveforms of the system. For low values of noise intensity the system stays in the vicinity of the equilibrium point. For low values of noise intensity the system stays in the vicinity of the equilibrium point.

Figure 6.9 shows the projection of the state plane time of the system in the planes (x_i, v_i), $(i = 1 \ldots 4)$ corresponding to each graphene membrane. For relatively high values of noise intensity, the trajectory can only evolve to an equilibrium point being the ambient vibration in this case not sufficient to make the system to oscillate between the two regions. When the noise intensity increases, as in the case of the single membrane, the system trajectory visits more the different equilibrium points and larger oscillation amplitude can be reached. All the four membranes are characterized by the same dynamics.

Figure 6.10 depicts the FFT spectra of the displacement corresponding to each graphene membrane. Since we have a system with four degrees of freedom, four different vibration modes of with different frequencies should appear. While this is the case for the low intensity noise case, the spectrum corresponding to a relatively high value of noise intensity is spread a broad range of frequencies. This occurs for all the membranes.

Figure 6.11 shows the probability density functions calculated for the different displacements of the membranes. For low values of noise intensity, the probability of finding the system in the vicinity of one of its equilibrium points is practically one. For relatively high values of noise intensity, the probability to visit a wider region increases being the maximum probability that corresponding to the vicinity of the equilibrium points.

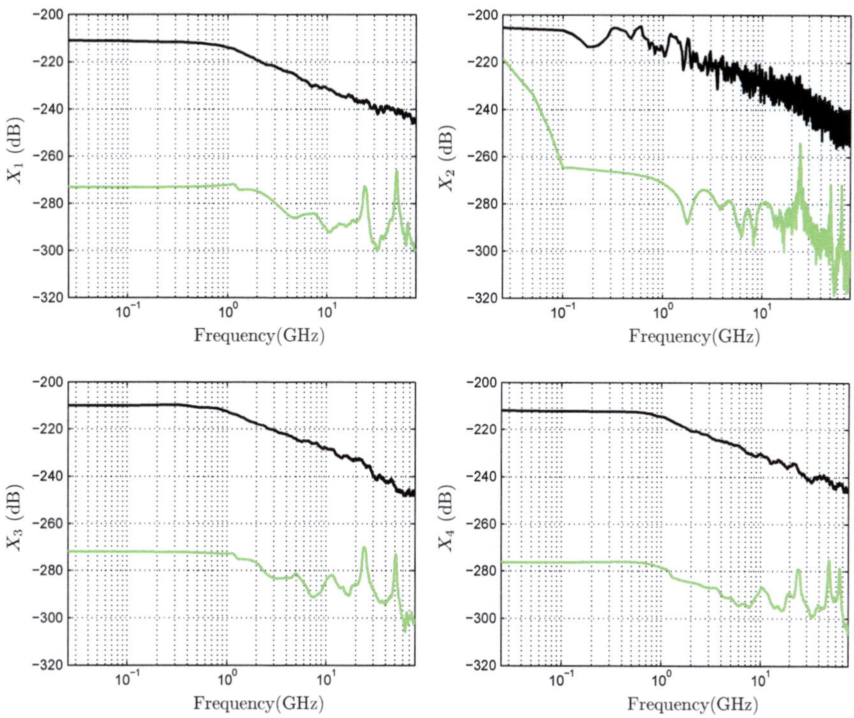

Fig. 6.10 FFT spectra of the displacements x_i $(i = 1 \ldots 4)$ corresponding to the four graphene membranes for two values of noise intensities

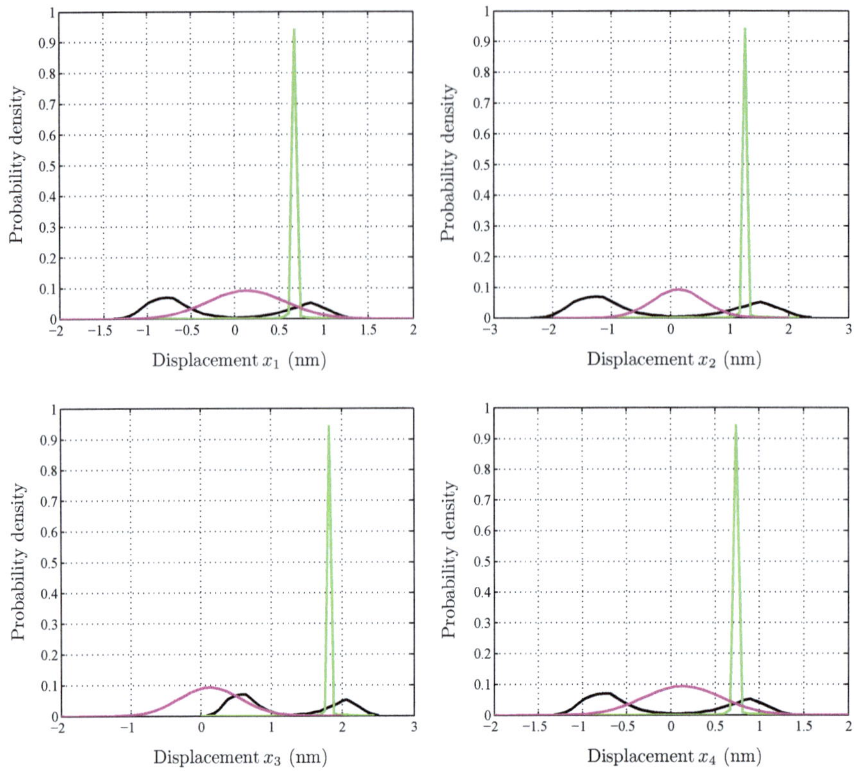

Fig. 6.11 Probability density function of the displacements x_i $(i = 1 \ldots 4)$ for two values of noise intensities. For reference, the probability density of the noise driving signal is also plotted

6.9 Conclusions

In this chapter, we have considered the nonlinear dynamics of four identical coupled nanostructured graphene vibrating membranes for ambient energy applications at the nanoscale. The compressed graphene presents a nonlinearity that has been shown to play an important role in increasing the efficiency of this energy harvesting device by increasing the RMS values of the displacement and the velocity. We presented a continuous-time nonlinear dynamical model of the coupled system. When random vibrations are considered as the main ambient energy source for the system, the performances of the system as an energy harvester are presented in the steady- state nonequilibrium regime when the noise level for a certain value of the compression ratio. Although the study can be carried out for the system with the above-mentioned eighth-order potential, a simpler double-well fourth-order potential can be considered in order to inherit many available analytical results on the behavior of the Duffing oscillator. Finally it should be noted that this structure of mechanically coupled graphene membranes is feasible and can physically fabricated. A possible way to get it is by suspending the membranes on a shared elastic anchors in a series configuration.

Acknowledgments This work was supported by the Spanish *ministerio de Economía y Competitividad* under grants DPI2013-47437-R and by the RUE CSD2009-00046 (Consolider-Ingenio 2010), FIS2009-12721-C04-03, FIS2012-37549-C05-05, ENE2009-14340-C02-02 and FP7-ICT-P.No.: 270005-ZEROPOWER.

References

1. Kaźmierski, T. J., & Beeby, S. (Eds.) (2011). *Energy harvesting systems. Principles, modeling and applications*. Springer.
2. Neri, I., Travasso, F., Vocca, H., & Gammaitoni, L. (2011). Nonlinear noise harvesters for nanosensors. *Nano Communication Networks*, *2*(4), 230–234.
3. Gammaitoni, L., Vocca, H., Neri, I., Travasso, F., & Orfei, F. (2007). Vibration energy harvestng: linear and nonlinear oscillator approaches. In *Sustainable energy harvesting technologies-past, present and future* (pp. 171–190).
4. Tang, L., Yang, Y., & Soh, K. (2010). Toward broadband vibration-based energy harvesting. *Journal of Intelligent Material Systems and Structures*, *21*, 1867–1896. Dec.
5. Ralman, R., Brennan, M. J., Mace, B. R., & Kovacic, I. (2010). Potential benefits of a non-linear stiffness in an energy harvesting device. *Nonlinear Dynamics*, *59*, 545–558.
6. Trigona, C., Dumas, N., Latorre, L., Andò, B., Bagliom, S., & Nouet, P. (2011). Exploiting benfits of a periodically-forced nonlinear oscillator for energy harvesting from ambient vibration. *Procedia Engineering*, 2–4.
7. Liu, H., Tay, C. J., Quan, C., Kobayashi, T., & Lee, C. (2011). Piezoelectric MEMS energy harvester for low-frequency vibrations with wideband operation range and steadily increased output power. *Journal of Microelectromechanical Systems*, *20*(5), 1131–1142.
8. Xu, S., Qin, Y., Xu, C., Wei, Y., Yang, R., & Wang, Z. L. (2010). Self-powered nanowire devices. *Nature Nanotechnology*, *5*(5).
9. Wang, Z. L., Wang, X., Song, J., Liu, J., & Gao, Y. (2008). Piezoelectric nanogenerators for self-powered nanodevices. *Pervasive Computing*, *7*(1), 48–55.
10. Quin, Y., Wang, X., & Wang, Z. L. (2008). Microfibre-nanowire hybrid structure for energy scavenging. *Nature*, *451*(14), 809–813.
11. Wang, Z. L., Wang, X., Song, J., Liu, J., & Gao, Y. (2008). Piezoelectric nanogenerators for self-powered nanodevices. *Pervasive Computing, IEEE*, *7*(1), 49–55.
12. Imam, S. A., Sabri, S., & Szkopek, T. (2010). Low-frequency noise and hysteresis in graphene field-effect transistors on oxide S.A. *Micro and Nano Letters, IET*, *5*(1), 37–41.
13. Geim, A. K., & Novoselov, K. S. (2011). The rise of graphene. *Nature*, *6*(3), 183–191.
14. Brownson, D. A. C., Kampouris, D. K., & Banks, C. E. (2011). An overview of graphene in energy production and storage applications. *Journal of Power Sources*, *196*, 4873–4885.
15. Grande, L., Chundi, V. T., Wei, D., Bower, C., Andrew, P., & Ryhänen, T. (2011). Graphene for energy harvesting/storage devices and printed electronics. *Particuology*, *10*, 1–8.
16. Pumera, M. (2011). Graphene in biosensing. *Materials Today*, *14*(7–8), 308–315.
17. Singh, V., Joung, D., Zhai, L., Das, S., Khondaker, S. I., & Seal, S. (2011). Graphene based materials: Past, present and future. *Progress in Materials Science*, *56*, 1178–1271.
18. López-Suárez, M., Rurali, R., Gammaitoni, L., & Abadal, G. (2011). Nanostructured graphene for energy harvesting. *Physical Review B*, *84*(161401), 1–5.
19. Kwon, J., Sharma, B. K., & Ahn, J. -H. (2013). Graphene based nanogenerator for energy harvesting. *Japanese Journal of Applied Physics*, *52*(6S) 06GA02.

20. Dhiman, P., Yavari, F., Mi, X., Gullapalli, H., Shi, Y., Ajayan, P. M., et al. (2011). Harvesting energy from water flow over graphene. *Nano Letters*, *11*(8), 3123–3127.
21. Que, R., Shao, Q., Li, Q., Shao, M., Cai, S., Wang, S., et al. (2012). Flexible nanogenerators based on graphene oxide films for acoustic energy harvesting. *Angewandte Chemie International Edition*, *51*(22), 5418–5422.
22. El Aroudi, A., Lopez-Suarez, M., Rurali, R., Alarcon, E., & Abadal, G. (2013). Nonlinear dynamics in a nanostructured graphene device for energy harvesting applications. In *IEEE international symposium on circuits and systems, 2013, ISCAS 2013, May 2013, Beijing, China* (pp. 2727–2730).
23. El Aroudi, A., Lopez-Suarez, M., Rurali, R., Alarcon, E., & Abadal, G. (2014). Nonlinear dynamics of an ambient noise driven array of coupled graphene nanostructured devices for energy harvesting. In *A: International conference on structural dynamics and diagnosis. MATEC web of conferences* (Vol. 16); *CSNDD 2014: International conference on structural nonlinear dynamics and diagnosis. Agadir: EDP Sciences* (pp. 01001-1–01001-4).
24. Soler, J. M., Artacho, E., Gale, J. D., Garcıa, A., Junquera, J., Ordejon, P., & Sánchez-Portal, D. (2002). The SIESTA method for ab initio order-N materials simulation. *Journal of Physics: Condensed Matter*, *14*(11), 2745.
25. Gammaitoni, L., Neri, I., & Vocca, H. (2009). Nonlinear oscillators for vibration energy harvesting. *Applied Physics Letters*, *94*(16), 164102-1–164102-3.

Chapter 7
End-Stop Nonlinearities in Vibration Energy Harvesters

Sukhdeep Kaur and Cuong Phu Le

7.1 Introduction

Vibration energy harvesters are gaining increase in popularity as a possible alternative to battery-based power sources. Vibration energy harvesters are typically resonant devices that produce appreciable output power when operating at resonance, but they often fail to deliver reliable power when they are made to vibrate at off-resonance frequencies. The energy sources of these devices are ambient vibrations, which are easily available and offer a wide variety of spectra, from narrowband to wideband. A wide range of vibration energy harvesters designed for various operating conditions has been proposed and demonstrated to date [1–8]. In particular, it is desirable to develop microscale energy harvesters that can perform efficiently when using excitation signals that are not only narrowbanded but are of variable frequency, either over time or from case to case in applications, or when using excitation signals that are relatively broadbanded [9, 10].

In that respect, nonlinear devices have emerged as a potential solution that can adapt to the various vibration spectra to increase both the bandwidth response and the output power. The nonlinearities may appear as part of the harvester design, such as the use of nonlinear springs that add nonlinearities to the device response at sufficiently large acceleration amplitudes and thus broaden the harvester's bandwidth [10, 11]. Another solution to widen the response of the harvester uses the impact of the proof mass with end-stops that add additional nonlinearities to the harvester responses [9, 12, 13]. Even if the proof mass vibrates linearly between the end-stops, the behavior changes abruptly when the mass hits the end-stops. The frequency

S. Kaur (✉) · C.P. Le
Faculty of Technology and Maritime Sciences, Department of Micro and Nano Systems Technology, University College of Southeast Norway, PO Box 2243, 3103 Tønsberg, Norway
e-mail: Sukhdeep.Kaur@hbv.no

C.P. Le
e-mail: Cuong.Le@hbv.no

E. Blokhina et al. (eds.), *Nonlinearity in Energy Harvesting Systems*,
DOI 10.1007/978-3-319-20355-3_7

response obtained in this manner depends on the nonlinear springs in the suspension or on the contact linearity between the proof mass and the end-stops. It should be possible to produce a wider response than that obtained from a linear resonant device through appropriate design of the device nonlinearities.

The motion of the proof mass is generally nonlinear on impact with the end-stop, depending on the system and end-stop parameters [14]. The end-stops in the vibration energy harvester not only add nonlinearities and enhance the bandwidth but also saturate the harvester's output power. Thus, the end-stops set an upper bound on the output power. This negative end-stop trait can be overcome by the introduction of sufficiently compliant end-stops to the device and collection of the lost energy at the end-stops on impact via integrated electronic circuits such as synchronous electric charge extraction (SECE) or synchronized switch harvesting on inductor (SSHI) circuits [15]. The energy that is otherwise lost in the end-stops can be used efficiently by introducing a transduction mechanism at the end-stops [13].

In this chapter, an overview of the use of end-stops in energy harvesters will be presented. The mathematical modeling of end-stop behavior in a harvester will be discussed in detail on the basis of phase space trajectories and eigenvalue problem. The effects of nonlinearities in the motion of the proof mass on variation of the coefficient of restitution and the device parameters will be demonstrated. The average output power from the harvester with end-stop nonlinearities for different coefficients of restitution will be shown in power graphs. Examples of devices with end-stop nonlinearities will be presented and discussed with respect to their experimental results, and conclusions will then be drawn.

General Aspects of Vibration Energy Harvesters

The schematic diagram in Fig. 7.1 shows a typical model of a vibration energy harvester. While this model, which was built up by William and Yates [16, 17] is mainly applicable to electromagnetic transducers, it can be applied to both piezoelectric and electrostatic transducers in appropriate ways. The model includes a proof mass m suspended from a mechanical spring with a stiffness k. The mechanical loss is represented by c_m, while the electrical loss c_e comes from the coupling between the mechanical and electrical domains in the system for energy conversion. The power is obtained from the motion of the proof mass relative to the device package when the harvester is subjected to vibration. Because of the finite size of the device package, the proof mass displacement is always constrained by a maximum amplitude $|x(t)| \leq X_{max}$. Under sinusoidal excitation with an angular frequency ω in the steady state [16, 17], the proof mass displacement amplitude and the power obtained are given by

$$|X| = \frac{A}{\sqrt{\left(\omega_0^2 - \omega^2\right)^2 + \left(c_m + c_e\right)^2 \omega^2/m^2}} \tag{7.1}$$

$$P = \frac{1}{2}\frac{C_e\omega^2A^2}{\left(\omega_0^2 - \omega^2\right)^2 + \left(c_m + c_e\right)^2 \omega^2/m^2} \tag{7.2}$$

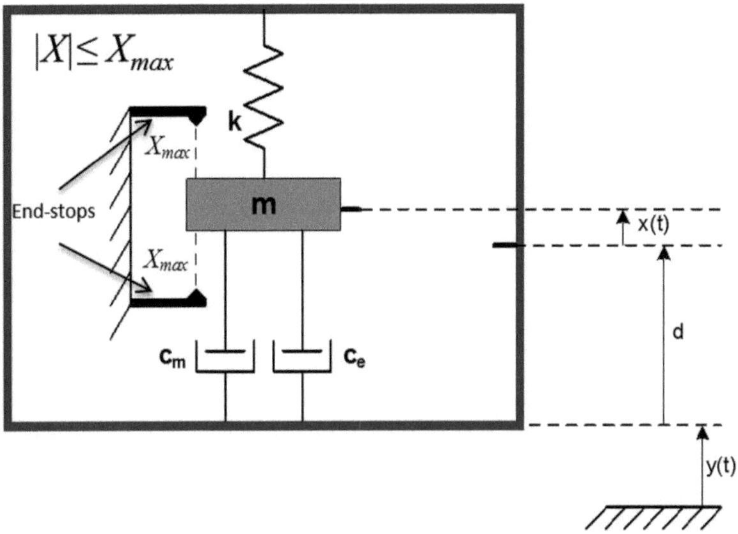

Fig. 7.1 Typical diagram of a vibration energy harvester model

where A is the acceleration amplitude and $\omega_0 = 2\pi f_0 = \sqrt{k/m}$. The optimum power is achieved when $c_m = c_e$. Additionally, the generated power is limited by $P_{lim} = (2/\pi)\, m\omega X_{max} A$ when $|X| = X_{max}$ [18, 19]. In addition, at resonance, for $|X| = X_{max}$, the power obtained is $P_{lim} = (1/2)\, m\omega X_{max} A$, which is approximately 78.1 % of P_{lim}.

Depending on their transduction mechanisms, vibration energy harvesters can be classified into electrostatic energy harvesters [1–3], piezoelectric energy harvesters [4–6] and electromagnetic energy harvesters [7, 8]. In an electrostatic energy harvester, the displacement of the movable proof mass electrode varies the capacitance between the movable proof mass electrode and the fixed electrode when an appropriate bias is placed on either the movable proof mass electrode or the fixed electrode. Piezoelectric conversion is a function of the material properties; strain in the material produces an electric field in that material, and vice versa. The principle of operation of an electromagnetic energy harvester is based on Faraday's law of induction. A variable electric field is generated in an electromagnetic energy harvester, creating an output voltage V, when the magnetic flux Φ passes through an electrical circuit. The harvested power can either be used directly for a specific application or can be managed efficiently using power management circuits such as SSHI or buck–boost converter circuits [15, 20].

The spectra of the vibration signals vary from narrowband to wideband [21–25]. The role of the energy harvester is to respond effectively to the wideband spectra but not to the single peaks. However, for applications where the response from a single frequency peak is the case of interest, the harvester should work efficiently for that frequency peak.

The harvesters are excited by ambient vibrations, which introduce the relative motion of the proof mass with respect to the device frame. The vibration amplitudes can be unpredictable in many cases. The harvester designs must be able to adapt to the vibration conditions to maximize the output power, while the harvester performance is required to be both robust and effective. In the following analysis, a sinusoidal vibration with angular frequency $\omega = 2\pi f$ and amplitude A is used to characterize the nonlinear behavior of the end-stops of vibration energy harvesters.

7.2 Modeling of the End-Stops

In conventional vibration energy harvesters, the end-stops are unavoidable in the real prototypes. The displacement is limited by the space constraints of the device package. The main function of the end-stops is to confine the proof mass motion under sufficiently large vibration conditions. In addition, they protect the beam structures and prevent degradation of the material properties. The end-stops set an upper bound on the output power at high acceleration amplitudes, i.e. they limit the harvested power, which would otherwise be infinitely high without the end-stops.

The end-stops introduce strong nonlinearities into the harvester response when the proof mass hits the end-stops under extreme accelerations. The positive and negative characteristics of the end-stops have been discussed in a number of papers. For example, the nonlinear behavior of proof mass motion in energy harvesters that include end-stops has been studied via both simulations and experiments [12, 15]. Mathematical modeling of the end-stops in an energy harvester will be discussed in detail in this section, followed by examples of working devices harvesting power from the active end-stops. Theoretical and experimental studies have been carried out to characterize the end-stops in the device design. In [4], a mechanical end-stop is applied only on one side of the cantilever beam of a piezoelectric energy harvester. The frequency response of this harvester is demonstrated by varying the end-stop parameters. In [12], the end-stops applied in an electromagnetic power generator increase the bandwidth of the frequency up-sweep while maintaining the same bandwidth for the down-sweep.

7.2.1 Mathematical Analysis

It is useful to have some tools that can provide an insight into the performance of an energy harvester with and without the end-stop nonlinearities using given parameters for the desired prototypes. A mathematical tool for nonlinearities related to the end-stops in vibration energy harvesters using an eigenvalue approach and time domain simulations is presented in this section. For this analysis, two end-stops that limit the motion of the proof mass are considered, where the motion is treated as a linear evolution between the impacts with velocity discontinuities at the end-stops.

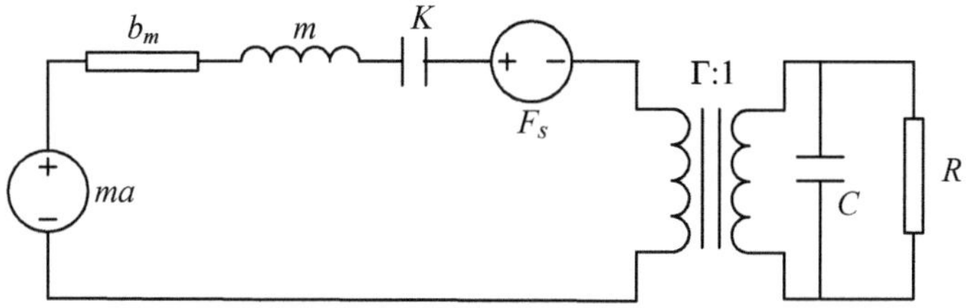

Fig. 7.2 Two-port linear transducer with end-stop effects

An equivalent circuit for an energy harvester model with end-stop effects, irrespective of the transduction mechanism, is shown in Fig. 7.2, where m is the proof mass, b_m is the mechanical damping, k is the electromechanical coupling factor, K is the effective spring stiffness, C is the transducer capacitance, $\Gamma = \sqrt{KC}$, and R is the load resistance. The excitation force is represented by $F_{ext} = ma$. The transducer model is similar to that of a velocity-damped resonant generator (VDRG). The end-stop effects are included as the impact force F_s on the mechanical domain circuit. The impact force F_s is activated when the proof mass displacement amplitude reaches its maximum.

At equilibrium, the total force on the proof mass and the voltage across the electrical load in the transducer are assumed to be zero. We observe deviations of the state variables from their equilibrium values when the harvester is excited, and these deviations depend on the strength of the excitation signal. The linear equations of proof mass motion under sinusoidal excitation can be formulated as Eqs. (7.3)–(7.7).

$$\frac{dx}{dt} = v \tag{7.3}$$

$$\frac{dv}{dt} = -\omega_0^2 x - \frac{\omega_0^2 k}{\Gamma} q - \frac{\omega_0}{Q} v + a \tag{7.4}$$

$$\frac{dq}{dt} = -\frac{\omega_0}{r} \Gamma k x - \frac{\omega_0}{r} q \tag{7.5}$$

$$\frac{da}{dt} = -\omega b \tag{7.6}$$

$$\frac{db}{dt} = \omega a \tag{7.7}$$

where x is the position of the proof mass, v is the velocity, q is the transducer charge, ω is the angular driving frequency, t is the time, Q is the open circuit quality factor of the device, k is the electromechanical coupling factor, ω_0 is the open circuit angular resonance frequency, $r = \omega_0 CR$, and acceleration $a = A\cos\omega t$, where A is the acceleration amplitude $A = \sqrt{a^2 + b^2}$. To make the system autonomous, an auxiliary quantity $b = A\sin\omega t$ is introduced into the state equations.

Equations (7.3)–(7.7) are translated into dimensionless forms with reference to the dimensionless time (phase angle) $\theta = \omega_0 t$, frequency $\varsigma = \omega/\omega_0$, and amplitude $\hat{A} = A/X_{max}\omega_0^2$. The dimensionless state variables then become

$$\hat{x}(\theta) = x(t)/X_{max} \tag{7.8}$$

$$\hat{v}(\theta) = v(t)/X_{max}\omega_0 \tag{7.9}$$

$$\hat{q}(\theta) = q(t)/X_{max}\Gamma \tag{7.10}$$

$$\hat{a}(\theta) = a(t)/X_{max}\omega_0^2 \tag{7.11}$$

$$\hat{b}(\theta) = b(t)/X_{max}\omega_0^2 \tag{7.12}$$

where $\pm X_{max}$ denotes the end-stop positions, i.e., $|x(t)| \leq X_{max}$. Typically, the motion of the proof mass before impact is linear, and thus the state vector is $\hat{u} = \left[\hat{x}\ \hat{v}\ \hat{q}\ \hat{a}\ \hat{b}\right]^T$ and Eqs. (7.3)–(7.7) then read:

$$\frac{d\hat{u}}{dt} = -\hat{L}\hat{u} \tag{7.13}$$

where

$$\hat{L} = \begin{bmatrix} 0 & -1 & 0 & 0 & 0 \\ 1 & \frac{1}{Q} & k & -1 & 0 \\ \frac{k}{r} & 0 & \frac{1}{r} & 0 & 0 \\ 0 & 0 & 0 & 0 & \varsigma \\ 0 & 0 & 0 & -\varsigma & 0 \end{bmatrix} \tag{7.14}$$

and the linear evolution of the system from, for example $\theta = \theta_1$ to $\theta = \theta_2$, is given by

$$\hat{u}(\theta_2) = \hat{U}(\theta_2 - \theta_1)\,\hat{u}(\theta_1) \tag{7.15}$$

where

$$\hat{U}(\theta) = \exp\left(-\theta\hat{L}\right) \tag{7.16}$$

If an impact occurs at the dimensionless time θ_1, then the change in velocity at the time of impact is modeled as $\hat{v}\left(\theta_1^+\right) = -e\hat{v}\left(\theta_1^-\right)$, where e is the coefficient of restitution and the superscript $\pm$ denotes a infinitesimally small time after/before θ_1. Thus, the change in the state vector at the time of impact is given by

$$\hat{u}\left(\theta_1^+\right) = S\hat{u}\left(\theta_1^-\right) \tag{7.17}$$

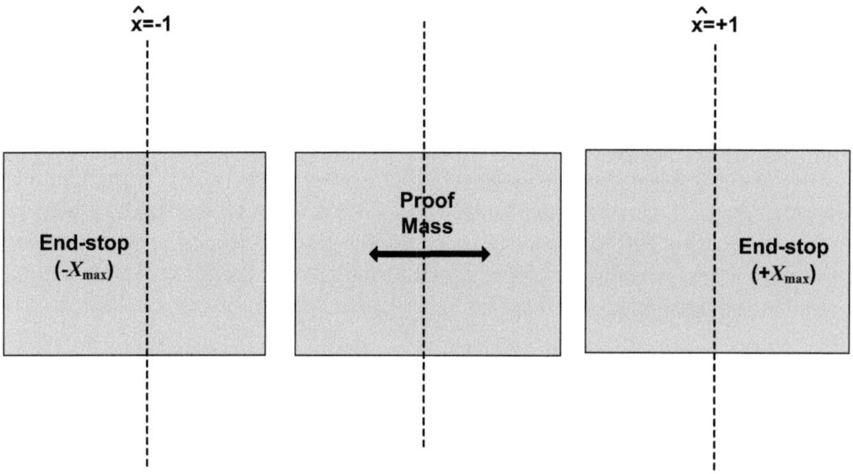

Fig. 7.3 Modeling of the impact at the end-stops

where

$$S = \begin{bmatrix} 1 & 0 & 0 & 0 & 0 \\ 0 & -e & 0 & 0 & 0 \\ 0 & 0 & 0 & 0 & 0 \\ 0 & 0 & 0 & 0 & 0 \\ 0 & 0 & 0 & 0 & 0 \end{bmatrix} \quad (7.18)$$

Figure 7.3 shows the end-stops model, where the end-stops are placed at the positions of $\pm X_{max}$. Thus, the impacts occur at the normalized positions $\hat{x} = -1$ and $\hat{x} = +1$. If the sequence of impacts per cycle of motion is known, then all possible solutions can be found from the solution to the eigenvalue problem.

Let $\theta = 0$ be the time of impact at $\hat{x} = -1$ and assume that, at some intermediate time θ_1, another impact occurs at $\hat{x} = +1$, and then the next impact occurs at $\hat{x} = -1$ at time $\theta_2 = 2\pi/\varsigma$, thus completing one whole cycle of the proof mass motion.

If the state vector is initially $\hat{u}\left(0^+\right) = u_0$, then the sequence of linear evolutions and impacts that occur up to the point in time just after the second impact at $\hat{x} = -1$ at time $\theta = \theta_2^+$ is given by

$$\hat{u}\left(\theta_1^-\right) = \hat{U}\left(\theta_1\right) u_0 \quad (7.19)$$

$$\hat{u}\left(\theta_1^+\right) = S\hat{u}(\theta_1^-) = S\hat{U}\left(\theta_1\right) u_0 \quad (7.20)$$

$$\hat{u}\left(\theta_2^-\right) = \hat{U}\left(\theta_2 - \theta_1\right) \hat{u}(\theta_1^+) = \hat{u}\left(\theta_2^-\right) = \hat{U}\left(\theta_2 - \theta_1\right) S\hat{U}\left(\theta_1\right) u_0 \quad (7.21)$$

$$\hat{u}\left(\theta_2^+\right) = S\hat{u}\left(\theta_2\right) = S\hat{U}\left(\theta_2 - \theta_1\right) S\hat{U}(\theta_1^+) u_0 \quad (7.22)$$

If the period of motion is equal to the period of the vibration, then $\hat{u}\left(\theta_2^+\right) = u_0$ in (7.22). Therefore, an admissible u_0 must be an eigenvector of the matrix

$S\hat{U}(\theta_2 - \theta_1)\, S\hat{U}(\theta_1)$ with an eigenvalue of 1. Thus, solution of the eigenvalue problem provides possible solutions for the motion of the proof mass for one whole period of the driving force. The final result must be checked against unphysical solutions where the proof mass motion extends beyond the limits of the end-stops. The state vectors give the package acceleration amplitude.

Similar analyses can be conducted for other types of motion, e.g., with more than one impact per hit. However, the analysis quickly becomes complex when there are several impacts per period of motion, because that leads to several unknown impact times and a new eigenvalue problem must be formulated for each case. This technique is similar to the technique used in [14].

7.2.2 *Analysis of the Numerical Results*

System parameters are required for solution of the eigenvalue problem. In this chapter, the eigenvalue problem is demonstrated using system parameters that are identical to those given in [15], with $\varsigma = 1$, $r = 1$, $Q = 350$, and $k^2 = 0.6\,\%$. The effects of changing the system parameters are also compared using the system parameters given in [13], with $\varsigma = 1$, $r = 1$, $Q = 203.5$, and $k^2 = 2.52\,\%$. The simplest case that can be used to check the nonlinear behavior at the end-stops is to assume that the coefficient of restitution $e = 1$, i.e., the end-stops are rigid and elastic collisions occur; no energy is gained or lost at the end-stops. If the end-stops are compliant by nature, then the coefficient of restitution is less than 1.

Figures 7.4, 7.5 and 7.6 show the physical and unphysical sets of solutions for one period of motion from the eigenvalue problem using Eqs. (7.19)–(7.22). Three different coefficients of restitution, i.e., $e = 1$, $e = 0$ and $e = 0.3$, are used for illustration purposes here. Figure 7.4 shows that there is a single solution in which the first

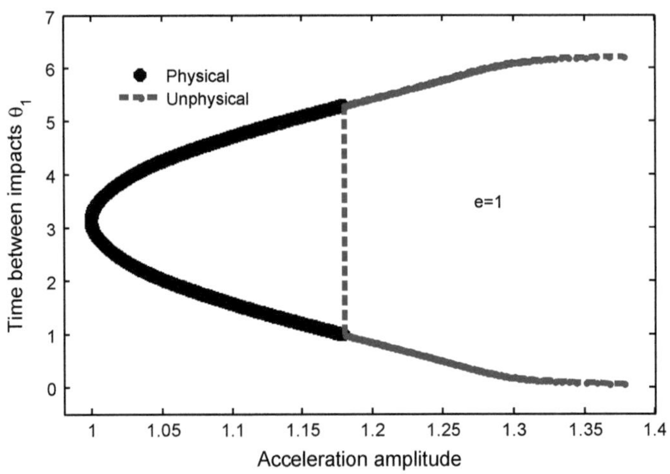

Fig. 7.4 Acceleration amplitude versus time between impacts for $e = 1$

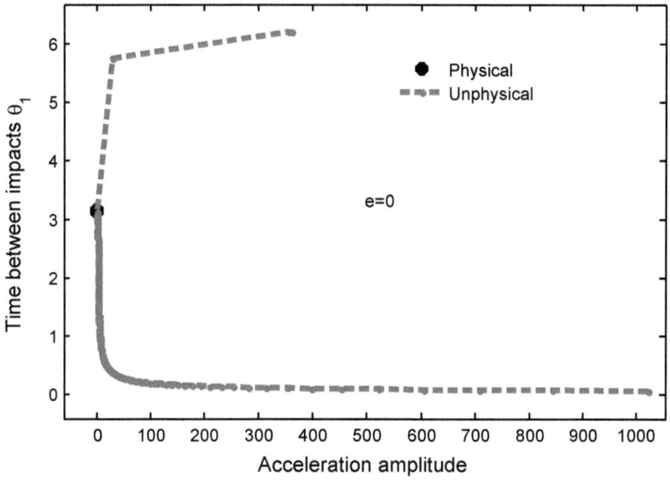

Fig. 7.5 Acceleration amplitude versus time between impacts for $e = 0$

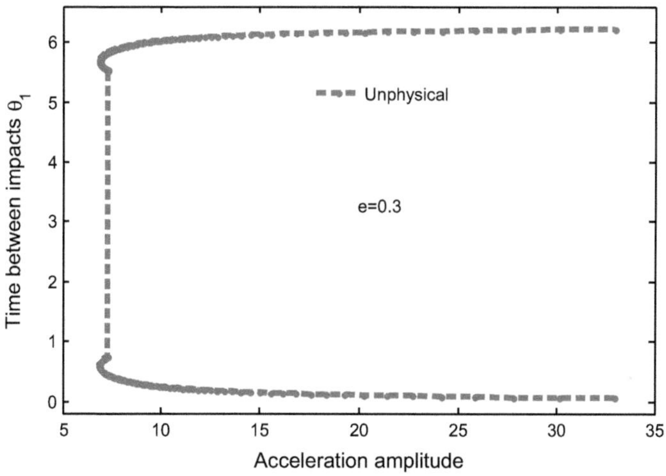

Fig. 7.6 Acceleration amplitude versus time between impacts for $e = 0.3$

impact occurs at the midpoint of the cycle for a normalized amplitude of 1. As the amplitude increases, this solution splits into two solutions with opposite asymmetry in their times between impacts. The motion of the proof mass changes dramatically for compliant end-stops. Figure 7.5 shows that the proof mass motion is unphysical for the wider set of acceleration amplitudes when compared with the set of physical solutions. The motion of the proof mass becomes more complicated for $e = 0.3$, as shown in Fig. 7.6. Not even single acceleration amplitude is detected for the physical motion from the eigenvalue problem. The eigenvalue approach used here cannot detect the motion of the proof mass other than as described in Eqs. (7.19)–(7.22).

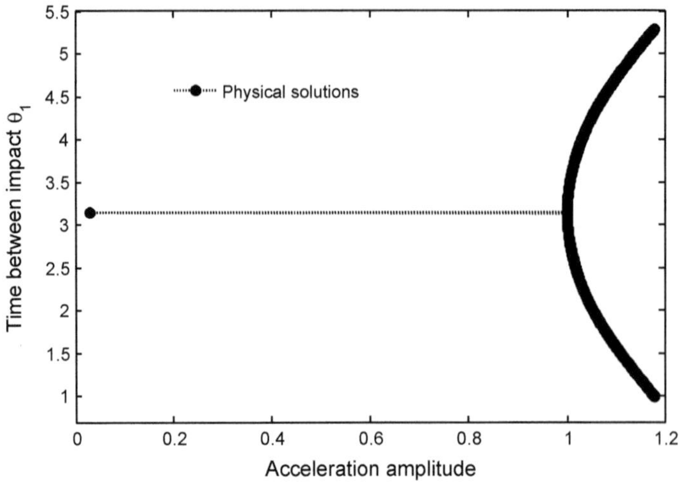

Fig. 7.7 A bare impact at the end-stops when $\hat{A} = 0.00658$ and $e = 1$

Figure 7.7 shows the acceleration amplitude where proof mass barely hits the end-stops.

The acceleration amplitude under the condition where the proof mass will barely hit the end-stops is given by

$$\hat{A} = \frac{\left(r + k^2 Q\right)^2 + 1}{\sqrt{\left(Q^2 + r^2 Q^2\right)\left(r + k^2 Q^2\right) + Q^2 + r^2 Q^2}} \tag{7.23}$$

To study the motion of the proof mass in detail, the acceleration amplitudes from each set of physical and unphysical solutions are selected for the given coefficient of restitution. Figures 7.8 and 7.9 show the motion of the proof mass for acceleration amplitudes $\hat{A} = 1.0023$ and $\hat{A} = 1.2271$ from the set of physical solutions and unphysical solutions for $e = 1$, respectively.

The motion of the proof mass shown in Fig. 7.9 is complex. The proof mass tends to go beyond the displacement limit that has been set by the end-stops, while in Fig. 7.8, the period of the proof mass motion is equal to the period of the driving force with one impact at each end-stop per cycle of the driving force. The jump in velocity at the point of impact can be seen clearly in these figures.

Figure 7.10 shows the evolution of the proof mass motion for the acceleration amplitudes from the set of physical solutions for $e = 1$. Figure 7.10 shows the different patterns of the proof mass motion for specific acceleration amplitudes. For $\hat{A} = 1.0945$, the period of the proof mass motion is longer than the period of the driving force. The eigenvalue problem requires further improvement to detect the different motion patterns as separate entities for the given set of parameters. Figure 7.11 shows the complexity of the proof mass motion pattern for $e = 0$. The acceleration

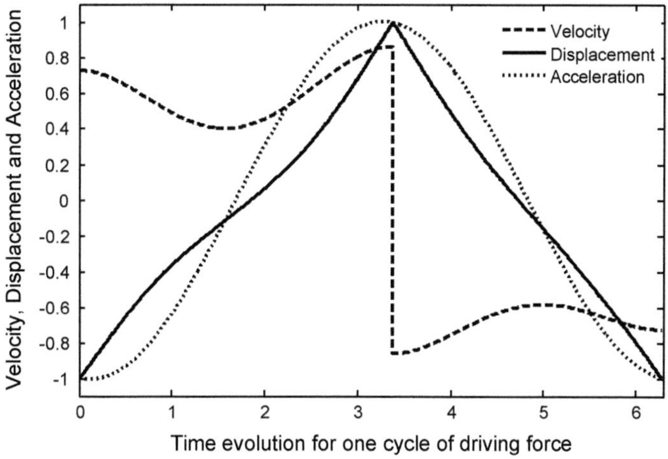

Fig. 7.8 Proof mass motion for $e = 1$ and $\hat{A} = 1.0023$

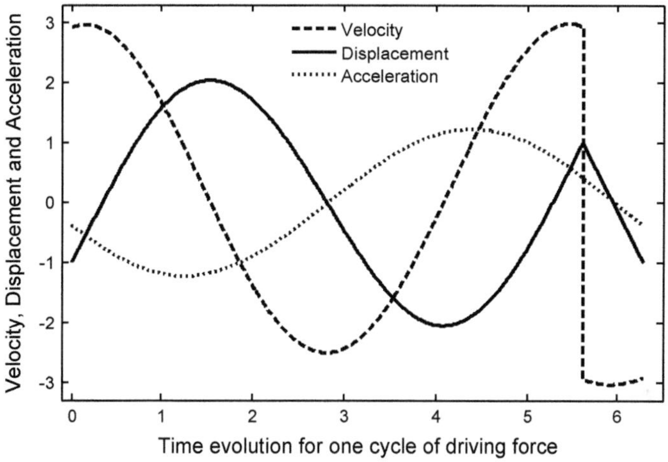

Fig. 7.9 Proof mass motion for $e = 1$ and $\hat{A} = 1.2271$

amplitude $\hat{A} = 1.0207\mathrm{e}+003$ in Fig. 7.11 corresponds to the set of unphysical solutions that was detected by the solution to the eigenvalue problem.

The coefficient of restitution $e = 0$ is closer to the conditions of real-world collisions. The proof mass will lose all energy at the end-stops on each impact. As shown in Fig. 7.11, the proof mass tends to go beyond the displacement limit when the acceleration amplitudes are sufficiently high. The restoring force from the end-stop then comes into play and limits the displacement of the proof mass to prevent it from passing beyond the end-stop. The proof mass will leave the end-stop when the restoring force from the end-stop reverses its direction. The restoring force phenomenon is not modeled in the eigenvalue analysis.

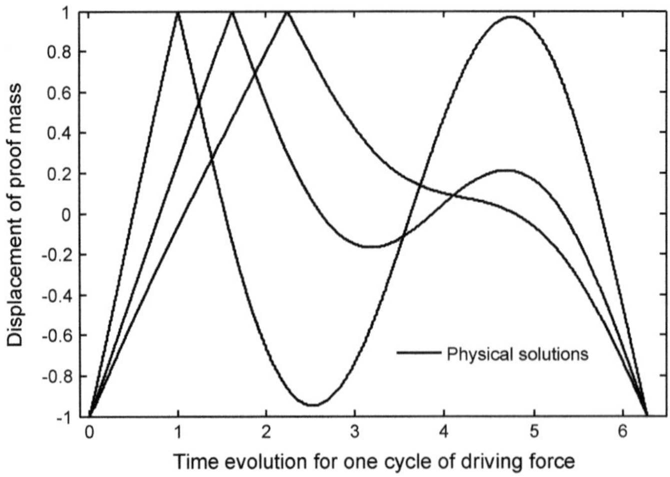

Fig. 7.10 Proof mass motion for $e = 1$ and acceleration amplitudes $\hat{A} = 1.1778, 1.0329, 1.0945$, from *left* to *right*

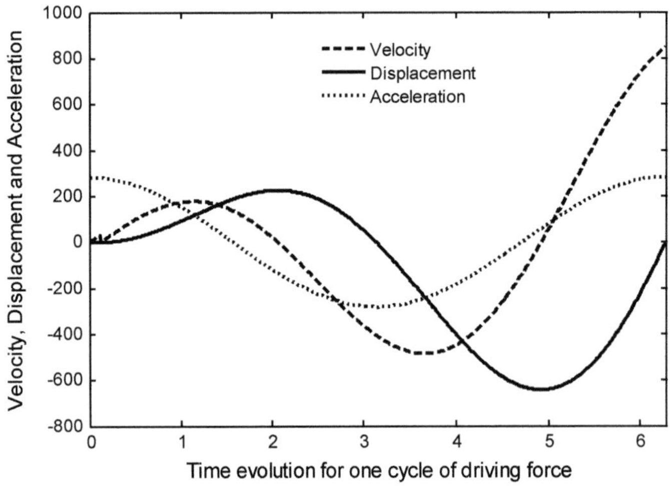

Fig. 7.11 Proof mass motion for $e = 0$ and $\hat{A} = 1.0207\mathrm{e} + 003$

The eigenvalue problem is a simple analysis that does not account for the initial transitions that occur in the system, which are otherwise always present during the experimental testing of the harvester. To characterize the system exclusively, it is important to simulate the system's behavior over a time period that is long enough for these initial transitions to die out.

To check whether simulation of the system for a large driving force cycle makes any difference to the proof mass motion patterns for given acceleration amplitudes, the phase space trajectories were studied. The phase space trajectories provide obvi-

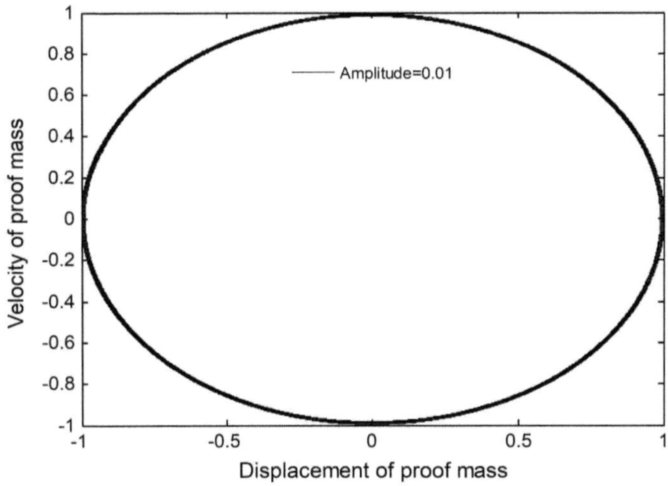

Fig. 7.12 Phase space trajectory for $e = 1$ and $\hat{A} = 0.01$

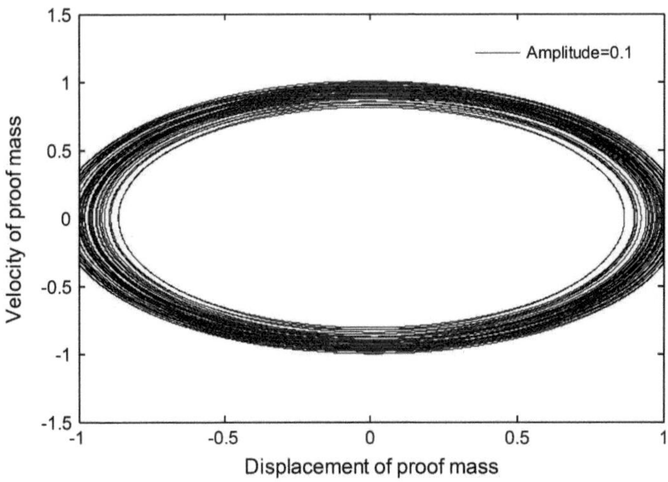

Fig. 7.13 Phase space trajectory for $e = 1$ and $\hat{A} = 0.1$

ous visualizations of the proof mass motion with changing acceleration amplitudes for a given coefficient of restitution. Figures 7.12, 7.13, 7.14 and 7.15 show the phase space trajectories projected into the $x - v$ plane for $e = 1$. Figure 7.12 shows the time evolution of the acceleration amplitude that is necessary to achieve the required impacts. The phase space trajectory shows the cyclic motion of the proof mass, but with the motion period being considerably longer than the period of the driving force. For slightly larger amplitudes, the phase space trajectory becomes very complex and chaotic, with no repeating patterns, as shown in Fig. 7.13. This may stem from the fact that eigenvalue analysis is unable to capture complex motion patterns.

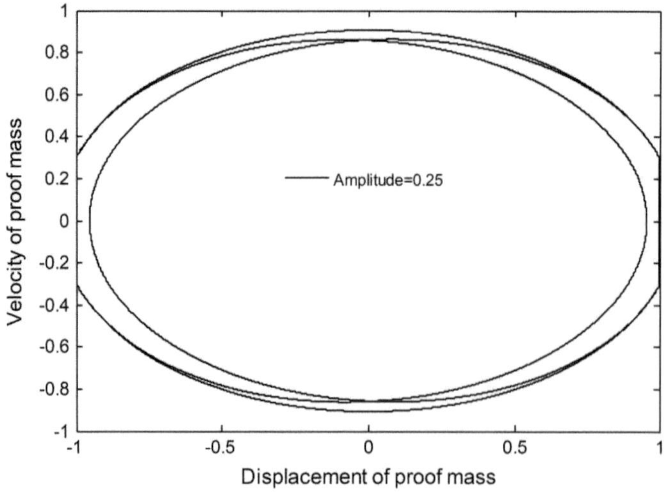

Fig. 7.14 Phase space trajectory for $e = 1$ and $\hat{A} = 0.25$

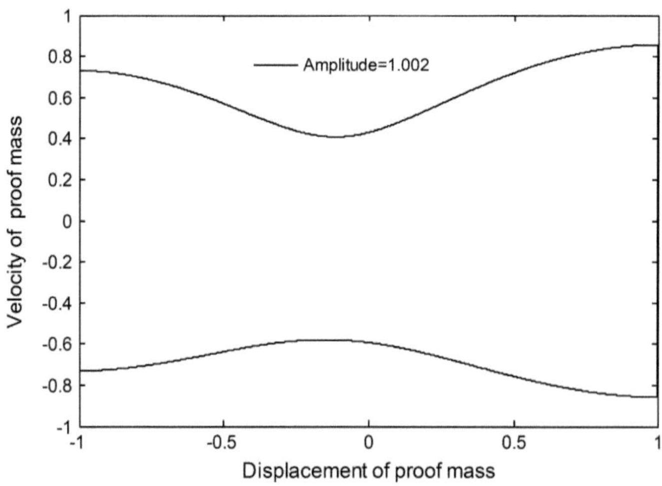

Fig. 7.15 Phase space trajectory for $e = 1$ and $\hat{A} = 1.002$

For further increases in the acceleration amplitude, the motion pattern becomes simpler, as shown in Figs. 7.14 and 7.15. Figure 7.14 shows that the motion period of the proof mass is double the period of the driving force. Upon a further increase, the acceleration amplitude then lies in the region where the period of the proof mass motion is exactly equal to the period of the driving force with one impact at each end-stop per cycle, as shown in Fig. 7.15.

The motion of the proof mass depends significantly on the system parameters. The proof mass motion therefore varies with different sets of parameters for the same

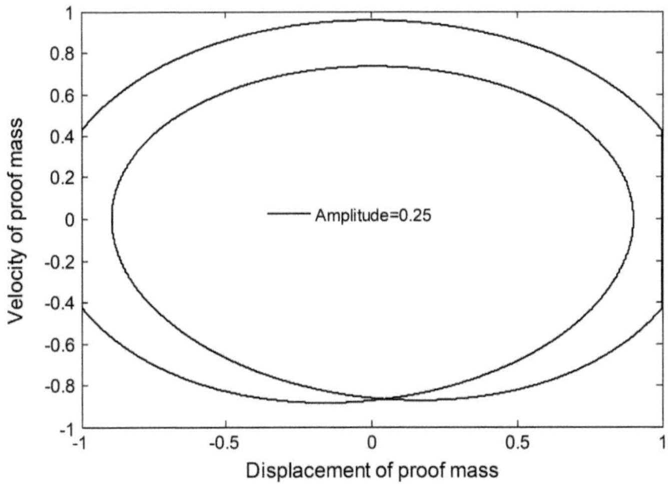

Fig. 7.16 Phase space trajectories for $e = 1$ and $\hat{A} = 0.25$ for the system parameters given in [13]

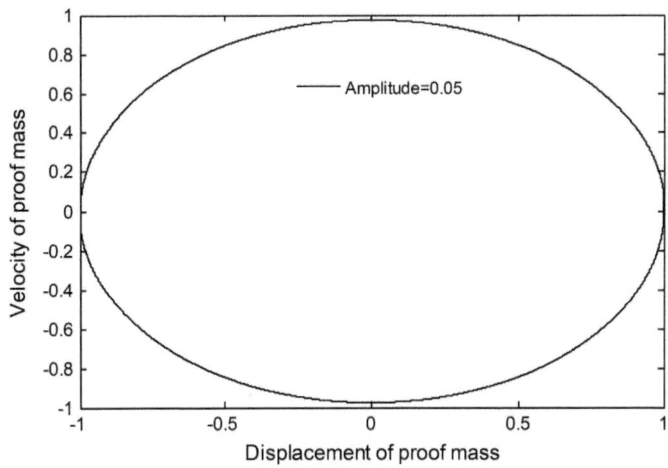

Fig. 7.17 Phase space trajectory for $e = 0$ and $\hat{A} = 0.05$

coefficient of restitution. Figure 7.16 shows the proof mass motion pattern based on the system parameters given in [13] for $e = 1$. Comparison of Fig. 7.14 with Fig. 7.16, where both are based on the same acceleration amplitude and coefficient of restitution, shows that the motion periods for the different sets of system parameters are dissimilar. The motion pattern in Fig. 7.16 is much simpler than that shown in Fig. 7.14. The appearance of the phase space trajectories thus varies with the different coefficients of restitution for the given system parameters, and vice versa.

It is interesting to observe the evolution of the phase space trajectories for $e = 0$ over time. Figures 7.17 and 7.18 show the trajectories for the acceleration amplitudes

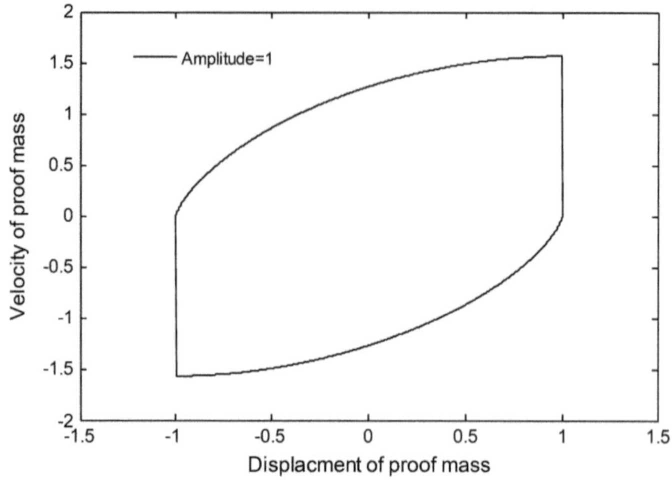

Fig. 7.18 Phase space trajectory for $e = 0$ and $\hat{A} = 1$

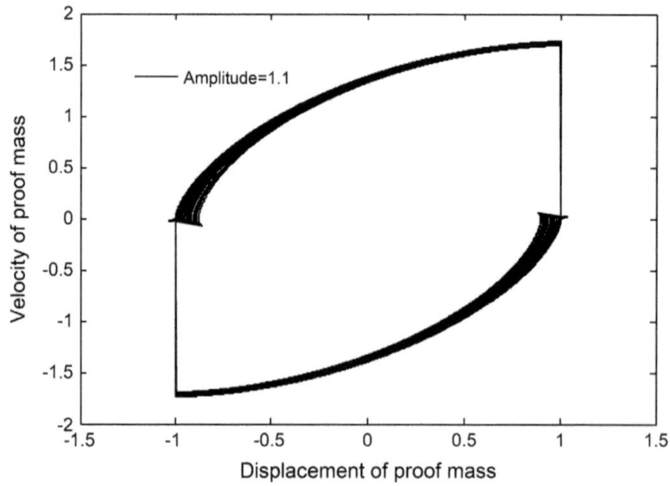

Fig. 7.19 Phase space trajectory for $e = 0$ and $\hat{A} = 1.1$ without a restoring force

where the motion period is equal to that of the driving force. For an acceleration amplitude of more than 1, the proof mass motion becomes complicated, because it tends to go beyond the displacement limit that was set by the end-stops, as shown in Fig. 7.19. Thus, modeling of the restoring force for the acceleration amplitude for which the proof mass tends to cross the displacement limit simplifies the proof mass motion, as shown in Fig. 7.20.

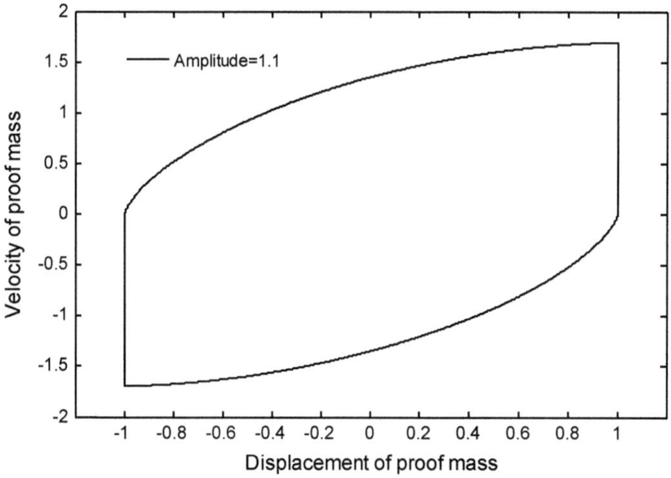

Fig. 7.20 Phase space trajectory for $e = 0$ and $\hat{A} = 1.1$ with restoring force

With the restoring force at the end-stops, the equation system from Eqs. (7.3)–(7.7) is reformulated in the form of

$$\frac{d\hat{u}}{dt} = -\hat{L}_1\hat{u} \tag{7.24}$$

where

$$\hat{L}_1 = \begin{bmatrix} 0 & 0 & 0 & 0 & 0 \\ 0 & 0 & 0 & 0 & 0 \\ \frac{k}{r} & 0 & \frac{1}{r} & 0 & 0 \\ 0 & 0 & 0 & 0 & \varsigma \\ 0 & 0 & 0 & -\varsigma & 0 \end{bmatrix} \tag{7.25}$$

For coefficients of restitution $0 < e < 1$, the analysis requires further investigation. For example, an impact with $e = 0.3$ is not perfectly inelastic, and thus the proof mass will have a tendency to bounce back and forth toward the end-stops several times before attaining continuous motion. The bouncing motion of the proof mass must be taken into account to study the complete motion pattern of the proof mass. The $0 < e < 1$ range has not been analyzed in detail in this chapter.

The average output power is calculated using Eq. (7.26) for a linear model with no end-stop impact. The output power with the end-stop effects is given in Eq. (7.28) by averaging the instantaneous power over the motion period.

$$\hat{P}_{linear} = \frac{P_{linear}}{mX_{max}^2\omega_0^3} = \frac{1}{2}r\hat{A}^2\left|\hat{h}(\varsigma)\right|^2 \tag{7.26}$$

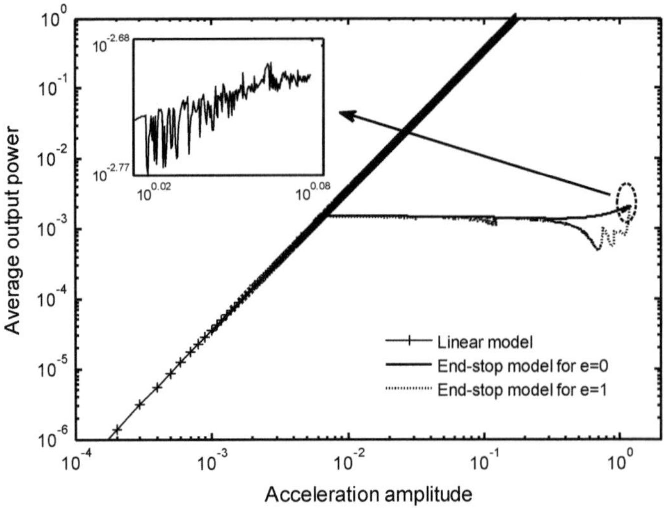

Fig. 7.21 Average output power versus acceleration amplitude

where

$$\hat{h}(\varsigma) = \frac{i\varsigma k}{\left\{\left(1-k^2\right) - \left(1+\frac{r}{Q}\right)\varsigma^2 + i\varsigma\left[\frac{1}{Q} + r\left(1-\varsigma^2\right)\right]\right\}} \quad (7.27)$$

$$\hat{P} = \frac{1}{r}\left(k\hat{x} + \hat{q}\right) \quad (7.28)$$

Figure 7.21 shows the average output power for the system with $e = 1$ and $e = 0$ that was simulated for large numbers of vibration cycles. The discontinuities shown in Fig. 7.21 for $e = 1$ correspond to the acceleration amplitudes that produce complex phase space trajectories with motion periods that differ from the period of the driving force, as shown above.

The inset image for $e = 0$ in Fig. 7.21 shows the complex pattern for an acceleration amplitude of more than 1, which corresponds to the phase space trajectories where the proof mass passes beyond the displacement limit, and thereby illustrates the need for the restoring force to be included in the model.

Figure 7.22 shows the average output power from the end-stop model when taking the restoring forces from the end-stops into account. The unevenness shown in Fig. 7.21 at acceleration amplitudes of more than 1 and for $e = 0$ is evened out in Fig. 7.22.

The results in Fig. 7.21 show that the power with $e = 0$ and $e = 1$ follows almost identical curves. However, this may or may not be the case for practical harvester prototypes, where the impacts are imperfectly elastic ($e = 0$). Additionally, many other factors affect the average output power, including the squeeze film damping

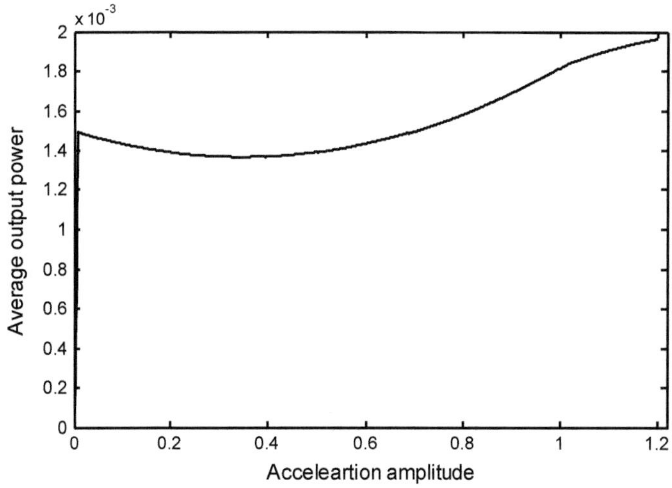

Fig. 7.22 Average output power versus acceleration amplitude with a restoring force

coefficient, the slide film damping coefficient, overcutting in the fabrication process, fringing fields, parasitic capacitance, and the end-stop configurations. One interesting phenomenon that is shown in Fig. 7.21 is that the average output power is weakly dependent on the acceleration amplitude.

7.2.3 Transducing End-Stops

The negative effects of power saturation for $e = 1$ and of the power loss for $0 < e < 1$ are obvious for the end-stops in vibration energy harvesters when the proof mass displacement reaches a maximum amplitude, as shown in Fig. 7.21. This effect has been demonstrated experimentally in many harvester prototypes [9, 11, 12, 26, 27]. The concept of replacement of passive end-stops with active end-stops to act as secondary transducers is illustrated in Fig. 7.23 [13, 28]. When the excitation levels are strong enough, the transducing end-stop is actuated by the force of the impact between the proof mass and the end-stop. The power from the end-stop transducer is added to that from the main transducer, and thus continuously increases the total power of the system when the acceleration level increases. As a result, the efficiency or effectiveness of the harvester is improved by this combination of the main transducer with the end-stop transducers.

The end-stop mechanism can be one of three types: electrostatic, electromagnetic, or piezoelectric, as stated earlier. The important function of the transducing end-stop is to efficiently convert the kinetic energy from the impact force into electrical power while maintaining the net stiffness to confine the primary transducer's motion. Figure 7.24 shows an example of a device that implements transducing end-stops

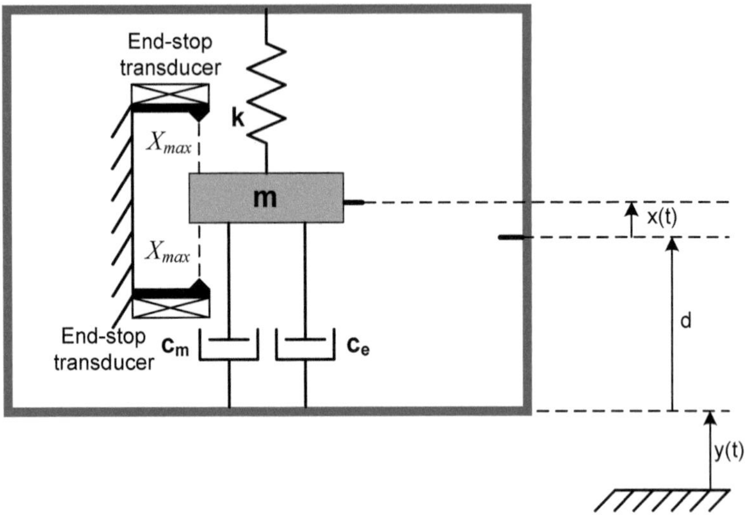

Fig. 7.23 Active end-stops acting as transducers included within the device package [13, 28]

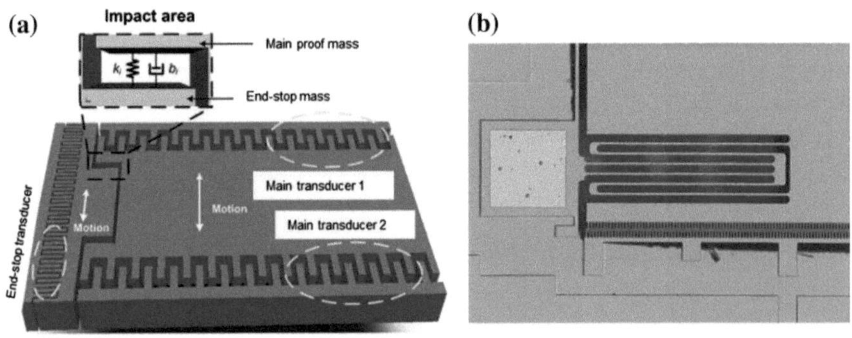

Fig. 7.24 **a** A schematic of the device prototype design using the active end-stop transducers, and **b** a close-up view of the fabricated device

based on electrostatic mechanisms. An overlap-varying comb-drive capacitor structure includes main transducers 1 and 2, which vary in anti-phase with each other. The end-stop transducer is a gap-closing capacitor structure. The masses of both the main and secondary transducers are suspended using linear folded springs. In the design, the maximum displacement amplitude of the main proof mass is $X_{max} = 10\,\mu\text{m}$. The end-stop transducer begins actuation when the relative displacement of the main proof mass passes beyond 6 μm.

Figure 7.24a shows the main characteristics of the in-plane harvester design. Figure 7.24b shows part of the microelectromechanical system (MEMS) device, which was fabricated using the silicon-on-insulator multi-user MEMS processing (SOI-MUMPS) method with a device layer thickness of 25 μm [29]. The total active

Table 7.1 Design parameters for the device prototype

Parameters	Main transducers 1 and 2	End-stop transducer
Finger length	25 μm	55 μm
Finger width	2 μm	2 μm
Nominal overlap	10 μm	50 μm
Nominal gap	2 μm	5 μm
Spring length	530 μm	335 μm
Spring width	6.2 μm	6.2 μm
Mass	1.15 mg	0.05 mg

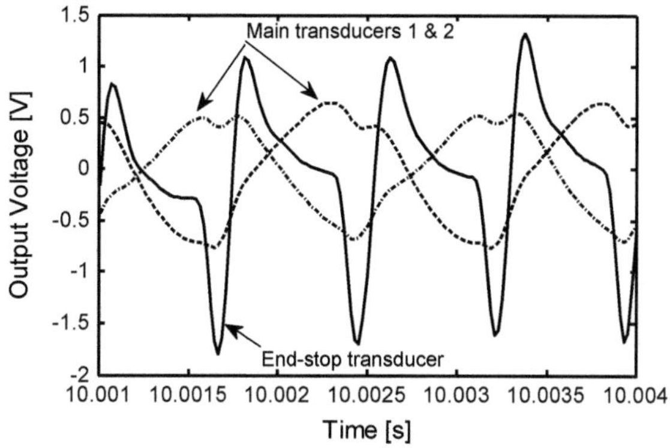

Fig. 7.25 Output voltage waveforms of the transducers for $A = 1.2$ g

area of the prototype is 4×5 mm^2. The design parameters of the device are listed in Table 7.1. In the experiments, all electrostatic transducers in the device prototype are biased using an external bias voltage $V_b = 12\,\text{V}$ in continuous mode. The output powers from the end-stop and main transducers are obtained by connecting the fixed electrodes of the transducers to an external load with component value $R_L = 18.5\,\text{M}\Omega$, which is the optimum load for the main transducers.

At low acceleration levels, the proof mass motion of the main transducers is less than the maximum displacement amplitude. Therefore, there is no internal impact between the main transducers and the end-stop transducer. Thus, the transducing end-stops are deactivated and produce almost no output power. The harvester outputs are linear and mainly come from the main transducers, 1 and 2, in opposite phase. Figure 7.25 shows the output voltage waveforms of the main transducers and the end-stop transducer for an acceleration $A = 1.2$ g. At this sufficiently large excitation level, the impact force between the masses is strong enough to drive the end-stop transduction significantly. The output voltage of the end-stop transducer then becomes comparable to that of the main transducers. This indicates that the energy conversion process is more effective with the addition of the end-stop transducer. The waveform of the end-stop transducer is characterized almost in transient time, which is restricted by the time interval between the impacts, as shown in Fig. 7.25.

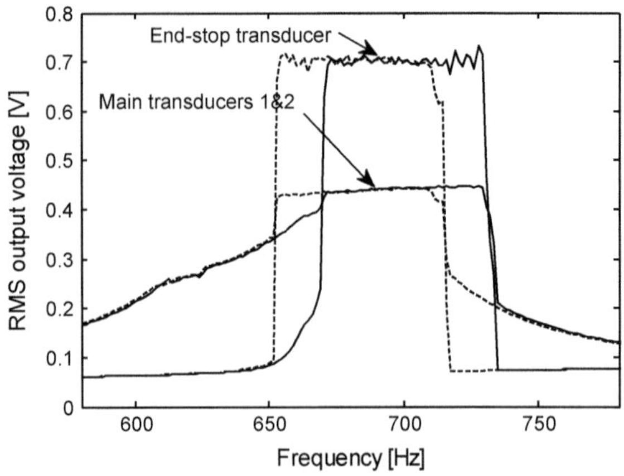

Fig. 7.26 Measured output voltages of the transducers versus frequency up-sweep (*solid line*) and down-sweep (*dashed line*) for $A = 1.2$ g

The gap-closing transduction in the end-stop transducer produces a motion period that is double that of the overlap-varying transduction in the main transducers.

Figure 7.26 shows the frequency responses of all measured output voltages for $A = 1.2$ g. In the linear regime, the main transducers give a resonant frequency of $f_0 = 665$ Hz and a 3-dB bandwidth of 6.9 Hz. The dynamic interaction due to the impact between the proof masses adds nonlinearities to the frequency responses of both the main transducers and the end-stop transducer. The frequency responses form a hysteresis pattern with a jump-down frequency $f_{down} = 653.1$ Hz and a jump-up frequency $f_{up} = 730.4$ Hz. The frequency band between the jump frequencies of the up-sweep and down-sweep is 77.3 Hz, which is approximately 11 times higher than the 3 dB bandwidth in the linear regime. Thus, the positive nonlinearities broaden the harvester bandwidth based on the end-stop transducer impact mechanism. The intermediate frequency range shows a high-amplitude revolution in their responses.

Additionally, the output voltage of the transducing end-stop is significant in the impact frequency range. During an impact, the main proof mass hits the end-stop and is then moved for an extra distance before it returns to the equilibrium position. The hysteresis observed in the low-frequency range is affected by the softening-spring nonlinearity [10, 30–32] because of the electrostatic pull of the end-stop transducer on the proof mass for small gap sizes. Therefore, the end-stop proof mass is driven toward a larger displacement amplitude, which leads to greater variation in the gap-closing capacitance in the end-stop transduction process. As shown in Fig. 7.26, the minimum gap for the end-stop transducer is achieved for $A = 1.2$ g. The outputs of both the main transducers and the end-stop transducer reach their maximum levels.

The benefits of the transducing end-stop can be seen by comparison with the reference harvester prototype under increasing acceleration amplitudes, as shown in Fig. 7.28. The reference transducers are designed to be identical to the main transducers in the same active area. Both prototypes have a maximum displacement amplitude $X_{max} = 10\,\mu$m. In the linear regime, the output power of the reference prototype is slightly higher than that of the impact prototype. This is because the reference har-

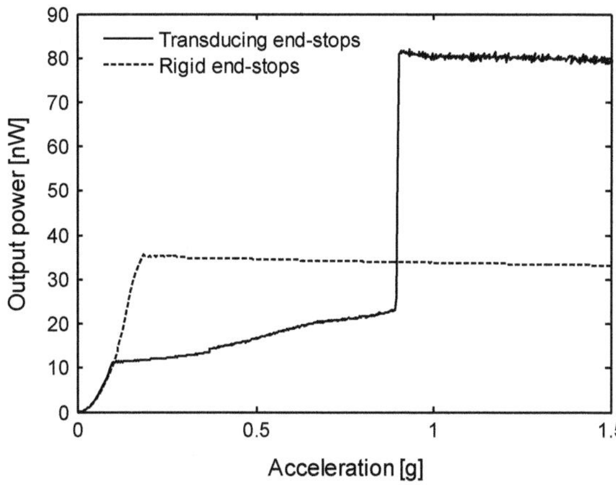

Fig. 7.27 Measured output power of the prototype with the transducing end-stops compared with that of the reference prototype with rigid end-stops at their resonant frequencies

vester has a larger proof mass and higher transduction designed using the same constraints. The proof mass motion of the reference device reaches its maximum displacement amplitude at an acceleration $A = 0.18$ g. A saturated output power of 35.0 nW is obtained for the reference device. With the additional power coming from the end-stop transducers, the total power of the impact prototype is higher than that of the reference harvester under the same conditions. The total power achieved is 81.5 nW at $A = 0.91$ g, which is 2.3 times higher than that of the reference prototype.

Figure 7.27 shows one drawback where a large acceleration gap from $A = 0.10$ g to $A = 0.91$ g must be covered to enable the end-stop transducer to be effective. This is because the designed mechanical stiffness of the end-stop transducer is rather too stiff. Therefore, strong acceleration forces are required to enhance the end-stop transduction of the gap-closing capacitance, which varies as $\sim 1/\left(g^2 - x_s^2\right)$. This problem can be overcome by designing the end-stop transducer to have a compliant stiffness, as shown in Fig. 7.28. With a compliant end-stop, the total output power increases almost linearly when the benefits of the end-stop transducer become recognizable at $A = 0.72$ g. The total power is higher than the saturated power of the impact device with the stiffer end-stop transducer for $A > 1.42$ g. The high-amplitude orbit of the end-stop proof mass is achieved quickly when the net stiffness of the end-stop transducer is reduced. The end-stop transduction is considerable even at small acceleration amplitudes, and becomes complicated at high acceleration levels. These complications can be explained with reference to the phase space trajectories from the mathematical analysis of the end-stop effects that was illustrated earlier. In the later prototype, the transducing end-stops are used on both the right and left sides of the main transducers [13].

Alternatively, the mechanical stiffness of the end-stop transducer can be further reduced by increasing the bias voltages applied to the electrostatic end-stop transducer. The electrostatic force from the gap-closing transducer cancels the spring force to produce a low net stiffness while maintaining sufficient strength to secure

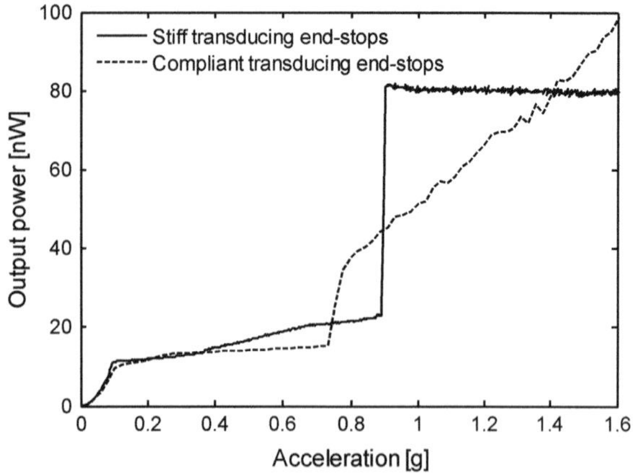

Fig. 7.28 Total output powers for two different end-stop designs

any unstable pull-in effects. Therefore, the end-stop transducer becomes more compliant. The transducing end-stops will need further improvements to improve the total system output power with given displacement constraints.

7.3 Conclusions

The numerical analysis that was carried out above is a useful tool for study of the nonlinearities of end-stop effects in vibration energy harvesters. Through simple modeling of the end-stops, it was found that this simple motion is atypical and that solutions with complicated trajectories exist in phase space. The periods of motion for these solutions can be very different from the period of the driving force, if indeed they are periodic at all. The main effect of the end-stops on the output power is to produce saturation behavior during continuous mode operation. The phase space trajectories show that the motion of the proof mass in the energy harvester can be complex, depending on the acceleration amplitude, but that the output power is weakly dependent on the acceleration amplitude in the impact regime. Therefore, the effects of the impacts on device performance during continuous operation are minor. The consideration of the restoring force at the impacts is demonstrated using the phase space trajectories and an output power graph.

The inelastic collisions with $e = 0$ will cause some energy to be lost in the end-stops on impact. This energy loss can be collected smartly using power conditioning circuits such as SECE or SSHI circuits. Alternatively, the lost energy on impact can be collected by introducing the transducing end-stops which not only increases the total output power of the harvester but also enhances its bandwidth. The compliant transducing end-stops harvest more power than rigid transducing end-stops, as demonstrated here using harvester prototypes. The harvester output can be further improved by efficient design of the stiffness of the end-stops.

The mathematical analysis described in this chapter can provide an insight into the device behavior for given system parameters. The tool can be used effectively and efficiently for energy harvester design. For example, when designing an energy harvester that uses the end-stops as switches for power conversion circuitry, the simulation approach illustrated in this chapter can be used to predict the switching on every cycle of the driving force, thus optimizing the power that is harvested.

Acknowledgments This work was supported by the Research Council of Norway under Grant no. 191282. We thank Prof. Einar Halvorsen for useful discussions and suggestions.

References

1. Naruse, Y., Matsubara, N., Mabuchi, K., Izumi, M., & Suzuki, S. (2009). Electrostatic micro power generation from low-frequency vibration such as human motion. *Journal of Micromechanics and Microengineering*, *19*, 094002.
2. Roundy, S. (2003). Energy scavenging for wireless sensor nodes with a focus on vibration to electricity conversion. Ph.D. thesis, The University of California, Berkeley, Spring
3. Halvorsen, E., Westby, E. R., Husa, S., Vogl, A., Østbø, N. P., Leonov, V., et al. (2009). An electrostatic energy harvester with electret bias. *Proceeding of Transducers*, *2009*, 1381–1384.
4. Blystad, L.-C. J., & Halvorsen, E. (2011). A piezoelectric energy harvester with a mechanical end stop on one side. *Microsystem Technologies*, *17*, 505–511.
5. Roundy, S., & Wright, P. K. (2004). A piezoelectric vibration based generator for wireless electronics. *Smart Materials and Structures*, *13*, 1131–1142.
6. Che, L., Halvorsen, E., Chen, X., & Yan, X. (2010). A micromachined piezoelectric PZT-based pantilever in d33 mode. In *Proceeding of the 5th IEEE International Conference on Nano/Micro Engineered and Molecular Systems* (pp. 785–788).
7. Amirtharajah, R., & Chandrakasan, A. P. (1998). Self-powered signal processing using vibration-based power generation. *IEEE Journal of Solid-State Circuits*, *33*, 687–695.
8. Cao, X., Chiang, W. J., King, Y. C., & Lee, Y. K. (2007). Electromagnetic energy harvesting circuit with feedforward and feedback DC-DC PWM boost converter for vibration power generator system. *IEEE Transactions on Power Electronics*, *22*, 679–685.
9. Le, C. P., & Halvorsen, E. (2012). MEMS electrostatic energy harvesters with end-stop effects. *Journal of Micromechanics and Microengineering*, *22*, 074013.
10. Tvedt, L. G. W., Nguyen, D. S., & Halvorsen, E. (2010). Nonlinear behavior of an electrostatic energy harvester with wide- and narrowband excitation. *Journal of Micromechanics and Microengineering*, *19*, 305–316.
11. Hoffmann, D., Folkmer, B., & Manoli, Y. (2009). Fabrication, characterization and modeling of electrostatic microgenerators. *Journal of Micromechanics and Microengineering*, *19*, 094001.
12. Soliman, M. S. M., Abdel-Rahman, E. M., El-Saadany, E. F., & Mansour, R. R. (2008). A wideband vibration based energy harvester. *Journal of Micromechanics and Microengineering*, *18*, 115021.
13. Le, C. P., Halvorsen, E., Søråsen, O., & Yeatman, E. (2012). Microscale electrostatic energy harvester using internal impacts. *Journal of Intelligent Material Systems and Structures*, *13*, 1409–1421.
14. Neubauer, M., Krack, M., & Wallaschek, J. (2010). Parametric studies on the harvested energy of piezoelectric switching techniques. *Smart Materials and Structures*, *19*, 025001.
15. Blystad, L.-C. J., Halvorsen, E., & Husa, S. (2010). Piezoelectric MEMS energy harvesting driven by harmonic and random vibrations. *IEEE Ultrasonics, Ferroelectrics and Frequency Control Society*, *57*, 908–919.

16. Williams, C. B., & Yates, R. B. (1995). Analysis of a micro-electric generator for microsystems. In *Proceeding of Transducers'95* (pp. 369–372).
17. Williams, C. B., & Yates, R. B. (1996). Analysis of a micro-electric generator for microsystems. *Sensors and Actuators A: Physical, 52*, 8–11.
18. Cantatore, E., & Ouwerkerk, M. (2006). Energy scavenging and power management in networks of autonomous microsensors. *Microelectronics Journal, 37*, 1584–1590.
19. Mitcheson, P. D., Yeatman, E. M., Rao, G. K., Holmes, A. S., & Green, T. C. (2008). Energy harvesting from human and machine motion for wireless electronic devices. *Proceedings of the IEEE, 96*, 1457–1486.
20. Meninger, S., JMur-Mirande, J. O., Amirtharajah, R., Chandrakasan, A. P., & Lang, J. H. (2001). Vibration to electric energy conversion. *IEEE Transactions on Very Large Scale Intergration (VLSI) Systems, 9*, 64–76.
21. Miller, L. M., Halvorsen, E., Dong, T., & Wright, P. K. (2011). Modeling and experimental verification of low-frequency MEMS energy harvesting from ambient vibrations. *Journal of Micromechanics and Microengineering, 21*, 045029.
22. Westby, E. R., & Halvorsen, E. (2012). Design and modeling of a patterned-electret based energy harvester for tire pressure monitoring systems. *IEEE/ASME Transactions on Mechatronics, 17*, 995–1005.
23. Vocca, H., Neri, I., Travasso, F., & Gammaitoni, L. (2012). Kinetic energy harvesting with bistable oscillators. *Applied Energy, 97*, 771–776.
24. Cottone, F., Basset, P., Guillemet, R., Galayko, D., Marty, F., & Bourouina, T. (2013). Nonlinear MEMS electrostatic kinetic energy harvester with a tunable multistable potential for stochastic vibrations. *Proceeding of Transducers, 2013*, 1336–1339.
25. Guillemet, R., Basset, P., Galayko, D., Cottone, F., Marty, F., & Bourounia, T. (2013). Wideband MEMS electrostatic vibration energy harvesters based on gap-closing interdigited combs with a trapezoidal cross section. *Proceeding of IEEE MEMS, 2013*, 817–820.
26. Hoffmann, D., Folkmer, B., & Manoli, Y. (2011). Analysis and characterization of triangular electrode structures for electrostatic energy harvesting. *Journal of Micromechanics and Microengineering, 21*, 104002.
27. Stanton, S. C., McGehee, C. C., & Mann, B. P. (2009). Reversible hysteresis for broadband magnetopiezoelastic energy harvesting. *Applied Physics Letters, 95*, 174103.
28. Le, C. P., Halvorsen, E., Søråsen, O., & Yeatman, E. M. (2012). Comparison of transducing end-stops with different stiffness in MEMS electrostatic energy harvesters. *Proceeding of PowerMEMS, 2012*, 444–447.
29. (2013). http://www.memscap.com/products/mumps/soimumps.
30. Mestrom, R. M. C., Fey, R. H. B., Phan, K. L., & Nijmeijer, H. (2010). Simulations and experiments of hardening and softening resonances in a clamped-clamped beam MEMS resonator. *Sensors and Actuators A: Physical, 162*, 225–234.
31. Amri, M., Basset, P., Cottone, F., Galayko, D., Najar, F., & Bourouina, T. (2013). Novel nonlinear spring design for wideband vibration energy harvesters. *Proceeding of PowerMEMS, 2011*, 189–192.
32. Elshurafa, A. M., Khirallah, K., Tawfik, H. H., Emira, A., Aziz, A. K. S. A., & Sedky, S. M. (2011). Nonlinear dynamics of spring softening and hardening in folded mems comb drive resonators. *Journal of Microelectromechanical Systems, 20*, 943–958.

Chapter 8
Conditioning Circuits for Capacitive Energy Harvesters

D. Galayko

8.1 Introduction

8.1.1 Generalities

Any sensor converting a physical quantity into electricity needs a minimal electrical conditioning: an appropriate biasing, a readout circuit with an optimal input impedance, etc. However, in many cases, the conditioning is more complex: for example resonant sensors need electrical oscillating circuits, zero displacement accelerometers require PWM or Sigma-Delta modulating feedback loops, etc. The capacitive transducer, being itself a passive device, requires a sophisticated electrical conditioning for electromechanical energy conversion. As was pointed out in Chap. 4, the primary goal of the conditioning is an implementation of a cyclic charge flow synchronized with the variation of the transducer capacitance. Different aspects of practical conditioning circuits are discussed in this chapter.

In Chap. 4, it was shown that the two equations linking the electrical and mechanical quantities (4.2) and (4.13) of an electrostatic transducer are nonlinear. It was also emphasized that an ideal scenario of the electromechanical energy conversion corresponding to a constant voltage QV cycle is implemented by a time variant, i.e., reconfigurable electrical network. As a consequence, a capacitive harvester as a whole can be seen as a nonlinear system. More sophisticated conditioning circuits, e.g., those implementing a rectangular QV cycle based on charge pumps using diodes, are generally nonlinear as well.

In this chapter, we propose a short review of conditioning circuits used for energy conversion with capacitive transducers.

D. Galayko (✉)
UPMC—Sorbonne Universities, Paris, France
e-mail: dimitri.galayko@lip6.fr

E. Blokhina et al. (eds.), *Nonlinearity in Energy Harvesting Systems*,
DOI 10.1007/978-3-319-20355-3_8

8.1.2 Classification of Conditioning Circuits for Capacitive Harvesters

As it was shown in Chap. 4, a capacitive transducer needs a dynamic biasing in order to generate electricity. For a capacitive transducer whose capacitance varies periodically following a time law $C_t(t)$, the energy conversion process is completely defined by the charge-voltage cycle Γ. Since the voltage, charge and capacitance of the transducer are bounded by the equation $Q_t = C_t V_t$, if $C_t(t)$ is defined, the energy conversion is defined by the voltage waveform applied to the transducer. A conditioning circuit for an electrostatic transducer can be seen as an electrical dipole, which is connected to the electrical terminals of the transducer, as presented in Fig. 1.3.

Since the past two decades, several architectures of conditioning circuits for eVEHs (electrostatic Vibration Energy Harvesters) were invented. We propose a classification of the conditioning circuits on the basis of the shape of the achieved charge-voltage cycles. Figure 8.1 presents all known basic QV cycles plotted in the same axes, under hypothesis that the maximum voltage V_t applied to the variable capacitor is the same for all cycles (V_{max}). We chose such a basis for the comparison, because in practice, any technology of electronic circuits always impose a maximum allowed voltage. We distinguish three families of conditioning circuits implementing three types of QV cycles

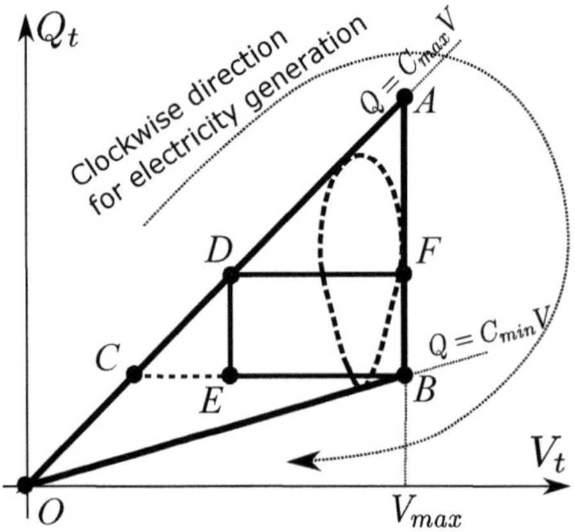

Fig. 8.1 Diagram showing all charge-voltage cycles which have been used to date for conditioning of capacitive transducers in VEHs: OCB (constant-charge), OAB (constant-voltage), DFBE (*rectangular*) and "tear drop" cycle corresponding to the continuous conditioning circuit or VEH with electret layer. All cycles are drawn for the same extreme values of the variable capacitance (C_{min} and C_{max}) and for the same maximum voltage V_{max}

1. Teardrop, oval, or egg-shaped QV cycle (an example is plotted with dotted lines)
2. Triangular QV cycle (OAB, OCB)
3. Rectangular QV cycle (DFBE)

In this chapter, we present practical topologies of conditioning circuits implementing these QV cycles.

8.1.3 Frame of the Analysis of Conditioning Circuit

The conditioning circuits are usually studied in electrical domain. It means, that the variation of the transducer capacitance is supposed to be defined and fully characterized by a function of time $C_t(t)$. A usual hypothesis is $C_t(t)$ is periodic with period T_e[1] and having only one local maximum and minimum (C_{max} and C_{min}) over a period. This is only a working hypothesis allowing to limit the complexity of the analysis. In reality, because of the electromechanical coupling, $C_t(t)$ depends also on the electrical operation of the conditioning circuit, as it will be shown in examples in Sect. 8.7. However, assuming a predetermined $C_t(t)$ is a necessary step in the study of the harvester as a whole.

In Sect. 8.7, we discuss how the presence of a transducer coupled with a mechanical resonator impacts the operation of the circuit.

8.2 Continuous Conditioning Circuit

The simplest conditioning circuit demonstrating a generation of electrical power out of variation of a capacitance is called "continuous conditioning circuit." Its topology is presented in Fig. 8.2. The circuit is composed of the variable capacitor C_t (the transducer), a resistive load R_L and an initially charged large reservoir capacitor C_{res}, or simply a DC voltage source, when only a laboratory test is aimed. The name of the circuit comes from the fact that all voltages and currents of the circuit are continuous functions of time; this is not the case of more sophisticated circuits which use switches or diodes and which will be considered later in this chapter.

The continuous conditioning circuit presented in Fig. 8.2 was first discussed in [17, 24]. It was proposed as an electric interface between a resistive load and the capacitive transducer provided with an electret layer [24], and later it was used with a passive capacitive transducer [1, 17] biased by a voltage source or by a fixed pre-charged capacitor. In spite of different nature of the transducer devices, these configurations are equivalent: a DC voltage source in series with a passive capacitive transducer is exactly an electrical model of a transducer biased by an electret layer

[1]The index "e" in T_e stands for "electrical." This is to emphasize that the variation of C_1 may have a different frequency that the mechanical vibrations, cf. [1].

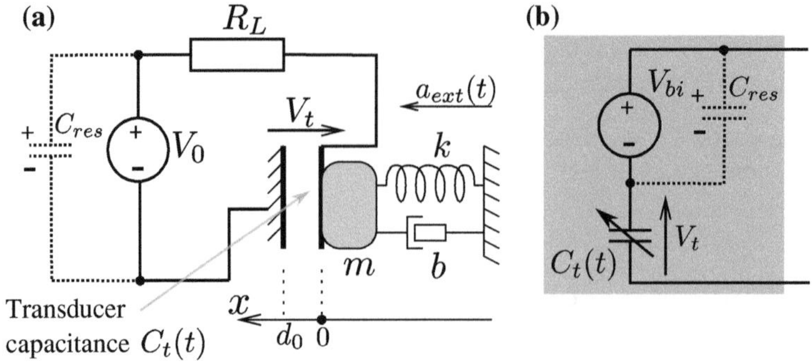

Fig. 8.2 Schematic of continuous conditioning circuit. In this study the continuous circuit represents all circuit configurations which employ a capacitive transducer, resistive load and some internal or external voltage to provide the initial bias on the transducer

[24]. Without major impact on prediction capability of the model, the DC voltage source may be replaced by a large pre-charged capacitor.

Because of the simplicity of its implementation, this circuit is often used in laboratory tests and characterization of capacitive transducers intended for the energy conversion. In this section we propose a brief analysis of this circuit and its main shortcomings.

8.2.1 *Qualitative Discussion on Operation of the Circuit*

In this subsection, we explain briefly the circuit operation on the example of a circuit where the biasing is provided by a large fixed capacitor C_{res} initially charged to a voltage V_0, other capacitors are initially discharged and where the transducer capacitance varies according to some known periodic time law $C_t(t)$ (Fig. 8.2a). We first consider the circuit operation in an extreme case, when R_L is close to zero [1]. It can be seen that as the transducer capacitance varies, there is a charge redistribution among C_t and C_{res}, which corresponds to a current $i(t)$ equal to

$$i(t) = V_0 \frac{\mathrm{d}C_t}{\mathrm{d}t} \tag{8.1}$$

This equation is valid if we can neglect the voltage on R_L comparing to V_0: that is, indeed the case if $R_L \approx 0$. It can be seen that this current dissipates power on R_L, whose instantaneous value is equal to

$$P_{RL} = i^2(t) R_L. \tag{8.2}$$

After each period of variation of $C_t(t)$, the circuit returns to the same electrical state, since the total electrical charge of C_t and C_{res} is constant, so is their total energy. One should conclude that the energy dissipated on the resistance comes from the conversion of energy from mechanical to electrical domain.

The power P_{RL} is proportional to the load resistance: it increases as R_L increases. However, for large R_L, the voltage on R_L cannot be neglected anymore, and the Eq. (8.1) is not valid.

We consider now the opposite extreme value of the load resistance: very large (infinite). If the power on the load resistance is expressed through the voltage on the load resistance V_L as

$$P_{RL} = V_L^2(t)/R_L, \tag{8.3}$$

we can show that the power tends to zero as $R_L \to \infty$. To prove this, it is enough to show that V_L has an upper bound. Indeed, since the current tends toward zero, the transducer keeps its charge constant. The value of this charge depends on the initial condition. Let it be Q_0, in this case the voltage on C_t is $Q_0/C_t(t)$, and the absolute value of voltage on the resistor is equal to $|V_0 - Q_0/C_t(t)|$. Evidently, if $C_{min} > 0$, this time function has an upper bound.

So, at zero and infinity values of the load resistance, the power is zero, but not for finite values of R_L. Hence, one should conclude that there is an optimal value of the load resistance, for which the converted power is maximum.

In the next two subsections, we propose a more detailed analysis of the circuit.

8.2.2 Analytical Model in the Electrical Domain

This section presents the mathematical model describing formally the operation of the conditioning circuit.

The governing equations describing the electrical behavior of the simple conditioning circuit are given by the Kirchhoff voltage law and the element equations:

$$R_\mathrm{L}\frac{dQ_t}{dt} + \frac{Q_t}{C_t(t)} = V_0, \tag{8.4}$$

where Q_t is the instantaneous charge on the transducer capacitor, $C_t(t)$ is the time evolution law of the transducer's capacitance. The electrical equation of the transducer is simply

$$V_t(t) = Q_t(t)/C_t(t), \tag{8.5}$$

where V_t is the voltage on the transducer.

The instantaneous power converted by the transducer is dissipated on the load resistance, and can easily be calculated with Eq. (8.2), where the current is given as

$$i = \frac{\mathrm{d}Q_t}{\mathrm{d}t} \tag{8.6}$$

The average power converted by the transducer is given by the average of (8.2) on one period of $C_t(t)$ variation.

In order to calculate the average converted power, the closed expression of $Q_t(t)$ should be found, and for that the differential Eq. 8.4 should be solved. The work [12] addressed the resolution of this equation. It was concluded that even for simple geometries of the transducer, this equation cannot be solved in closed form, and the analytical solution for $Q_t(t)$ can be expressed as infinite Fourier series, in which, however, the terms above third or forth can be neglected in the most practical cases. The coefficients of the series are expressed through Bessel functions. For this reason, a comprehensive analytical expression relating the converted power to the parameters of the system is difficult to establish.

We present here an example of practical study of the continuous conditioning circuit in the electrical domain. For that, we need to define the function defining the variation of the transducer capacitance over time, $C_t(t)$. In this example, we study the case of a gap closing transducer (see Chap. 4) whose mobile plane moves according to a sinusoidal law. We have:

$$C_t(t) = C_t(x(t)) = \varepsilon_0 \frac{S}{d_0 - x(t)} = \varepsilon_0 \frac{S}{d_0 - X_0 \sin \omega t}, \tag{8.7}$$

where S and d_0 are the overlapping area and the initial transducer gap, respectively, X_0 and ω are the amplitude and the angular velocity of mobile electrode motion.

This function is submitted to Eq. (8.4). The steady-state solution of the linear parametric equation (8.4) describes a periodic variation of the charge $Q_t(t)$ and of the current $\dot{Q}_t(t)$. The solution can be obtained by any tool for EDT solving, or by a Spice simulator able to model a variable capacitance. Here, we present the result of numerical resolution of the equation for the values of the parameters presented in Table 8.1.

Table 8.1 Values of parameters used for simulation of continuous conditioning circuit with gap closing capacitive transducer

Physical parameters						
	Fixed parameters				Variable parameters	
Parameter name	d (m)	S (m^2)	ω (rad $\cdot$ s^{-1})	V_0 (V)	X_0 (m)	R_L (Ω)
Parameter value	50×10^{-6}	1×10^{-4}	$2 \cdot \pi \cdot 100$	10	$(30, 40, 45) \times 10^{-6}$	$(1 \ldots 100) \times 10^6$

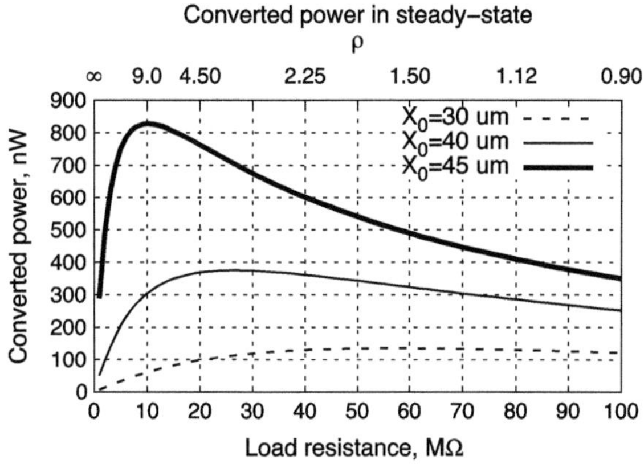

Fig. 8.3 Average power converted in a steady-state by the continuous conditioning circuit used with a gap closing transducer against the load resistance, for three different amplitudes of the mobile electrode motion

Figure 8.3 presents the plot relating the average power generated on the load resistance in the steady-state mode (after the transient process) to the value of the load resistance. The plot is given for three amplitudes of the mobile mass vibrations. It can be seen that the curves are similar to what is obtained for the power-load characteristic of a real voltage source having some internal resistance R_s. In this case, the power is maximum when the load resistance is equal to the internal resistance of the source. This allows us to consider the load resistance value at which the power on the plot of Fig. 8.3 is at its maximum as the equivalent internal resistance of the power source represented by the transducer and the biasing source. Note, that the internal resistance of such a source depends on the amplitude of the mobile electrode motion, as shows the plot. This highlights nonlinearity of the system: indeed, in a linear system, the impedance matching condition for the load does not depend on the amplitude of the input signal.

Figure 8.4 presents three families of QV cycles in steady-state mode (after the end of the transient process), corresponding to three different amplitudes of the mass displacement X_0. Each plot contains several QV cycles obtained for different values of the load resistance. The QV cycles drawn with a thick plain line correspond to the values of the load resistance at which the converted power is maximum (cf. Fig. 8.3): this QV cycle has the largest area over all cycles of the family.

Let us have a look on the evolution in the shape of the QV cycle as the load resistance increases. For low R_L, the QV cycles are more "vertical," and as $R_L \to 0$, the cycle is degenerated to a vertical line (meaning that there is no voltage across the load resistance). On the contrary, for large R_L, the QV cycle is more horizontal, and in the limit case when $R_L \to \infty$, the QV cycle is a horizontal line (meaning that there is no current flowing through the load resistance). For these two extreme cases, the energy converted by a cycle is zero (cf. discussion in Sect. 8.2.1). The QV cycle has a "tear-drop" shape, which, as we discussed before, can not be expressed by any analytical function written in closed form.

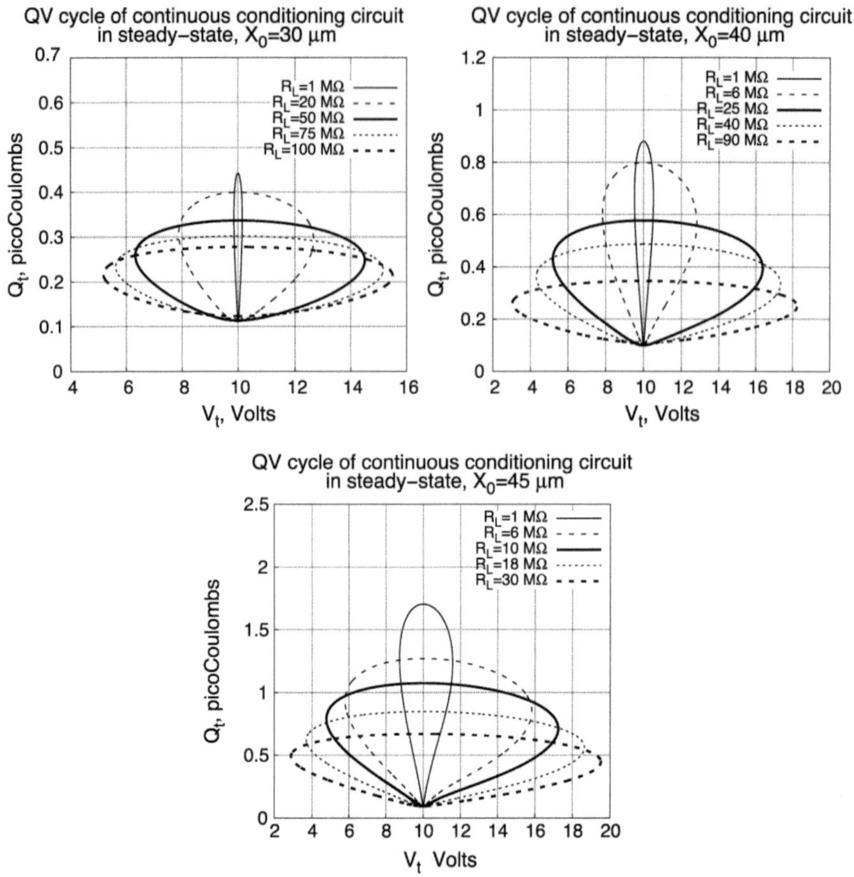

Fig. 8.4 Three families of QV cycles corresponding to steady-state operation of the continuous conditioning circuit used with a gap closing transducer, each family is plotted for a value of the amplitude of mobile electrode motion. The QV cycles on each plot correspond to different load resistances. The optimal QV cycle (those maximizing the converted power and hence having a maximum area) is plotted with *thick solid line*

The plots of Fig. 8.5 presents the time evolution of the transducer capacitance, the transducer voltage and current at the configuration where $X_0 = 45\,\mu\text{m}$, $R_L = 10\,\text{M}\Omega$. The transient process lasting for less than one period is observable on these plots. The nonlinear nature of the system can clearly be seen from these curves.

Shortcomings of the Continuous Conditioning Circuit. Auto-Increasing of the Biasing

The main advantage of the continuous conditioning circuit is its simplicity of implementation. Unfortunately, the drawbacks are numerous, and this is the main motivation for study of alternative but more complex solutions. In this conclusion, we summarize the drawbacks of this conditioning circuit.

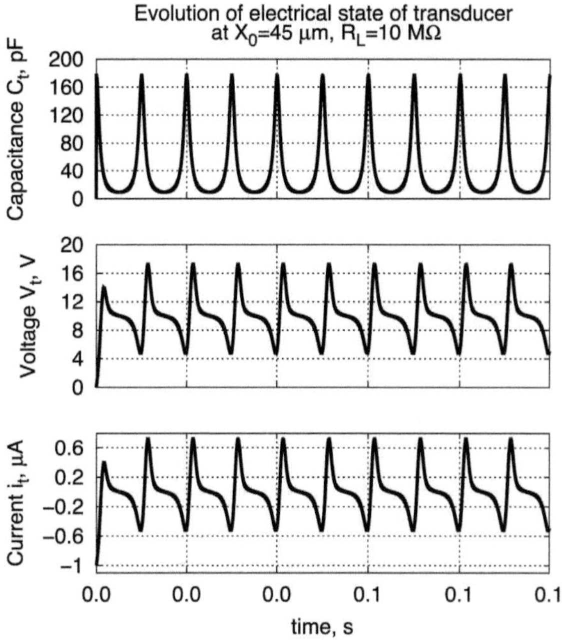

Fig. 8.5 Time evolution of the electrical quantities in the continuos conditioning circuit used with a gap closing transducer, obtained by simulation. The transient and the steady-state behavior are clearly observed

- The voltage on the load resistance is AC; it can be seen from the current flowing through the transducer in Fig. 8.5. An AC–DC conversion is required; in the case of the electret transducer this may be done by conventional rectifying networks [20].
- The order of magnitude of the voltage on the load and the bias voltage may be the same. In the case of the electret layer, the bias voltage may be of several tens of volts, and a downscale of the voltage is necessary. This is related to generally very large internal impedance (resistance) of the energy source provided by the continuous conditioning circuit (cf. discussion in Sect. 8.2.2). An active impedance matching is required in order to optimally supply an eventually low impedance load.
- However, the main shortcoming of the continuous conditioning circuit is in the case when the available bias voltage is, for some reasons, low. Practical studies highlight that up to tens of volts of bias may be needed for optimal operation [1]. However, the initial charge can only be obtained by a low voltage battery existing in the system, or by a complementary piezoelectric VEH as proposed in [11] (or maybe, by a solar cell), and the generated initial voltage will be not greater than few volts. A low bias voltage means a low output power, since the latter is proportional to the square of the voltage. A similar problem is when the electret layer is weakly biased, for example, because of the depolarization due to the aging.

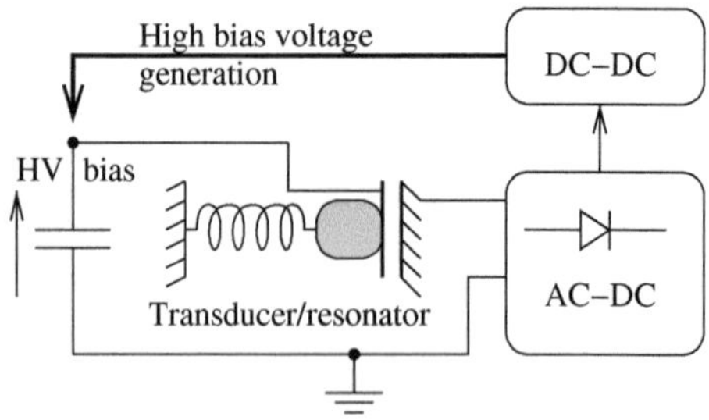

Fig. 8.6 Functional diagram of a conditioning circuit allowing an accumulation of the converted energy on the reservoir capacitor

- If the bias voltage is generated by a large pre-charged capacitor, the leakage reduces its charges with time, and the circuit becomes inactive. Similar problems have been reported with electret layers, which tend to depolarize with time.

The last two points are related to the impossibility of the continuous conditioning circuit to accumulate the energy converted from the mechanical domain, in order to increase its biasing. One of the possible functional diagrams required for a conditioning circuit is presented in Fig. 8.6: there should be a mechanism allowing a generation of a high voltage by using the energy converted from the mechanical domain. This voltage should then be used for biasing of the transducer. Since the biasing does not consume energy (under hypothesis of low leakage of capacitors), the circuit will be able to operate at high biasing, while the start-up may be done from a low voltage. Obviously, at the initial stage, the energy converted from the mechanical domain should be used for generation of the high voltage biasing, and only after the biasing is established, the converted energy can be provided to the load with a high power due to a high voltage biasing. We call the mode in which the conditioning circuit accumulate its energy in order to increase its biasing "the auto-accumulative" or "self-increasing" mode.

The circuits presented in the next sections are free from the drawback of the continuous conditioning circuit, and all provide the possibility to accumulate the converted energy on a storage capacitor, which then can be used for the load supply.

8.3 Conditioning Circuits Implementing Triangular QV Cycles

Two triangular QV cycles are known for conditioning of capacitive transducers in energy converters: a constant voltage QV cycle (OAB in Fig. 8.1), and a constant charge QV cycle (OCB in Fig. 8.1). The two cycles have a very important common property: a discontinuity in time of the voltage on the variable capacitors. It can be obvious from their QV diagrams: the latter include paths on segments corresponding to a fixed value of the transducer capacitances (the lines OA, OC, and OB). Since the transducer capacitance varies in a continuous way (for instance, following the law given by (8.7)), these lines correspond to instantaneous, and hence, discontinuous variation of the charge and of the voltage. That is physically impossible; in practice, an implementation of such an operation supposes a *fast* variation of the transducer voltage, in comparison with the evolution speed of $C_t(t)$. It can be shown that only inductive DC–DC conversion allows a very fast voltage change on a capacitor without substantial loss of energy. As a consequence, conditioning circuits implementing triangular QV cycles need inductive DC–DC conversion stages synchronized with the variation of the variable capacitance.

8.3.1 Constant Voltage Conditioning Circuit

The constant voltage QV cycle was discussed in Sect. 4.1.5. We showed that the constant voltage QV cycle provides the best energy yield achievable with a given transducer having a given dynamics, supporting a given maximum voltage.

In this subsection, we propose a discussion about how to implement practically a circuit achieving a constant voltage QV cycle on a capacitive transducer. We highlight fundamental difficulties of implementation, and explain the reasons of very limited use of the constant voltage scenario and as a consequence, the motivation for exploring alternative suboptimal solutions.

We summarize the scenario corresponding to the constant voltage QV cycle, and propose a corresponding electrical network (Fig. 8.7). We suppose that a voltage source V_{max} is available. This can be a large fixed capacitance pre-charged to this voltage. At the beginning of the cycle, C_t is discharged, SW is OFF.

(1) When C_t is at its maximum, the transducer is pre-charged to a voltage V_{max} and immediately after that the transducer is connected to the source of voltage V_{max}. Note that the transducer cannot be directly connected to the voltage source, if their voltages are not equal. The only way to avoid losses when charging C_t from 0 to V_{max} is to use an inductive DC–DC converter. This conversion is denoted "Conversion 1" in Fig. 8.7

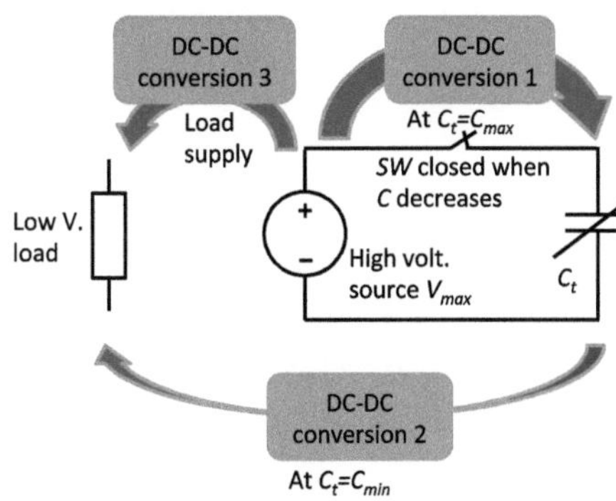

Fig. 8.7 Diagram illustrating the principle of operation of a conditioning circuit implementing a constant voltage QV cycle

(2) When C_t decreases from C_{max} to C_{min}, the transducer is connected to the voltage source (SW is ON). During this time, the voltage source receives charges and energy from C_t. Indeed, the variable capacitor loses charges $\Delta Q = (C_{max} - C_{min})V_{max}$, and the source receives the energy $\Delta Q V_{max}$.
(3) When C_t is at its minimum, the switch becomes OFF, the remaining charges of C_t are transferred to V_{max}, so that the voltage of C_t is zero. The transfer is achieved by an inductive DC–DC conversion denoted as "Conversion 2."
(4) Since at each cycle V_{res} receives more energy that it delivers, the average energy of V_{res} increases with time. If the voltage source is implemented as a large capacitor, its voltage increases with time: that is not desirable, since, by initial hypothesis, V_{max} is the maximum voltage allowed in the given technology. The circuit should contain a mechanism removing energy from V_{max}, for example, by supplying a load. Since the load is supplied by a low voltage, an inductive DC–DC converter should be implemented, which is denoted as "Conversion 3."

We note that this operation must be precisely synchronized with the variation of C_t. In this description we do not consider how to implement such a synchronization.

Summarizing, an implementation of a constant voltage scenario requires:

- a high voltage source,
- three inductive DC–DC conversions at each cycle,
- a switch connecting the transducer and the V_{max} source,
- precise synchronization of the DC–DC conversions with the variation of C_t.

The challenges of implementation of such a circuit are obvious, especially because of very small available energy for the operation of the control electronics.

An example of a successful implementation of a constant voltage conditioning circuit has been reported in [25], by a research group of Georgia Tech University.

8.3.2 Constant Charge Conditioning Circuits

The constant charge energy conversion scenario is similar to the constant voltage scenario, except when the transducer capacitance decreases, the circuit fixes the charge of the transducer, and not the voltage. The corresponding aspect of QV cycle is given in Fig. 8.1, cycle OCB. The energy converted by the cycle is given by the formula:

$$\Delta W = \frac{1}{2}V_0^2 C_{max}\left(\frac{C_{max}}{C_{min}} - 1\right) = \frac{1}{2}V_{max}^2 \frac{C_{min}}{C_{max}}(C_{max} - C_{min}). \tag{8.8}$$

where V_0 is the voltage on C_t when $C_t = C_{max}$.

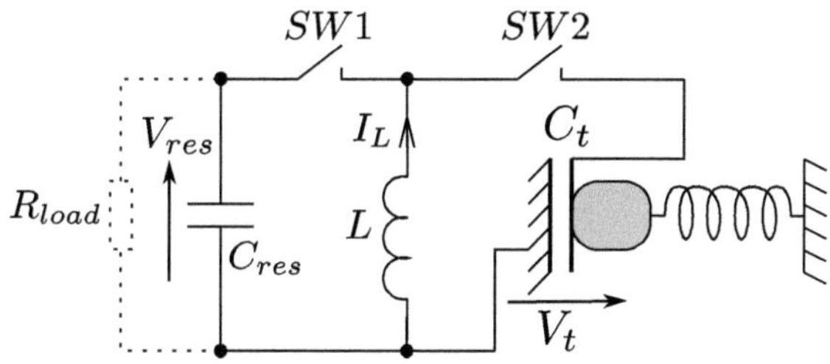

Fig. 8.8 Circuit implementing a constant-charge QV cycle

If one compares this formula with Eq. (4.19), it is evident that at the same V_{max}, C_{max} and C_{min}, the constant charge QV cycle converts C_{max}/C_{min} times less energy than the constant voltage QV cycle. This can also be seen from the plot of Fig. 8.1.

However, a conditioning circuit implementing a constant charge QV cycle is simpler than for a constant voltage QV cycle. A schematic of such a circuit is given in Fig. 8.8. The circuit is composed of a large capacitor C_{res} initially pre-charged to the voltage V_{res}, or of a voltage source generating the same voltage, of an inductor and of two externally controlled switches. The operation scenario is the following:

- when C_t is at its maximum and discharged, an inductive DC–DC converter transfers charges from a reservoir capacitor C_{res} to C_t, so that C_t has a charge Q_0 and a corresponding voltage $V_0 = Q_0 C_{max}$. Note that the initial voltage V_0 is determined by the timing of the DC–DC conversion (namely, by the ON time of the switch SW1), and not by the voltage V_{res}. An appropriate command of the switches SW1 and SW2 is required, as described in [18].
- both switches are off, the transducer capacitance reduces while the charge of the transducer remains constant. The internal electrical energy of C_t increases.
- when C_t is at its minimum, the DC-DC converter transfers the charge back from C_t to C_{res}, without energy losses. The capacitor C_{res} receives a greater energy that it initially has given to C_t.
- the transducer capacitance increases while its charge is zero (the switches are off), until it reaches the maximum, and the cycle repeats.

In this way, the conditioning circuit operates as a bidirectional DC–DC converter. Two advantages over the constant voltage circuit should be noted:

(1) The voltage V_{res} may be a low voltage, easy to generate. Nevertheless, the used DC–DC converter is able to generate a higher initial voltage V_0 on C_t, if it is required for optimal energy conversion by the transducer.
(2) The voltage V_{res} being low, it is compatible with the requirement of the load supply. In this way, one DC–DC conversion is removed (the Conversion 3 in Fig. 8.7).

A need of a precise synchronization of the DC–DC conversion with the motion of the mobile mass is a strong negative point of the constant-charge conditioning scenario, which makes the implementation of the control electronics complex and energy consuming. Study of practical implementation of this solution can be found in [5, 6, 16, 18].

8.4 Circuits Implementing Rectangular QV Cycles

In this section, we propose to study a class of circuits implementing rectangular QV cycle (DFBE in Fig. 8.1). We present here two subfamilies of such circuits: one based on a charge pump firstly proposed by Roundy et al. [22], and one based on the Bennet's doubler [3]. Other circuits which implement such a QV cycle are full wave and half wave diode rectifiers loaded by a reservoir capacitor. They are used with transducer biased by electret layer [20], and are not discussed in this book.

Priorly, to study the circuit topologies, we propose a discussion on the common properties of these circuits, which are due to the geometry of the implemented QV cycle.

8.4.1 Study of the Rectangular QV Cycle

Let us consider a rectangular QV cycle having the following parameters related to the cycle geometry (Fig. 8.9):

- The extreme voltages of the cycle which we call V_{res}, V_{st}, such as $V_{res} < V_{st}$
- The extreme charges of the cycle which we call Q_1, Q_2, such as $Q_1 > Q_2$

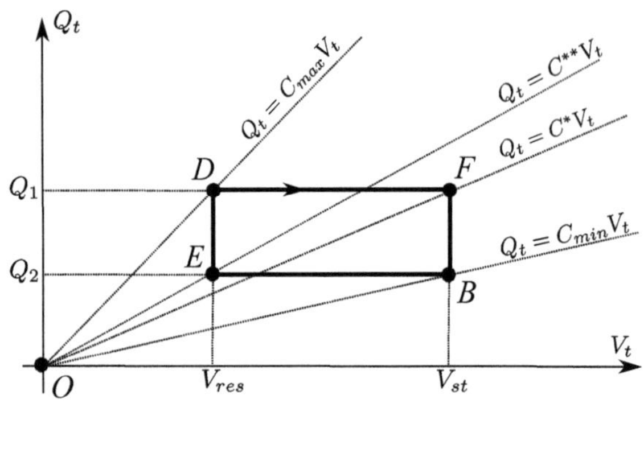

Fig. 8.9 Geometry of an ideal rectangular QV cycle

- The four values of the transducer capacitance corresponding to four points of the rectangle[2]:
  $C_{max} = Q_1/V_{res}$ (corresponds to the point D)
  $C_{min} = Q_2/V_{st}$ (corresponds to the point B)
  $C^* = Q_1/V_{st}$ (corresponds to the point F)
  $C^{**} = Q_2/V_{res}$ (corresponds to the point E)

If a capacitive transducer is biased so as to follow the electrical state defined by such a cycle, the energy converted in one cycle is easily calculated as the area of the cycle

$$\Delta W = (V_{st} - V_{res})(Q_1 - Q_2) = V_{res}^2 C_{max}\left(\frac{V_{st}}{V_{res}} - 1\right)\left(1 - \frac{V_{st}/V_{res}}{C_{max}/C_{min}}\right) \quad (8.9)$$

To have a physical meaning, this quantity must be positive. Hence, it is required that

$$\frac{C_{max}}{C_{min}} \geq \frac{V_{st}}{V_{res}}. \quad (8.10)$$

The expression ΔW is a quadratic form of the variables V_{res}, V_{st}. As a consequence, if one of these voltages is fixed, there is an optimal value of other voltage maximizing ΔW. Also, if there is some limitation applied on the voltages V_{res} and V_{st} (for instance, a maximum allowed value), there is an optimal ratio V_{res}/V_{st}, which is a function of the C_{max}/C_{min} ratio [13].

How practically a rectangular QV cycle can be implemented? It was said in Chap. 1, that a conditioning circuit can be seen as a dipole (Fig. 1.3). We now propose a discussion on main properties of the dipole implementing a conditioning cycle with a rectangular QV cycle, and we define its current-voltage characteristic.

By analyzing the QV diagram (8.9), it can be seen that when $V_{res} < V_t < V_{st}$, the transducer current is zero (since $Q_t = const$). We conclude, that at this operation stage, the conditioning circuit presents a high impedance (an open circuit).

As the transducer voltage reaches V_{st} *when the transducer capacitance decreases* (the segment DF), the voltage of the transducer is fixed to V_{st} (the segment FB), i.e., the conditioning circuit presents a voltage source V_{st}. Note that the current of transducer is

$$i_t = \frac{dQ_t}{dt} = V_{st}\frac{dC_t}{dt} < 0 \quad (8.11)$$

since $C_t(t)$ decreases. As a consequence, the transducer gives its charges (and hence its energy) to the conditioning circuit.

---

[2]Each point (V, Q) on a QV diagram defines an unique value of the variable capacitor given by Q/V, otherwise, by the slope of the line connecting the point (V, Q) and the origin.

A similar process happens when the transducer capacitance increases and its voltage reduces to V_{res} (the segment BE), while the capacitance continues to increase. In this case, the conditioning circuit behaves as a voltage source V_{res} (the segment ED). In this case

$$i_t = \frac{\mathrm{d}Q_t}{\mathrm{d}t} = V_{res}\frac{\mathrm{d}C_t}{\mathrm{d}t} > 0 \tag{8.12}$$

since C_t increases, and as a consequence, the transducer takes charges and energy from the conditioning circuit.

We can define the current-voltage characteristic (relating i_t and V_t) of the conditioning circuit as follows:

- if $V_{res} < V_t < V_{st}$, $i_t = 0$; (open circuit)
- if $i_t > 0$, $V_t = V_{res}$; (voltage source V_{res})
- if $i_t < 0$, $V_t = V_{st}$; (voltage source V_{st})

Figure 8.10a presents the plot of the corresponding current-voltage characteristic. It can be seen that such a characteristic corresponds to a voltage limiter implemented by the network given in Fig. 8.10b, often used for the ESD protection of integrated circuits, and known as "voltage limiter." The use of this circuit for implementation of a rectangular QV cycle in capacitive energy harvesters was first proposed by Roundy [22]. In the literature addressing the energy harvesting, this circuit is usually drawn as in Fig. 8.11a, and is called "charge pump." Indeed, as follows from our discussion, when this circuit is connected to a variable capacitor with periodic variation of the

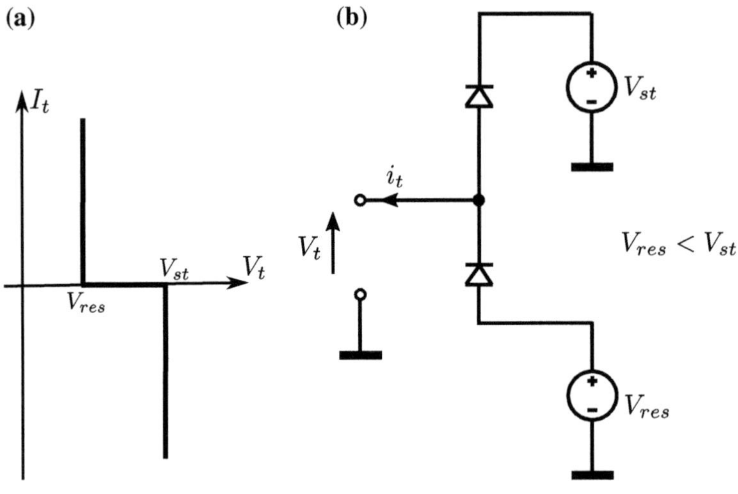

Fig. 8.10 Implementation of an ideal rectangular QV cycle: **a** Current-voltage characteristic of a dipole which, when connected to a variable capacitor, implements a rectangular QV cycle, **b** An electrical network having such a characteristic

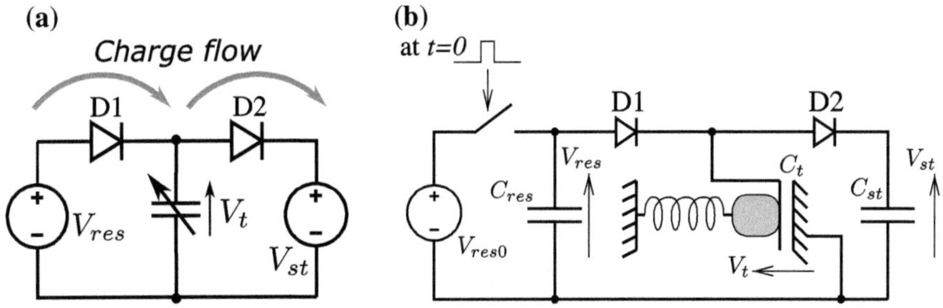

Fig. 8.11 Schematic of a charge pump implementing a rectangular QV cycle: **a** ideal charge pump, **b** practical implementation

capacitance, the low voltage source V_{res} gives charges, whereas the high voltage source V_{st} receives charges. In such a way, the circuit creates a charge flow from a low potential (V_{res}) to a high potential (V_{st}). The additional energy necessary for such an operation comes from the mechanical domain, thanks to the variation of the transducer capacitor.

The advantage of this circuit with comparison to a constant voltage or constant charge circuit is obvious: the automatic synchronization of the phases of the biasing with the variation of the transducer capacitance, without any need of external control. However, practical use of the charge pump in capacitive energy harvesters requires to answer the following questions:

- How to generate the voltages V_{res} and V_{st},
- How to transfer periodically the energy gained by the source V_{st} to some low voltage storage device (a large capacitor or a battery).

In this section, we consider two families of circuit implementing rectangular QV cycle solving these problems.

8.4.2 Practical Implementation of the Charge Pump

An implementation of the charge pump close to the described idealized circuit is obtained with the network of Fig. 8.11b. The voltage sources are implemented by charged fixed capacitances of large values. The values of C_{res} and C_{st} should be chosen so that

$$C_{max} \ll C_{st} \ll C_{res}. \tag{8.13}$$

C_{res} capacitor is initially charged to a voltage $V_{res} = V_{res0}$, and so are C_{st} and C_t capacitors, because of the charge sharing through the diodes D1 and D2.

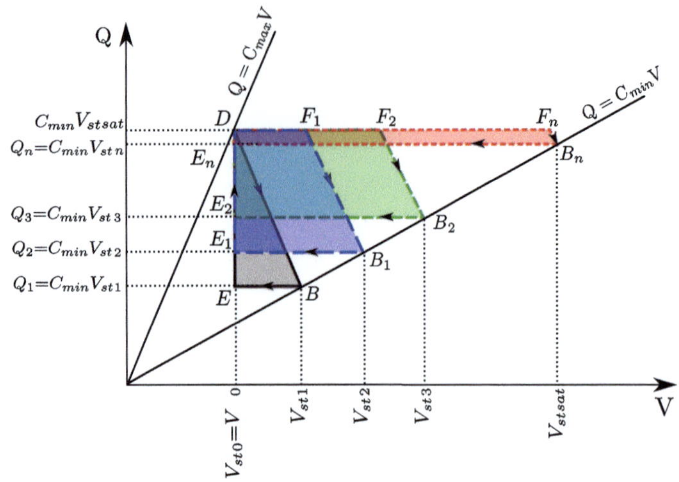

Fig. 8.12 Succession of charge-voltage cycles as the charge pump operates starting from the state when $V_{st} = V_{res}$: the first cycle DBE, the subsequent cycles $DF_1B_1E_1, DF_2B_2E_2, \ldots, DF_nB_nE_n$. If $n \to \infty$, the cycle is degenerated to a line passing through the points D, F_1, $F_2 \ldots F_n$

By initially precharging the circuit to a voltage V_{res0}, we implement the configuration of Fig. 8.11a with $V_{res} = V_{st}$, in which the converted energy per cycle is zero, according to (8.9). However, since the V_{st} voltage source is implemented by a capacitor C_{st}, the charge transfer from C_{res} to C_{st} during the first cycle slightly increases the voltage across C_{st}. Hence, the corresponding QV cycle is actually triangular, as shown in Fig. 8.12, cycle DBE. The next cycle starts in a configuration with a slightly larger V_{st}, yielding a trapezoidal QV cycle. After each cycle, V_{st} increases, so that all further cycles are different. It can be shown that the slope of the line DB, $F_1B_1, \ldots, F_nB_n$ is $-C_{st}$ (steeper than C_{max}, but exaggerated in the figure): the shape of the cycles is close to rectangular, as $C_{st} \gg C_{max}$. The QV trajectory follows a succession of cycles, starting from $V_{st} \approx V_{res}$ and asymptotically approaching a degenerated cycle represented by a trapeze with a zero area (a horizontal line) corresponding to the V_{st} voltage given by:

$$V_{st\ sat} = V_{res}C_{max}/C_{min}. \tag{8.14}$$

This value is called the "saturation voltage" of the charge pump.

As a result of the operation of this circuit, the capacitor C_{st} receives energy from C_{res}, and in addition, it receives energy converted from the mechanical domain (the harvested energy). At every QV cycle of the charge pump operation, the overall electrical energy of the circuit increases. In what follows, we present a quantitative analysis of this process.

The voltage V_{st} at the end of the nth cycle evolves following the recursive equation [26]:

$$V_{st\,n} = \alpha V_{st\,n-1} + \beta, \tag{8.15}$$

where

$$\alpha = \frac{C_{st}}{C_{st} + C_{min}}, \text{ and } \beta = V_{res}\frac{C_{max}}{C_{st} + C_{min}}. \tag{8.16}$$

The solution of the discrete equation (8.15) is

$$V_{st\,n} = \left(V_{st0} - V_{res}\frac{C_{max}}{C_{min}}\right)\left(\frac{C_{st}}{C_{st} + C_{min}}\right)^n + V_{res}\frac{C_{max}}{C_{min}}, \tag{8.17}$$

where V_{st0} is the voltage on C_{st} capacitor at the zeroth cycle. In this example, it is assumed $V_{st0} = V_{res0}$.

Note that if C_t variations are periodic, the variable n is the operating time of the charge pump divided by the period of C_t variation, T_e. From (8.17) it can be derived that the asymptotic value of $V_{st\,n}$ as $n \to \infty$ is $V_{st\,sat}$ is given by (8.14).

As V_{st} increases, the voltage V_{res} slowly decreases, as the amount of charges on C_{res} and C_{st} is constant (as one neglects the charges on C_t). The evolution of V_n is given by:

$$V_{res\,n} \approx [\text{if } C_{st} \gg C_{max}] \approx V_{res0} - \frac{C_{st}}{C_{res}} \cdot (V_{st\,n} - V_{res0}) \approx [\text{if } C_{res} \gg C_{st}] \approx V_{res0}. \tag{8.18}$$

An example of the evolution of V_{res} and V_{st} is given in Fig. 8.13a, where the charge pump starts from the state $V_{res} = V_{st} = V_t = V_{res0}$. From formula (8.15) and (8.16) it can be seen that if $C_{st} \gg C_{max}$, the increment of V_{st} during every cycle is small, and the charge pump runs through cycles for all possible V_{st}, from V_{res0} to the saturation value $V_{st\,sat}$.

8.4.3 Evolution of the Harvested Energy

Neglecting the energy in the transducer capacitance since it is small, and supposing that the C_{st} voltage is initially V_{res0}, the energy harvested during the n first C_t variation periods is stored in the capacitors C_{res} and C_{st}, and is given as

$$W_n = \frac{1}{2}C_{st}V_{st\,n}^2 + \frac{1}{2}C_{res}V_{res\,n}^2 - \frac{1}{2}(C_{res} + C_{st})V_{res0}^2, \tag{8.19}$$

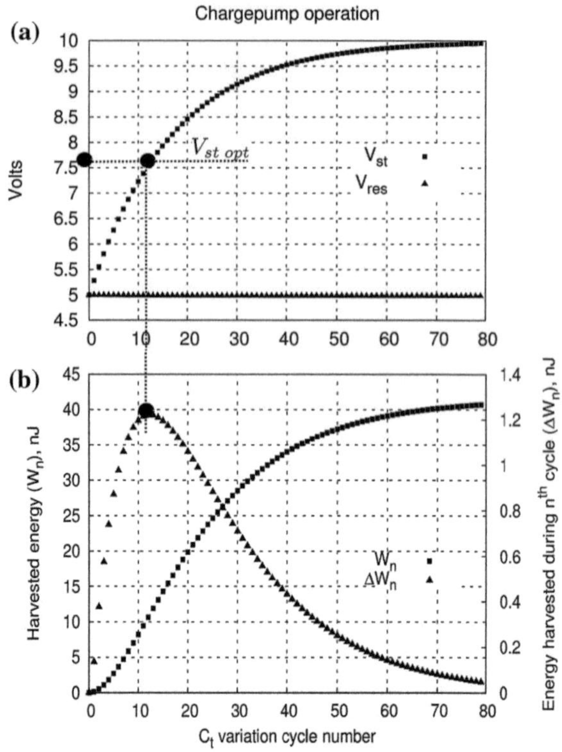

Fig. 8.13 Evolution of the electrical state of the charge pump without flyback. **a** Evolution of V_{st} and V_{res} voltages, **b** evolution of the total converted energy (W_n) and of the energy converted at one cycle (ΔW_n). The simulation was done with $C_{min} = 200\,\text{pF}$, $C_{max} = 400\,\text{pF}$, $C_{st} = 3.3\,\text{nF}$, $C_{res} = 10\,\mu\text{F}$, frequency of C_t variation is 300 Hz

where $V_{res\,n}$ is the C_{res} voltage after n vibration cycles and V_{res0} is the initial voltage of C_{res}. At the same time, the sum of charges of C_{st} and C_{res} is constant, so, the following relation holds (neglecting the charge shared with C_t):

$$C_{st}V_{res0} + C_{res}V_{res0} = C_{st}V_{st\,n} + C_{res}V_{res\,n}. \tag{8.20}$$

Using (8.20), Eq. (8.19) can be simplified

$$W_n = \frac{C_{st}C_{res}}{C_{st} + C_{res}}(V_{st\,n} - V_{res\,n})^2/2. \tag{8.21}$$

This equation suggests that the harvested energy is stored in the capacitor composed of C_{st} and C_{res} connected in series. This formula can be further simplified if $C_{st} \ll C_{res}$ and if C_{res} is large, so that $V_{res\,n} \approx V_{res0}$. In this case, we can write

$$W_n \approx C_{st}(V_{st\,n} - V_{res0})^2/2. \tag{8.22}$$

This equation suggests that the converted energy is mainly stored in the C_{st} capacitor.

The value

$$\Delta W_n = W_n - W_{n-1}, \tag{8.23}$$

gives the energy converted during one cycle. For the case when $C_{st} \gg C_{max}$, ΔW_n is equal to ΔW given by Eq. (8.9), if in (8.9) V_{st} and V_{res} are replaced by V_{res0} and $V_{st\,n}$ corresponding to the cycle.

In the formula (8.22) expressing the energy converted from the state where $V_{st} = V_{res0}$, there is no explicit dependence on discrete time n (the cycle number). If $C_{st} \gg C_{max}$, the variation of the voltage V_{st} is negligible at the time scale of one cycle. For this reason, we can state that V_{st} evolves continuously, and consider it as "macroscopic state parameters" defining the energy of the charge pump. It allows us to remove the indices n in the Eq. (8.22) and to define the total harvested energy W as

$$W(V_{res0}, V_{st}) = C_{st}(V_{st} - V_{res0})^2/2. \tag{8.24}$$

If, during the operation of the charge pump, the V_{st} voltage increases from some V_1 to some V_2, the corresponding converted energy is given by

$$\begin{aligned} \Delta W(V_{res0}, V_1, V_2) &= W(V_{res0}, V_2) - W(V_{res0}, V_1) = \\ &C_{st}(V_2 - V_1)(V_2 + V_1 - V_{res0})/2. \end{aligned} \tag{8.25}$$

Figure 8.13b presents the plots characterizing the evolution of the energy in the charge pump as a function of the cycle number. During each cycle n, the total energy (square dots) increases by the value given by (8.9) calculated for the V_{st} resulting the previous cycle (triangular dots). Only a few cycles convert a significant amount of energy: those corresponding to V_{st} values close to the optimal $V_{st\,opt}$, (cf. the analysis of Sect. 8.4.1 for fixed V_{res}). This value is situated in-between the extreme values of V_{st}, and a freely running charge pump finishes by entering into the saturation mode, in which $V_{st} \approx V_{st\,sat}$ and $\Delta W_n \approx 0$.

8.4.4 Shortcomings of the Single Charge Pump

Because of a dynamic evolution of V_{st} voltage leading to a saturation, the charge pump alone is not useful for capacitive VEHs. Practical conditioning circuits should be able to (i) ensure a sustainable energy conversion by fixing the V_{st}/V_{res} ratio to the optimal value as mentioned in Sect. 8.4.1, (ii) accumulate the converted energy. These two very important points are discussed in this subsection.

Need for a Flyback

The plot ΔW_n in Fig. 8.13b can be seen as the average power converted in a cycle. The goal of a harvester is to permanently maintain the converted power close to a

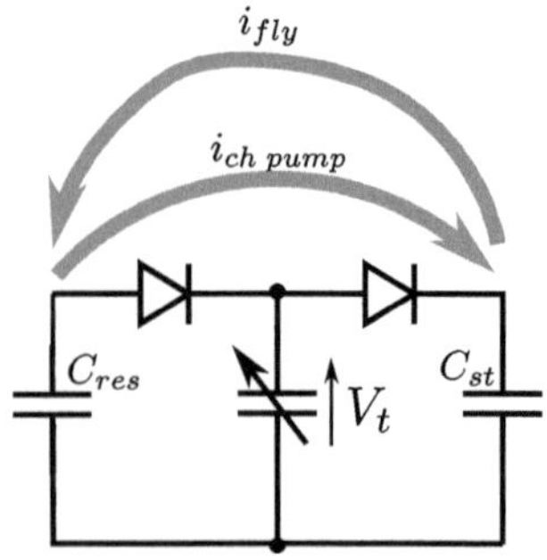

Fig. 8.14 Principle of operation of a charge pump with a mechanism of return of charges from C_{res} to C_{st}, called "flyback"

maximum level. For this reason, a charge pump needs periodic *flyback*: a mechanism returning some of charges from C_{st} to C_{res} so as to keep V_{st} close to the optimal value and to prevent the charge pump from saturation. Possible implementations of the flyback will be discussed in Sect. 8.5.

The action of the charge pump can be seen as the generation of an average current flowing from a low potential (V_{res}) to a higher potential V_{st} (Fig. 8.14). This current is roughly equal to

$$i_{ch\ pump} = \frac{Q_1 - Q_2}{T_e} = \frac{V_{res}C_{max} - V_{st}C_{min}}{T_e}. \tag{8.26}$$

The flyback must create an opposite current from C_{st} to C_{res} i_{fly}, preventing the accumulation of charges on C_{st}, and so, avoiding the charge pump saturation. This current, flowing from the high-to-low potential, *removes* the electrical energy from the charge pump, and in a practical application, this energy should be transferred to a useful load. In the steady-state mode, the rate at which the energy is removed from the charge pump is equal to the power of the electromechanical conversion.

Auto-Increasing of the Internal Energy

From (8.9) and further analysis of Sect. 8.4.2, it can be seen that the energy converted by a charge pump is proportional to the square of the voltage V_{st} and V_{res}, i.e., proportional to the internal energy of the circuit in Fig. 8.11b. Practical studies highlight that up to tens of volts may be needed for optimal operation [1]. However, the initial charge can only be obtained by a low voltage battery existing in the system, or by a complementary piezoelectric VEH as proposed in [11]. In both cases, the initial voltage generated in this way will certainly be low (few volts).

As a consequence, the conditioning circuit should be able to use a part of the converted energy in order to *increase* its biasing: its internal energy represented by the voltage on the largest capacitor, V_{res}. This is the "accumulative" or "self-increasing" mode discussed in Sect. 8.4.4.

The auto-increasing capability is not offered by the basic charge pump.

8.5 Circuits Derived from the Primitive Charge Pump

In this section, we discuss more sophisticated conditioning circuits which are derived from the architecture of the basic charge pump considered above. All of them implement a rectangular QV cycle. In all these circuits, the basic topology of the charge pump presented in Fig. 8.11 is complemented by networks allowing a return of the charges, so to allow a steady operation providing a fixed ratio V_{st}/V_{res}. Here, we present a short discussion of these solutions.

8.5.1 *Resistive Flyback*

The simplest way to implement a flyback is to connect a load between C_{res} and C_{st} (Fig. 8.15), originally proposed in [11]. Here, the load is represented with a resistance R_{load} whose current is

$$i_{fly} = \frac{V_{st} - V_{res}}{R_{load}}. \tag{8.27}$$

A simplified analysis may be done by analyzing independently the charge flows of the charge pump and of the resistor, and then by superimposing them. The point at which both currents are equal to each other ($i_{ch\ pump} = i_{fly}$) corresponds to steady-state operation. Although it is an approximation based on the averaging method [19], it gives good analytical results.

In the steady state, the average V_{st} voltage on a period T_e is the same for all cycles. The stability of this mode can be easily proven by supposing a perturbation which yields, for example, $i_{ch\ pump} < i_{fly}$. In this case, the load current consumption outweighs the current due to the charge generation by the pump, and V_{st} decreases. This results in decreasing of i_{fly}, and so in reducing the initial perturbation.

Under these considerations, the average steady-state voltage V_{st} with resistive flyback can be predetermined by equating (8.26) and (8.27). Conversely, the value of R_{load} yielding a desired V_{st} (for example, that yielding a maximum ΔW) can be calculated.

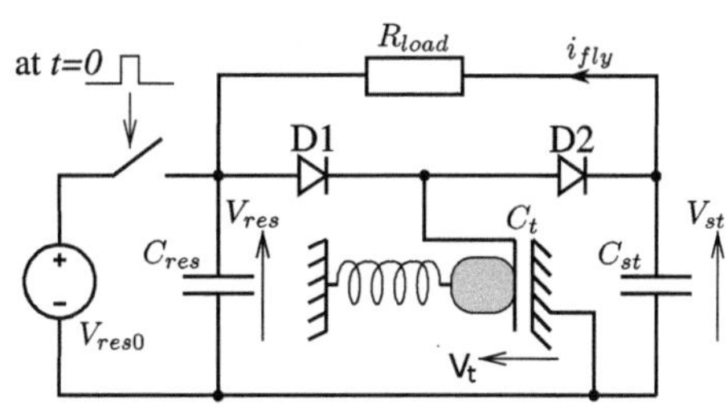

Fig. 8.15 Charge pump with resistive flyback

The resistive flyback configuration provides a sustainable energy conversion by preventing the charge pump from the saturation. This is an efficient solution from the point of view of the simplicity of implementation. However, it does not allow an auto-increasing of its internal energy (cf. Sect. 8.4.4), and for this reason, is not usable in practical applications.

8.5.2 *Inductive Flyback*

The inductive flyback is a Buck DC–DC converter, transferring energy from C_{st} to C_{res} (Fig. 8.16a). The flyback operates in two steps

(i) When V_{st} is high so that the efficiency of the charge pump decreases, some external control device (cf. the explanation below) activates the switch SW, and the current in the inductor L starts to increase. This results in: (i) charges being transferred from C_{st} to C_{res}, so that V_{st} decreases, (ii) the energy previously generated by the charge pump on C_{st} is transferred to L.
(ii) As the magnitude of V_{st} decreases, it becomes close to V_{res}, and the same external device cuts the switch, so that the inductor discharges through the flyback diode $D3$ onto C_{res}. This results in a transfer of the inductor energy to C_{res}, and an *increase* of the C_{res} voltage, since the diode $D3$ is connected to the ground. In this way, the converted energy is used to extract charges from the ground and to inject new charges into the charge pump.

It can be seen that C_{res} receives charges twice: in the first phase, it receives the charges previously given to C_{st} during the pump operation, and in a second phase, it receives new charges from the ground. In this way, C_{res} receives the energy it has given to C_{st} during the operation of the charge pump, and it receives the energy converted from the mechanical domain. If such an operation is cyclic, the average voltage and energy of C_{res} will grow. In this way, two problems are solved: (i) even if

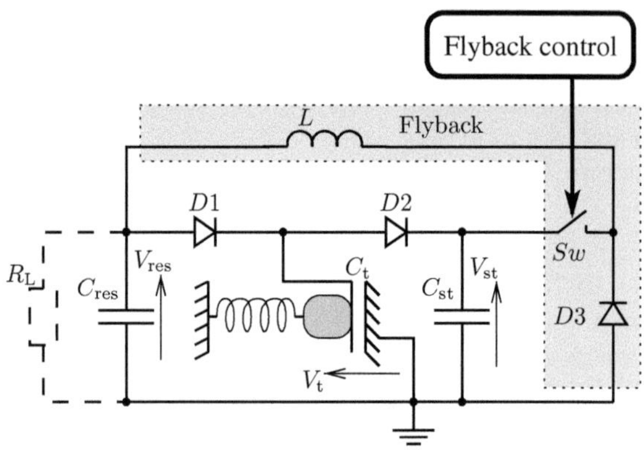

Fig. 8.16 Conditioning circuit with inductive flyback. **a** The architecture of the circuit: the basic charge pump (as presented in Sect. 8.4.2) and the inductive flyback (*grey* background)

the capacitors are leaky, the losses are compensated by the charges extracted from the ground, (ii) a large capacitor C_{res} contains the harvested energy available for the load (represented by dotted lines in Fig. 8.16a). If no load is present, C_{res} accumulates the converted energy, and so implements an "auto-increasing" or "accumulative" mode.

The advantage of the charge pump with inductive flyback is the possibility of a precise control of the QV cycle corresponding to the energy conversion. This is achieved by a modulation of the average V_{st} voltage so as to remain in the optimal mode for energy conversion.

A drawback of the circuit is the need of an external command for the switch. However, this command is synchronized not with the mobile mass vibrations, but with the evolution speed of the voltage V_{st}: first, a voltage can easily be measured, and second, the evolution is low compared to the frequency of the capacitance variation. Hence, the cost of such a command is smaller than for circuits implementing constant voltage or constant charge QV cycles (see Sects. 8.3.1 and 8.3.2).

Practical implementation of adaptive architecture in Fig. 8.16 is challenged by the relatively low power available for the control circuitry, and by the need to manage high voltages in the conditioning circuit. Most studies have used an "old" CMOS technology supporting high voltages (e.g., 0.7 μm CMOS in [14]). The work [9] used a mixed high-voltage/low-voltage 0.35 μm CMOS technology, where high voltage circuits are used for the interface with the charge pump, and the processing is done by low voltage parts. This allows a minimization of the power overhead of the control electronics. Implementation of a high-side high voltage MOS switch for the flyback control with a low voltage control interface represented a particular difficulty. An advanced study of a fully integrated CMOS implementation of the control architecture, as in Fig. 8.16, was presented in [9, 10]. To date, implementation of a working IC prototype of this architecture is still a subject of ongoing work in several research groups.

8.6 Conditioning Circuits Based on the Bennet's Doubler

8.6.1 Introduction of the Principle

Introduced at the end of the eighteenth century [2], the Bennet's electricity doubler is one of the first devices allowing a measure of a voltage, by amplifying the induced charge. It has recently been proposed by de Queiroz [3] for capacitive kinetic energy harvesters. Since then, several works have further developed this concept, adapting it to microscale VEH [7, 8, 15, 21].

The Bennet's doubler is a switched capacitor network, whose goal is a steady separation of the electrical charges and an accumulation of the separated charges in the capacitors. A steady accumulation of the charges may lead to very high (the-

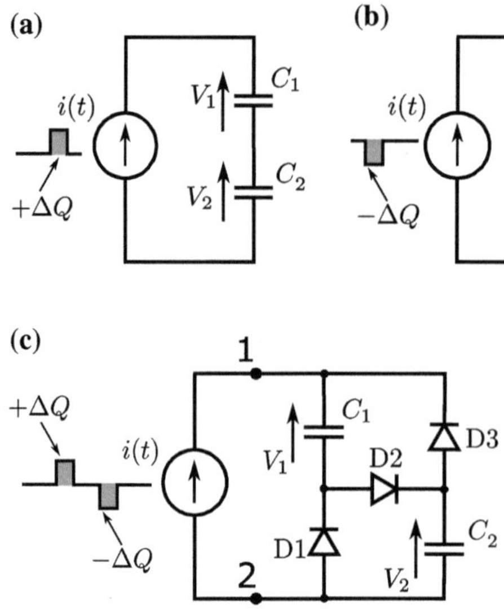

Fig. 8.17 Principle of the charge doubling in Bennet's doubler: **a** series configuration of capacitors when the current is positive, **b** parallel configuration of capacitors when the current is negative, **c** a diode network allowing an automatic switching between the configurations, depending on the sign of the current

oretically unlimited) voltages. Before analyzing conditioning circuit based on the Bennet's doubler, we propose an analysis of a more simple network widely used as a serial-parallel switched capacitors DC–DC converters. The purpose of the proposed discussion is to help the reader to acquire an intuitive understanding of the operation of the Bennet's doubler based conditioning circuit, whose more detailed analysis will be given in Sects. 8.6.2 and 8.6.3.

Consider an AC current source providing at the first half period a charge ΔQ, and pulling at the second half period the same charge. If during the first half period such a source is connected to a series capacitive network composed of two identical capacitors initially charged to identical voltages (Fig. 8.17a, $V_1 = V_2$), each capacitor receives a charge ΔQ. However, if during the second half period, the network topology is reconfigured and the previously charged capacitors are connected in parallel (Fig. 8.17b), the current source takes a charge $\Delta Q/2$ from each capacitors. At the end, each capacitor receives a charge $\Delta Q/2$, which is added to the preexisting charges of the capacitors. The cyclic reconfiguration of the circuit topology from series to parallel is usually achieved by externally controlled switches. In this case, the circuit can be seen as an AC-DC converter, since the AC current with finite maximum and minimum values is converted to a DC voltage, whose value is controlled by the number of operation cycles.

In order to achieve an automatic cyclic reconfiguration of the circuit between the topologies of Fig. 8.17a and b (i.e., without use of an external control), one may benefit from the fact that in each topology the current has a specific direction. Figure 8.17c presents a topology in which this reconfiguration is automatic thanks to a diode network. When the current is positive, the diode D2 is ON (series configuration), and when the current is negative, the diodes D1 and D3 are ON (parallel configuration). The use of the diodes introduces a new feature to the network: the voltages across C_1 and C_2 may be different (for example, because of the initial charging). In this case, when the current is negative, all the current passes through the branch with the maximum voltage, and the current source generates a voltage $max(V_1, V_2)$.

The requirement applied to the current source is to be able to generate a voltage[3] $V_1 + V_2$ during the positive current phase, and a voltage $\max(V_1, V_2)$ during the negative current phase. This is always possible for an *ideal* current source, which, by definition, can generates any voltage necessary to ensure the required current value. This is not the case of a real current source, and it leads to significant consequences when the current source is implemented by a transducer, as it will be discussed later in this chapter.

It can be noticed that on the time scale of one period, the network in Fig. 8.17c implements a voltage limiter with the characteristic as in Fig. 8.10a if C_1 and C_2 are large, and if ΔQ is small comparing to the charges of these capacitors. Indeed, in this case the variation of the charges and hence of the voltages on C_1 and C_2 are negligible at the time scale of one period, and the current source voltage is $\max(V_1, V_2)$ when the current is negative, and $V_1 + V_2$ when the current is positive. If, for some reason, the voltage applied to the terminals 1 and 2 of the network in Fig. 8.17c is in-between these limits, all diodes are blocked and the current is zero.

A similarity of the considered network (Fig. 8.17c) with a voltage limiter suggests that it can be used for implementation of a conditioning circuit achieving a rectangular QV cycle. Such a configuration is shown in Fig. 8.18, and was originally proposed by de Queiroz [3]. Its operation principle can be understood if one consider a functional similarity between a variable capacitor $C_t(t)$ and a current source. Consider a variable capacitor whose capacitance $C_t(t)$ reduces in time. When connected to a voltage source V, it generates a positive current $i = V\mathrm{d}C_t/\mathrm{d}t$: the similarity with a positive current source is obvious. When such a variable capacitor is connected to a open circuit, it generates an increasing voltage $Q_t/C(t)$, where Q_t is the charge of the transducer (which is constant in time, since the current is zero). If the C_t decrease is bounded by some C_{min}, the voltage increase is bounded by Q_t/C_{min}, and if $C_{min} \to \infty$, the voltage tends to become infinite. A current source connected to a open circuit generates an infinite voltage if the source is ideal, or a voltage limited by IR_S if the source is real, i.e., having an internal resistance R_S. There is hence a similarity in the

[3] We remind that an ideal current source generates in the external network connected to it a current of a given intensity. For that, it generates a *voltage* necessary to fix such a current. The value of this voltage is determined by the external circuits. For instance, if a resistance R is connected to a current source generating a current I, the source generates a voltage RI.

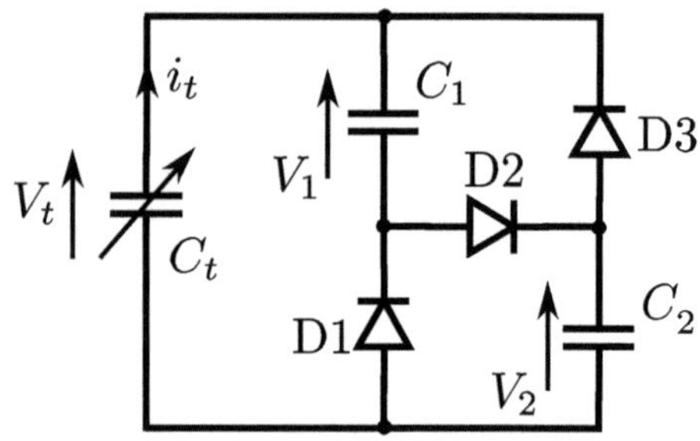

Fig. 8.18 Bennet's doubler used as a conditioning circuit for a capacitive VEH, as proposed in [4]

electrical behavior between a capacitor having an decreasing/increasing capacitance and a positive/negative current source correspondingly.

As a conclusion, a Bennett's doubler conditioning circuit is a series-parallel switched capacitors AC–DC converter, where the input AC current is generated by a variable capacitor. As the analysis presented in the next section will show, this circuit does not experience a saturation like the charge pump in Fig. 8.11b, and as a consequence, doesn't need any external control for steady generation of electricity.

8.6.2 Operation of a Bennet's Doubler in the Electrical Domain

We propose to consider the operation of the Bennet's doubler starting from a state at which one of the capacitors is initially charged to a voltage V_0, and the other capacitors are discharged. The Bennet's doubler will first experience a transient process during which the voltage of the second fixed capacitor will rise till V_0, and then, the circuit enters into the steady-state, in which the both voltages increase exponentially with time.

In order to illustrate the operation of the circuit, we performed an Eldo simulation of the circuit with the following configuration: $C_1 = 10\,\text{nF}$, $C_2 = 1\,\text{nF}$, the transducer is a gap closing transducer as presented in Sect. 4.1.1 with $d_0 = 50\,\mu\text{m}$ and $S = 10^{-4}\,\text{m}^2$. Capacitor C_1 is initially charged to $V_0 = 5\,\text{V}$. The mobile electrode of the transducer moves according to sinusoidal law, with amplitude of $X_0 = 30\,\mu\text{m}$ and frequency $f = 100$ Hz. In that way, the transducer capacitance C_t varies according to Eq. (8.7). From this equation we calculate $C_{min} = 11.05$ pF and $C_{max} = 44.2$ pF. The listing of the used Eldo model is given in Listing 8.1, and the time evolutions of V_1 and V_2 obtained by simulation are given in plots of Fig. 8.19. The plot Fig. 8.19a presents a long-term evolution of the voltages on capacitors C_1 and C_2, and the plot Fig. 8.19b presents a zoom of the time interval covering the transient and the beginning of the steady-state mode.

Listing 8.1 Bennet's doubler with variable capacitance: Eldo netlist

```
***** The first line is ignored in spice netlists******

.model diode1 d level=1

**** The parameter definition *****
*The transducer gap
.param d=50e-6
*The transducer area
.param S=1e-4
*Permittivity of vacuum
.param eps0=8.85e-12
*Simulation time
.param simtime=30
*Amplitude of the mobile plane motion
.param X0=30e-6
*Frequency of the mobile plane motion
.param f=100

*The voltage source defining the trajectory of the mobile electrode
Vx x_value 0 sin 0 X0 f
*The voltage controlled voltage source
*defining the variation of the transducer capacitance
ECt Ct_value 0 value={eps0*S/(d-v(x_value))}

*The transducer
Ct b3 0 value={v(Ct_value)}

*The two fixed capacitances C1 and C2, initially charged
*to 10V and to 0V respectively
C1 b1 0 10e-9 ic=10
C2 b3 b2 1e-9 ic=0

*The three diodes
db1 0 b2 diode1
db2 b2 b1 diode1
db3 b1 b3 diode1

*Transient simulation command,
*with option uic (Use Initial Conditions)
.tran {simtime} {simtime} uic
*Plotting commands
.plot V(b1) V(b2) V(b3)
.plot V(b3,b2)
.plot V(b1,0)
.plot V(Ct_value)
*Writing of waveforms to text files, for the plot drawing
.printfile tran v(b1) file=V2.txt start=0 stop={simtime} step=5e-6
.printfile tran v(b3, b2) file=V1.txt start=0 stop={simtime} step=5e-6
.printfile tran v(Ct_value) file=Ct.txt start=0 stop={simtime} step=5e
    -6
.printfile tran v(b3) file=Vt.txt start=0 stop={simtime} step=5e-6

*simulator options
.options hmax=10e-6
```

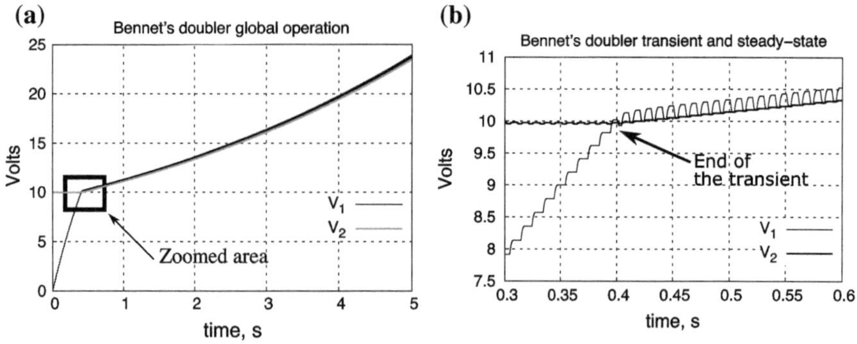

Fig. 8.19 Evolution of voltages in the Bennet's doubler: **a** long-term evolution of the voltages V_1 and V_2, **b** zoom on the time interval covering the end of the transient process and the beginning of the steady-state

Note that in this configuration the values of the fixed capacitors are much larger that the value of the maximum transducer capacitance. It means that the variation of the voltages on C_1 and C_2 at the time scale of one period are small comparing to the voltage variation on C_t.

The analysis of starts from $C_t = C_{max}$ at $t = 0$. We consider that thanks to the diode $D1$, the transducer is pre-charged to the same voltage as C_1. As C_t decreases, diode D_2 is on and both C_1 and C_2 receive charges from the transducer. In this way, the voltage on C_2 increases slightly. As C_t stars to decrease, the capacitor having the highest voltage becomes in parallel with the transducer. Since C_1 was initially charged to 5 V and C_2 experienced a small voltage increase, C_1 is the one becoming in parallel to C_1 through the diode D1, and C_1 gives the charges it previously received from C_t. At the end of this cycle, C_1 has the same voltage as in the beginning, and C_2 voltage increased.

The same process continues in the next cycles, till the voltages on the both fixed capacitors become equal. In the presented simulation, it happens at $t = 0.4$ s. Starting from that cycle, the network enters in the steady-state operation, in which the both voltages increase exponentially, and in average, they are equal. Since the diode network switch so that the voltage applied on the transducer is $max(V_1, V_2)$ and $V_1 + V_2$, the max-to-min ratio of the voltage on the transducer in the steady-state mode is 2. The voltage on C_1 displays a very low ripple, comparing with the voltage on C_2. This is because of the difference between the value of the capacitances C_1 and C_2: the charge variation is the approximately the same on both, but the voltage variation given by Q/C is greater on the smaller capacitor.

8.6.3 QV Cycle of the Bennet's Doubler and Approximated Analysis in Steady State

Figure 8.20a present the simulated plot of QV cycle that the Bennet's doubler with the presented configuration achieves in the steady state. The QV diagram is drawn for the cycle corresponding to the time instant 0.8 s (the simulation shown in Fig. 8.19). The vertical segment of the QV cycle correspond to the phases at which the transducer is connected to

- The both fixed capacitors in series (the right segment, D2 is on)
- The smallest of the fixed capacitor of the fixed capacitors (the lower left segment, only diode D3 is on)
- The both fixed capacitors in parallel (the upper left segment, both diodes D1 and D3 are on)

Each of this segment has a slope $1/C$, where C is the total equivalent capacitance of the capacitive network connected to the transducer by the diodes in each case. It can be noticed a non-infinite slope of the vertical segments of the QV cycle, however, the shape is very close to be rectangular. This is because the fixed capacitors are very large comparing to C_{max}. In order to provide an insight into the operation of the circuit, the same network was simulated with smaller capacitors C_1 and C_2 (Fig. 8.20b), whose values are now of the same order of magnitude as the transducer capacitance. It can clearly be seen that

- The three non-horizontal sections have different and non-infinite slopes,
- The cycle is not closed: the values of Q_t and V_t at $C_t = C_{max}$ at the beginning of the cycle are smaller that at the end of the cycle; this highlights the increase of the internal energy of the network.

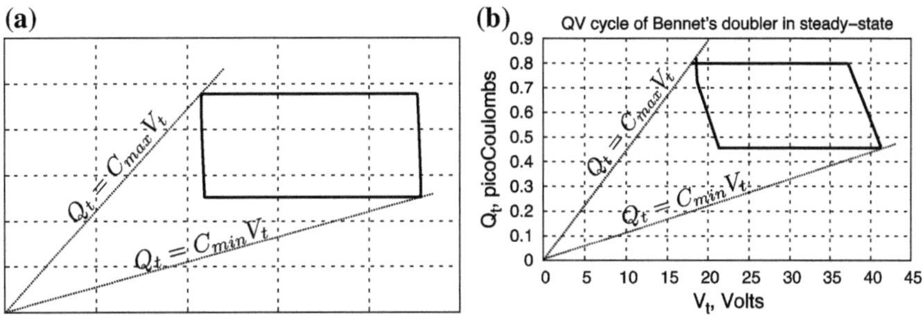

Fig. 8.20 QV diagram of Bennet's doubler obtained by simulation with the netlist given in Listing 8.1, with different values of capacitors C_1 and C_2: **a** $C_1 = 10\,\text{nF}$, $C_2 = 1\,\text{nF}$, as in the netlist given in Listing 8.1, **b** $C_1 = 500\,\text{pF}$, $C_2 = 100\,\text{pF}$. The corresponding C_{min} and C_{max} are 11.05 pF and 44.2 pF, respectively. It can be seen that when $\min(C_1, C_2) \gg C_{max}$, the cycle is very close to be rectangular (**a**). As the fixed capacitances are of the same order of magnitude as the transducer capacitance, the cycle is not exactly rectangular and highlight different stages of the circuit operation corresponding to different states of the diodes

Considering the QV cycle of the Bennet's doubler, it is possible to give an approximated expression of the energy converted in one cycle in the steady-state mode. If $C_{max} \ll C_1$ and $C_{max} \ll C_2$, the voltage on C_1 and C_2 may be considered fixed at the time scale of one cycle, and according to (8.9), we have:

$$\Delta W_{ss} \approx (2V - V)(VC_{max} - 2VC_{min}) = V^2 (C_{max} - 2C_{min}) \quad (8.28)$$

Here, V is the voltage on one of the capacitors C_1 and C_2: in the steady state these voltages are roughly equal.

The energy ΔW_{ss} is positive if $C_{max} - 2C_{min} > 0$. This is a necessary and sufficient condition allowing the Bennet's doubler to highlight an accumulative mode with exponentially increasing voltages on the fixed capacitors. If this condition is not fulfilled, the Bennet's doubler is still able to generate electricity (i.e., to convert energy) in the transient mode, as far as the following condition holds:

$$\frac{C_{max}}{C_{min}} > \frac{V_1 + V_2}{max(V_1 + V_2)} \quad (8.29)$$

In this case, the energy converted at each cycle is

$$\Delta W_{tran} \approx (V_1 + V_2 - max(V_1, V_2))(max(V_1, V_2)C_{max} - (V_1 + V_2)C_{min}) \quad (8.30)$$

As shows the plot Fig. 8.19a, in all this energy is accumulated in the capacitor with the smallest voltage, so that at some moment the condition (8.29) is not fulfilled since the smaller voltage increases. Then, the Bennet's doubler saturates, similarly with a classical charge pump of Fig. 8.11b, the voltages on C_1 and C_2 do not increase, the QV cycle is degenerated into a line and no energy conversion is achieved.

8.7 Dynamic behavior and Electromechanical Coupling of Rectangular QV Cycle Conditioning Circuits

Analysis of a conditioning circuit in the electrical domain assumes that the variation of the transducer capacitance is independent from the electrical state of the transducer. As we mentioned in Chap. 4, in practice, this does not hold for real electrostatic transducers connected to microscale mechanical resonators. Indeed, as it was said in Sect. 4.1.6, the electromechanical coupling impacts the dynamic of the mobile mass vibrations. As a consequence, the extreme values of the transducer capacitance C_{max} and C_{min} can also vary with time. As show on the example of charge pump and Bennet's doubler conditioning circuits, the electrical dynamic of the circuits depends on C_{max} and C_{min} which, in turn, depends on the electrical force generated by a dynamically biased transducer. Such a coupled behavior, to be described formally, requires advanced mathematical tools such as nonlinear differential equations, perturbation techniques, etc., which were introduced in Chap. 3. In this paragraph, we only present an example of how electromechanical coupling impacts the operation of a conditioning circuit, by studying the case of the Bennet's doubler.

We chose the example of Bennet's doubler since the difference between its operation in electrical domain and in a context where a real transducer/resonator are connected is flagrant. Indeed, without mechanical coupling, the output voltage of the Bennet's doubler (e.g., the voltage V_1 or V_2, Fig. 8.18) increases exponentially without any limit. If the variation of the capacitance is due to the motion of a mechanical system, one cannot expect an exponential evolution of the voltages and the energy converted at one cycle, because of the fundamental limit of the power which can be absorbed from the external vibrations, given by Eq. (3.92) from Chap. 3. The mechanism practically limiting the increase of the voltage is the presence of the electrostatic force generated transducer, which impacts the amplitude of the mobile mass vibrations so to reduce C_{max}/C_{min} and to limit the power converted by the transducer. Simulation and experiments highlight two possible behaviors of the system

- a "smooth" saturation. In this case, as the energy of the Bennet's doubler increases, the amplitude of the mobile mass decreases smoothly, and the ratio C_{max}/C_{min} asymptotically approaches 2. That stops the energy conversion by the circuit. Such a behavior was experimentally observed in study [7]. Figure 8.21 presents simulation curves highlighting such a behavior. As the voltages V_1 and V_2 grow, the amplitude of the mobile mass vibration changes in a sophisticated manner: it increases first, reaches a peak and decreases. This mechanical dynamics impacts the variation of the transducer's capacitance. The evolution of the voltages is, in turn, impacted by the C_{max}/C_{min} ratio evolution: the rate is slow at the beginning, increases near the peak of the amplitude, decreases after the peak, and becomes zero as C_{max}/C_{min} approaches 2 after time $t = 10$ s. The presence of a peak of amplitude at 2.4 s reminds a resonance phenomenon: indeed, we obtain a similar amplitude envelope if a linear resonator is excited by a sinusoidal signal with a frequency sweep. In our model, the frequency of the external acceleration is fixed, but the resonance frequency of the resonator is continuously modified by the voltages V_1 and V_2 responsible for the QV cycle definition. The drift of the resonance frequency is clearly observed in the plot for the displacement x.
- an "abrupt" saturation. In this case, the increase of the output voltage does not produce the reduction of the amplitude of the mobile mass, but creates conditions for a pull-in phenomenon proper to the gap closing transducer (cf. [23]). In this case, as the bias voltage of the circuit increases, the mobile mass sticks to the stoppers, and does not display any significant motion, so that C_{max}/C_{min} ratio is close to 1. This abruptly stops the increase of the output voltage. The resulting evolution is given in Fig. 8.22.

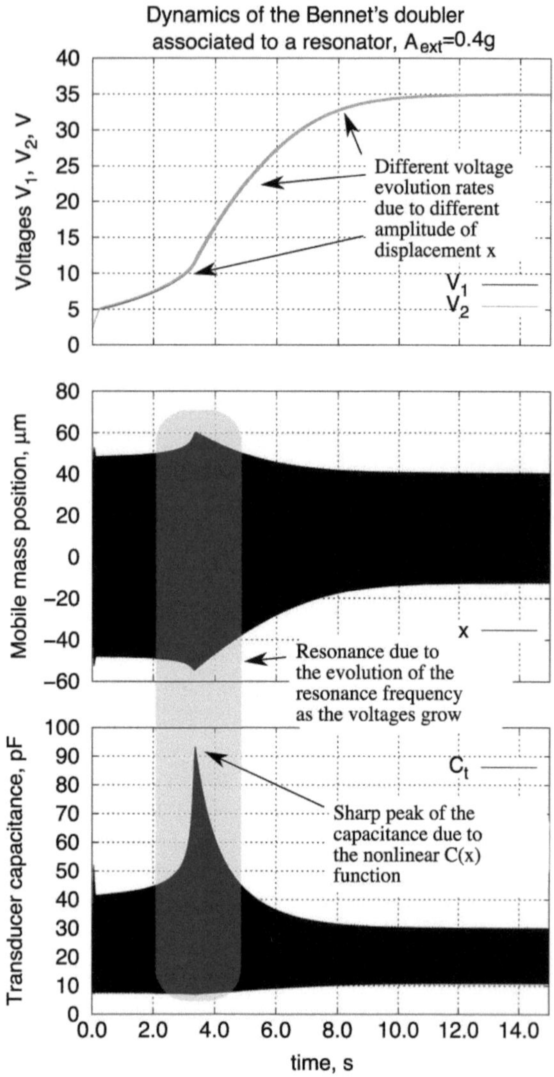

Fig. 8.21 Simulation results demonstrating a smooth saturation of a Bennet's doubler (Fig. 8.18), where the variation of the transducer capacitance is done by a gap closing transducer connected to a resonator. The variation of the transducer capacitance is achieved by a gap closing transducer with $S = 1 \times 10^{-4}$ m^2, $d_0 = 70$ µm, a resonator with $m = 57 \times 10^{-6}$ kg, $k = 30.8$ Nm^{-1} and $Q = 10$, frequency of the external vibrations is 110 Hz, acceleration amplitude is 0.4g, $C_1 = 10$ nF, $C_2 = 1$ nF, initial voltage V_1 is 5 V

As a consequence, in practice, the Bennet's doubler based conditioning circuit highlights a saturation of the output voltage, similarly with the charge pump considered in Sect. 8.4.2. Hence, for all circuits is it possible to find the value of the output voltage yielding a maximum energy conversion rate. A realistic design requires to take into account the mechanical coupling, which modify the optimal values of the output voltages calculated by the analysis in the electrical domain. This is discussed in the next section.

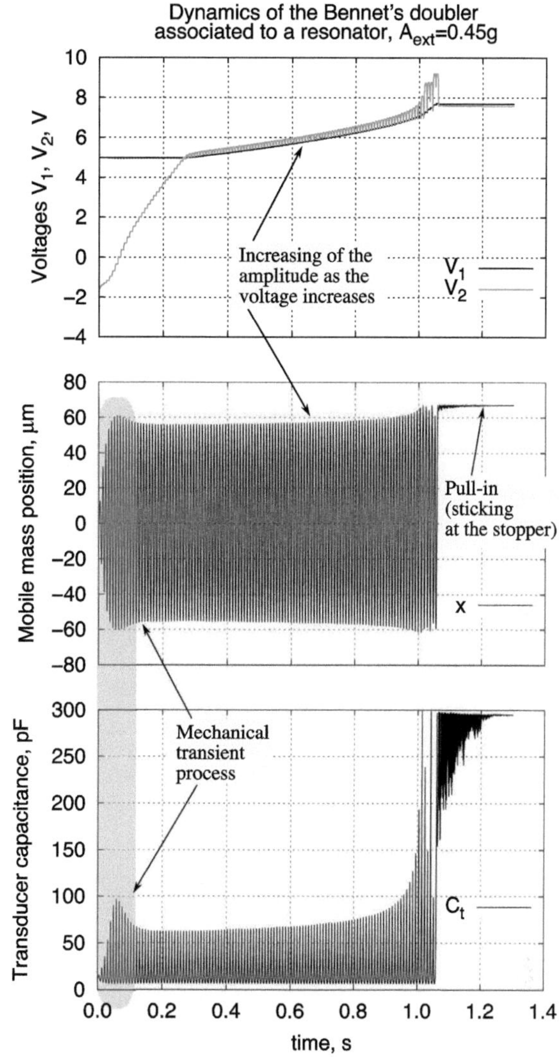

Fig. 8.22 Simulation results demonstrating an abrupt saturation of a Bennet's doubler (Fig. 8.18), where the variation of the transducer capacitance is done by a gap closing transducer connected to a resonator, with the same parameters as for the plots of Fig. 8.21, except the acceleration amplitude is 0.45 g, a stopper is placed at 3 µm from the fixed electrode

8.8 Practical Use of Conditioning Circuits with Rectangular QV Cycle

The last question we would like to discuss is the following: how it is possible to maintain the output voltage of a conditioning circuit at the optimal level? The most straightforward and generic technique is presented in Fig. 8.23.

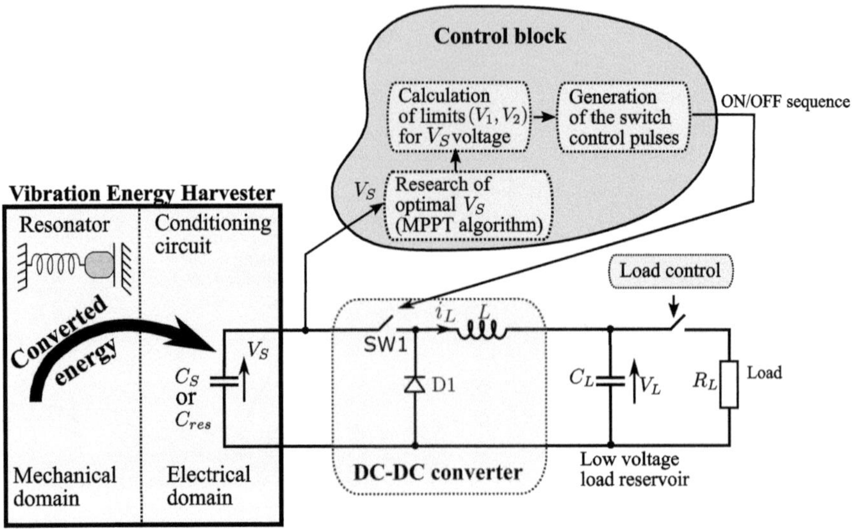

Fig. 8.23 Architecture implementing the control of the internal energy (the output voltage) of the vibration energy harvester in order to maximize the converted power

This architecture is based on a DC–DC converter which interfaces the output capacitance of the conditioning circuit (C_S or C_{res} in the case of the charge pump with flyback) with a low voltage energy reservoir available for the load supply. The output voltage of the harvester is defined by two energy flows: (i) the energy converted from the mechanical domain which tends to increase the output voltage, (ii) the DC–DC conversion removing energy from the conditioning circuit to the load reservoir. By controlling the second flow, it is possible to control the average level of the output voltage, and to guarantee that it corresponds to the optimal value for the maximum power conversion. The technique is very similar with what we proposed for the control of the voltage V_{st} of the charge pump in Sect. 8.5.2 [9]: the definition of the optimal voltage $V_{st\ opt}$, the definition of the acceptable interval for the output voltage variation, the control of a DC–DC converter in order to remove the extra energy from the output capacitor, when necessary. The energy is accumulated on the load capacitor C_L and when there is enough energy for accomplishing some useful operation, the load is supplied. The most critical block is the one implementing the control of the DC–DC converter (represented as a gray cloud in the diagram), since an complex analog information processing is required, as discussed in Sect. 8.5.2. A simplified version of such an interface with the load is proposed in [3].

The implementation of this technique is a very challenging task, mainly because of the low level of power available for the implementation of the control algorithm. Ultra-low power analog integrated circuits is required to make it possible. This issue is currently being studied in several research groups.

8.9 Conclusion on Conditioning Circuits for eVEHs

This chapter presented essential information about the electrical conditioning of electrostatic transducers for vibration energy conversion. Such use of the capacitive transducer is relatively recent (not more than 10–15 years), and the associated practical knowledge is relatively small. To the date of this book's writing, most of the presented concepts have been essentially validated by prototypes issued from research projects and by simulation. Substantial efforts from the community of researchers and engineers are still necessary in order to elevate the maturity level of these techniques so to employ them into commercial applications.

In many existing studies, the conditioning circuits of vibration energy harvesters have been analyzed without accounting for the mechanical coupling. However, both experiment and the theory emphasize that the electromechanical coupling deeply modifies the behavior of the circuit. In particular, whereas the electrical analysis claims that the bias voltage must be the largest possible in order to maximize the converted power (cf. for instance formulae (8.9) and (8.28)), analysis of the full system, simulations and experiment highlights the existence of an optimal value of the bias voltage [7]. Chapter 3 introduces analytical tools which can be used for the study of the coupling phenomena in capacitive energy harvesters, and provide some examples. The used mathematical tools are quite involved even for simple configurations, and to date, more investigation is required to have a comprehensive and handful representation of the effects of the electromechanical coupling.

Implementation of smart and adaptive behavior of capacitive VEHs like represented in diagrams Figs. 8.16a and 8.23, is mainly impeded by the low amount of power available from a microscale capacitive energy harvester. However, this limitation is specific to the technologies of integrated circuits (IC) which have been available for the studies carried out to date. Future evolutions of the IC technologies will allow the implementation of the control circuitry with an acceptable power overhead. From this standpoint, electrostatic vibration energy harvesting may be seen as an emerging technology, having a strong potential for tomorrow's applications.

References

1. Basset, P., Galayko, D., Cottone, F., Guillemet, R., Blokhina, E., Marty, F., et al. (2014). Electrostatic vibration energy harvester with combined effect of electrical nonlinearities and mechanical impact. *Journal of Micromechanics and Microengineering, 24*(3), 035,001.
2. Bennet, A., & Kaye, R. (1787). An account of a doubler of electricity, or a machine by which the least conceivable quantity of positive or negative electricity may be continually doubled, till it becomes perceptible by common electrometers, or visible in sparks. *Philosophical Transactions of the Royal Society of London, 77*, 288–296.
3. de Queiroz, A. C. M., & Domingues, M. (2011). The doubler of electricity used as battery charger. *IEEE Transactions on Circuits and Systems II: Express Briefs*, *58*(12), 797–801.
4. de Queiroz, A. C. M. (2010). Electrostatic vibrational energy harvesting using a variation of Bennet's doubler. In *2010 53rd IEEE International Midwest Symposium on Circuits and Systems (MWSCAS)* (pp. 404–407). IEEE.

5. Despesse, G. (2005). Etude des phénomènes physiques utilisables pour alimenter en énergie électrique des micro-systèmes communicants.
6. Despesse, G., Jager, T., Jean-Jacques, C., Léger, J. M., Vassilev, A., Basrour, S., et al. (2005). Fabrication and characterization of high damping electrostatic micro devices for vibration energy scavenging. In *Proceedings of the Design, Test, Integration and Packaging of MEMS and MOEMS* (pp. 386–390).
7. Dorzhiev, V., Karami, A., Basset, P., Marty, F., Dragunov, V., & Galayko, D. (2014). Electret-free micromachined silicon electrostatic vibration energy harvester with the bennet's doubler as conditioning circuit. *Electron Device Letters, 36*(2), 183–135.
8. Dragunov, V., & Dorzhiev, V. (2013). Electrostatic vibration energy harvester with increased charging current. In *Journal of Physics: Conference Series* (Vol. 476, p. 012115). IOP Publishing.
9. Dudka, A., Galayko, D., & Basset, P. (2012). Design of controller IC for asynchronous conditioning circuit of an electrostatic vibration energy harvester. In *IEEE International Conference on Internet of Things, 2012 Workshop on energy and Wireless Sensors*.
10. Dudka, A., Galayko, D., Blokhina, E., & Basset, P. (2014). Smart integrated conditioning electronics for electrostatic vibration energy harvesters. In *2014 IEEE International Symposium on Circuits and Systems (ISCAS)* (pp. 2600–2603). IEEE.
11. Florentino, H. R., Freire, R. C. S., Sá, A. V. S., Florentino, C., & Galayko, D. (2011). Electrostatic vibration energy harvester with piezoelectric start-up generator. In *2011 IEEE International Symposium on Circuits and Systems (ISCAS)* (pp. 1343–1346). IEEE.
12. Galayko, D., Blokhina, E., Basset, P., Cottone, F., Dudka, A., O'Riordan, E., et al. (2013). Tools for analytical and numerical analysis of electrostatic vibration energy harvesters: Application to a continuous mode conditioning circuit. In *Journal of Physics: Conference Series* (Vol. 476, p. 012076).
13. Galayko, D., Dudka, A., Karami, A., O'Riordan, E., Blokhina, E., Feely, O., et al. (2015). Capacitive energy conversion with circuits implementing a rectangular charge-voltage cycle—part 1: Analysis of the electrical domain. *IEEE Transactions on Circuits and System I, 62*(11), 2652–2663.
14. Kempitiya, A., Borca-Tasciuc, D. A., & Hella, M. M. (2013). Low-power interface IC for triplate electrostatic energy converters. *IEEE Transactions on Power Electronics, 28*(2), 609–614. doi:10.1109/TPEL.2012.2213676.
15. Lefeuvre, E., Wei, J., Mathias, H., & Costa, F. (2015). Single-switch inductorless power management circuit for electrostatic vibration energy harvesters. In *Proceeding of IEEE NEWCAS 2015 Conference*.
16. Meninger, S., Mur-Miranda, J., Amirtharajah, R., Chandrakasan, A., & Lang, J. (2001). Vibration-to-electric energy conversion. *IEEE Transactions on Very Large Scale Integration (VLSI) Systems, 9*(1), 64–76 (2001).
17. Mitcheson, P. D., Sterken, T., He, C., Kiziroglou, M., Yeatman, E., & Puers, R. (2008). Electrostatic microgenerators. *Measurement and Control, 41*(4), 114–119.
18. Mur-Miranda, J. (2004). Electrostatic Vibration-to-Electric Energy Conversion. Ph.D. thesis. MIT.
19. Nayfeh, A. (1993). *Introduction to perturbation techniques*. Wiley.
20. Okamoto, H., Suzuki, T., Mori, K., & Kuwano, H. (2009). *A concept of an electret power generator integrated with a rectifier*. Washington DC: PowerMEMS.
21. Risquez, S., Wei, J., Woytasik, M., Parrain, F., & Lefeuvre, E. (2014). Self-biased inductorless interface circuit for electret-free electrostatic energy harvesters. In *Journal of Physics: Conference Series* (Vol. 557, pp. 12,052–12,056). IOP Publishing.
22. Roundy, S., Wright, P., & Pister, K. (2002). Micro-electrostatic vibration-to-electricity converters. In *Proceedings of 2002 ASME International Mechanical Engineering Congress*.
23. Senturia, S. D. (2001). *Microsystem design* (Vol. 3). Boston: Kluwer Academic Publishers.
24. Sterken, T., Fiorini, P., Baert, K., Puers, R., & Borghs, G. (2003). An electret-based electrostatic/spl mu/-generator. In *12th International Conference on TRANSDUCERS, Solid-State Sensors, Actuators and Microsystems* (Vol. 2, pp. 1291–1294). IEEE.

25. Torres, E. O., & Rincon-Mora, G. A. (2009). Electrostatic energy-harvesting and battery-charging CMOS system prototype. *IEEE Transactions on Circuits and Systems*, *I*(56), 1938–1948.
26. Yen, B. C., & Lang, J. H. (2006). A variable-capacitance vibration-to-electric energy harvester. *IEEE Transaction on Circuits and Systems-I: Regular papers IEEE*, *53*(2), 288–295.

Chapter 9
Analysis and Modelling of Nonlinearties in Vibration Energy Harvesters

Peter Harte, Dimitri Galayko, Orla Feely and Elena Blokhina

9.1 Introduction

There are many ways of modelling, simulating and analysing the behaviour of electrostatic vibration energy harvesters (eVEHs) and this is the focus of this chapter. A common way of studying many engineering, biological and financial systems (amongst others) is to develop a mathematical model for them. These mathematical models can take any form, from very simple one-dimensional maps to highly complex and nonlinear differential equations. To simulate such complex equations can be very time consuming and computationally intensive. They also give very little information about the stability of systems and their bifurcation parameters. This is why we need analytical and semi-analytical techniques similar to the ones outlined in the following sections.

9.2 Mathematical Model of an eVEH

9.2.1 *Basis of Mathematical Model*

An eVEH, as seen in Fig. 9.1, consists of a high-Q linear or nonlinear oscillator, a capacitive transducer and a conditioning circuit that implements an energy conversion cycle [1]. Capacitive (or electrostatic) transducers consist of a parallel plate capacitor

P. Harte · D. Galayko · O. Feely · E. Blokhina (✉)
University College Dublin, Dublin, Ireland
e-mail: elena.blokhina@ucd.ie

P. Harte
e-mail: peter.harte@ucdconnect.ie

E. Blokhina et al. (eds.), *Nonlinearity in Energy Harvesting Systems*,
DOI 10.1007/978-3-319-20355-3_9

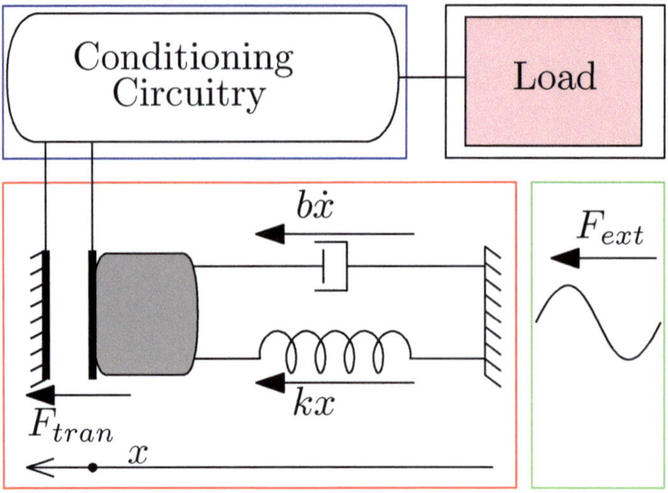

Fig. 9.1 Generic schematic for an eVEH. The mechanical resonator can be described by the mass-spring-damper equation. The external driving force, from which the mechanical energy is harvested, is assumed to be sinusoidal. The transducer force f_{tran} depends on the architecture of the conditioning circuitry. The direction of the forces F_{tran}, kx and $b\dot{x}$ depend on x

with mobile electrodes. These transducers can have different geometries, e.g. Gap Closing or Area Overlap. The capacitance, $C(x)$, of such a transducer depends upon its mobile displacement and the transducer dynamics depends upon the state of the conditioning circuit. The operation of the conditioning circuit depends on the state of the transducer (i.e. the transducer displacement $x(t)$ and/or velocity $v(t)$). As a result, the two domains (electrical and mechanical) are coupled.

To mathematically model an eVEH in the electromechanical domain, all mechanical forces acting upon the system must be taken into account, i.e. the force of the nonlinear resonator, the external driving force and the transducer force. The mechanical nonlinear resonator, including the mechanical nonlinearity terms, can be represented as a mass-spring-damper system. The resonator is driven by some form of external mechanical vibrations, from which the system is harvesting energy. In all examples throughout this chapter this external driving force, F_{ext}, is assumed to be sinusoidal with amplitude A_{ext}. Thus, the effect of the nonlinear resonator, the driving force and the transducer force F_{tran} can be written mathematically as follows

$$\underbrace{\ddot{x} + \frac{b}{m}\dot{x} + \omega_0^2 x + \sum_{\text{n}=2}^{\text{N}} \frac{k_{\text{n}}}{m} x^{\text{n}}}_{\text{Nonlinear resonator}} = \underbrace{A_{\text{ext}} \cos(\omega_{\text{ext}} t)}_{\text{Mechanical vibrations}} + \underbrace{F_{\text{tran}}(x, \dot{x})/m}_{\text{Transducer force}} \tag{9.1}$$

where the meanings of the various variables can be found in Table 9.1. This is a standard mathematical model of an eVEH operated with a nonlinear resonator. This can

Table 9.1 Definitions of dimensional and normalised parameters

Dimensional	Meaning	Normalised
x	Displacement (m)	$y = x/d$
t	Time (s)	$\tau = \omega_0 t$
$v = \dot{x}$	Velocity (ms^{-1})	$y' = v/(d\omega_0)$
m	Resonator mass (kg)	–
d	Transducer gap at rest (m)	–
k_1	Linear spring stiffness (Nm^{-1})	–
k_n	nth-order mechanical nonlinearity ($\text{Nm}^{-\text{n}}$)	$\kappa_\text{n} = (k_\text{n} d^{\text{n}-1})/(m\omega_0^2)$
b	Damping factor (Nsm^{-1})	$\beta = b/(2m\omega_0)$
$\omega_0 = \sqrt{k_1/m}$	The natural frequency of the resonator ($\text{rad}\,\text{s}^{-1}$)	–
ω_ext	Frequency of external vibrations ($\text{rad}\,\text{s}^{-1}$)	$\Omega = \omega_\text{ext}/\omega_0 = 1 + \sigma$
A_ext	Acceleration of external vibrations (ms^{-2})	$\alpha = A_\text{ext}/(d\omega_0^2)$
F_tran	Transducer force (N)	$f_\text{tran} = F_\text{tran}/(d\,m\,\omega_0^2)$
W_0	Energy placed on transducer (J)	$\nu_\text{W} = W_0/(d^2 m\omega_0^2)$

also be used to model an eVEH with a linear resonator by setting all of the mechanical nonlinearity coefficients, k_n, to zero. The only expression that will differ depending on the topology of the eVEH is that of F_tran since the transducer force depends on the geometry of the transducer and how the conditioning circuitry operates. Thus, to complete the mathematical model an expression for F_tran must be found.

9.2.2 The Transducer Force

The mechanical force of the transducer, F_tran, is generated by the voltage/current in the electrical domain. The instantaneous electrical power flowing to the transducer is

$$P_e = V_\text{tran}(t) I_\text{tran}(t) \tag{9.2}$$

where V_tran is the voltage across the transducer and I_tran is the current flowing into the transducer. The instantaneous mechanical power at the output is

$$P_m = F_\text{tran} \frac{dx}{dt} \tag{9.3}$$

where F_tran is the mechanical force generated by the transducer and dx/dt is the velocity $v(t)$. Electrical power P_e at the input of the transducer is spent on changing the energy of the transducer, W_tran, and on generating mechanical power P_m at the output of the transducer:

$$P_e = \frac{dW_{\text{tran}}}{dt} + P_m \tag{9.4}$$

Using the relationship between transducer charge and voltage

$$Q_{\text{tran}}(t) = C_{\text{tran}}(x(t))V_{\text{tran}}(t) \tag{9.5}$$

where C_{tran} is the transducer capacitance, a function of the transducer displacement and defined by its geometry, the current flowing through the transducer can be defined as

$$I_{\text{tran}} = \frac{dQ_{\text{tran}}}{dt} = \frac{d(C_{\text{tran}}(x(t))V_{\text{tran}}(t))}{dt} = C_{\text{tran}}(x)\frac{dV_{\text{tran}}}{dt} + V_{\text{tran}}\frac{dC_{\text{tran}}}{dx}\frac{dx}{dt} \tag{9.6}$$

Combining (9.2) and (9.6), the instantaneous electrical power can now be written as

$$\begin{aligned} P_e &= C_{\text{tran}}(x)V_{\text{tran}}\frac{dV_{\text{tran}}}{dt} + V_{\text{tran}}^2\frac{dC_{\text{tran}}}{dx}\frac{dx}{dt} \\ &= C_{\text{tran}}(x)V_{\text{tran}}\frac{dV_{\text{tran}}}{dt} + \frac{1}{2}V_{\text{tran}}^2\frac{dC_{\text{tran}}}{dx}\frac{dx}{dt} + \frac{1}{2}V_{\text{tran}}^2\frac{dC_{\text{tran}}}{dx}\frac{dx}{dt} \\ &= \frac{d}{dt}\left[\frac{C_{\text{tran}}(x)V_{\text{tran}}^2}{2}\right] + \frac{1}{2}V_{\text{tran}}^2\frac{dC_{\text{tran}}}{dx}\frac{dx}{dt} \end{aligned} \tag{9.7}$$

Since we assume that the system is lossless and following on from (9.4) it is clear to see that

$$P_m(t) = \frac{V_{\text{tran}}^2}{2}\frac{dC_{\text{tran}}}{dx}\frac{dx}{dt} = \frac{V_{\text{tran}}^2}{2}\frac{dC_{\text{tran}}}{dx}v \tag{9.8}$$

and

$$W_{\text{tran}} = \frac{C_{\text{tran}}(x)V_{\text{tran}}^2}{2} \tag{9.9}$$

By definition from (9.3)

$$P_m = F_{\text{tran}}\frac{dx}{dt} = F_{\text{tran}}v \tag{9.10}$$

Comparing (9.8) and (9.10) it is clear that the mechanical force generated by the electromechanical transducer is

$$F_{\text{tran}} = \frac{1}{2}V_{\text{tran}}^2\frac{dC_{\text{tran}}}{dx} \tag{9.11}$$

Thus, the voltage generated by the conditioning electronics causes an additional force F_{tran} on the mechanical resonator and the magnitude of this force is given by (9.11). This derivation methodology can be applied to any VEH transducer.

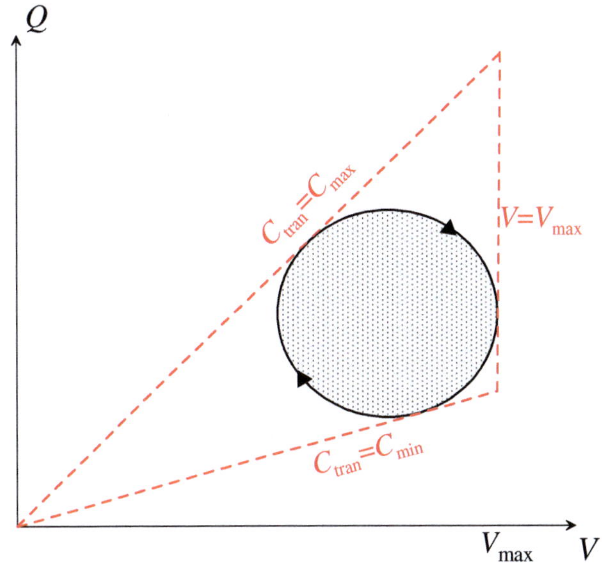

Fig. 9.2 This is a generic QV cycle for an eVEH. Here the transducer capacitance varies between two known values, C_{max} and C_{min}. V_{max} is the maximal voltage allowed by the technology. The area of the cycle is equal to the net energy converted by the transducer during the cycle

9.2.3 The Constant Charge Energy Conversion Cycle

Even in the absence of electromechanical coupling, i.e. when the mobile mass vibration amplitude of the resonator does not depend on the electrical processes, the analysis of even the simplest conditioning circuits is very difficult. Suchsystems can be described by a linear ordinary differential equation with time variable coefficients, which cannot be integrated in closed form even for a simple $C_{tran}(x)$ function. Charge–voltage (QV) diagrams are used to give a graphical representation of the state evolution of a variable capacitor in the (V, Q) plane. For the study of an energy harvester, QV diagrams are plotted in the steady-state mode when the mobile electrode of a transducer is assumed to oscillate with the given sinusoidal law: $x(t) = X\cos(\Omega t)$. A typical QV diagram is a closed cycle (See Fig. 9.2). The area of the cycle is equal to the net energy converted by the transducer during that cycle.

If one assumes a small displacement, dx, of the mobile electrodes, then the work done, dW_{tran}, by the transducer force F_{tran} from (9.11) is

$$dW_{tran} = F_{tran}dx = \frac{1}{2}V_{tran}^2\frac{dC_{tran}}{dx}dx = \frac{1}{2}V_{tran}^2 dC_{tran} \tag{9.12}$$

If dW_{tran} is positive, energy is converted from the electrical domain to the mechanical domain. If dW_{tran} is negative then energy is converted from the mechanical domain to the electrical domain. Therefore, eVEHs generate electricity when C_{tran} is decreasing. Thus, in order to optimally convert mechanical energy to electrical energy using an eVEH, the voltage V_{tran} should be maximised when C_{tran} is decreasing and minimised when C_{tran} is increasing. Therefore the ideal scenario would set $V_{tran} = 0$ V when C_{tran} is increasing and $V_{tran} = V_{max}$ when C_{tran} is decreasing (where V_{max} is the maximal voltage allowed by the technology). This is the basis of the constant voltage conversion cycle, more details of which can be found in [1]. For the remainder of this chapter, we will present and analyse a mathematical model of an eVEH with a gap closing transducer and conditioning circuitry that implements the constant charge energy conversion cycle [1]. The reason for focusing on this particular eVEH implementation is because it displays every type of nonlinear behaviour that can be seen in VEHs, e.g. multistability, period-doubling bifurcations, chaos and sliding phenomena, and as a result is a very good general example. The nonlinear behaviour present in the gap closing, constant charge circuit (GCT-CQ) is typical of both continuous and discontinuous circuits (sliding phenomena can only be seen in discontinuous systems). The capacitance of a gap closing transducer, with a hyperbolic dependence on displacement, is defined as

$$C_{tran}(x) = \frac{C_0}{(1 - x/d)} \tag{9.13}$$

where C_0 is the rest capacitance of the transducer and d is its rest gap. The constant charge energy conversion cycle for an eVEH was first presented in [1]. The QV diagram for the conversion cycle can be seen in Fig. 9.3. When an eVEH is operating in this "constant charge" mode, the conditioning circuit discharges the transducer to zero when the transducer capacitance is at a local minimum and charges it to a charge Q_0 when its capacitance is at a local maximum. The conversion from mechanical energy to electrical energy happens when the transducer capacitance decreases, due to the movement of the capacitor plates, from C_{max} to C_{min}, whilst during this movement the charge Q_0 on the capacitor remains constant. During this process, the transducer essentially acts as a damper in the mechanical domain. Realising a peak detector for a system such as this is quite difficult since there should be no charge on the transducer whilst its capacitance varies from C_{min} to C_{max} and thus one cannot just simply measure the voltage V_{tran} to determine when these local maxima and minima occur. In practice though, techniques such as optically detecting the resonator position or measuring V_{tran} by keeping some residual charge on the transducer are used. A schematic of the system can be seen in Fig. 9.4 and a detailed description of the conversion cycle of the system can be found in [2].

Thus, it is clear from the description and from (9.11) that the dimensional transducer force F_{tran} for an eVEH operating in constant charge mode is defined as

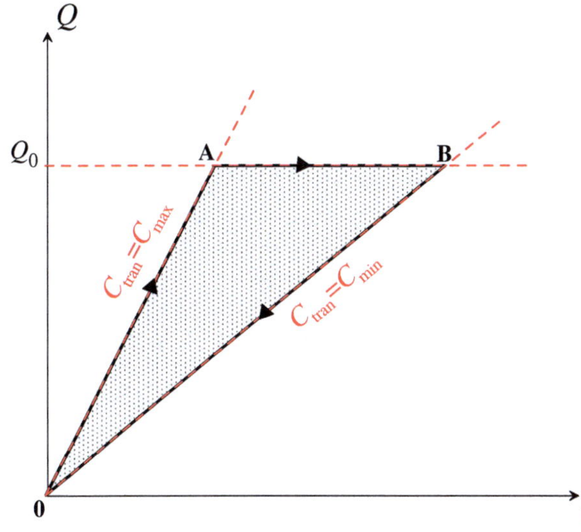

Fig. 9.3 QV cycle for the constant charge energy conversion cycle. The conditioning circuit discharges the transducer to zero when the transducer capacitance is at a local minimum (**B→0**) and charges it to a charge Q_0 when its capacitance is at a local maximum (**0→A**). As the capacitance decreases from C_{max} to C_{min} (**A→B**), the charge is held constant on the capacitor at Q_0. It is from **A→B** that the mechanical energy is converted to electricity

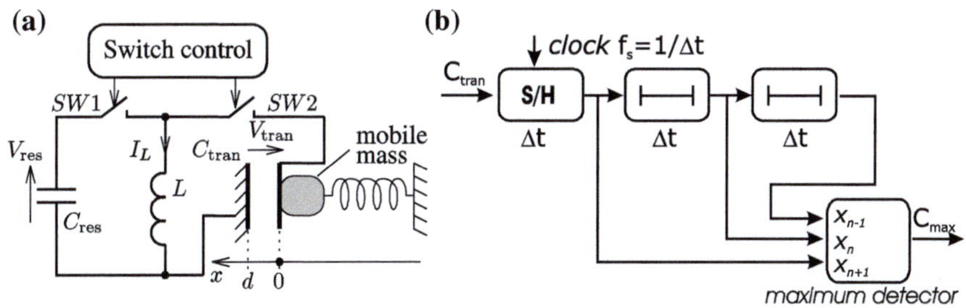

Fig. 9.4 **a** Schematic of the electronic oscillator which models a vibration electrostatic energy harvester. Note the mass-spring-damper resonator which makes up part of the capacitive transducer (also note the direction of the displacement axis). The switches SW1 and SW2 along with inductor L control the transfer of the harvested energy from capacitor C_{tran} to the reservoir capacitor C_{res}. **b** Schematic example of a digital detector of local maxima. It samples a signal with a very high frequency and holds three values. By comparing the middle value with the others, it detects an event (peak of the signal). Similarly a local minima detector compares three values and detects a trough of the signal

$$F_{tran}(x, v) = \begin{cases} \frac{V_{tran}^2}{2} \frac{dC_{tran}(x)}{dx} & v < 0 \\ 0 & v > 0 \end{cases} \tag{9.14}$$

As discussed earlier, the formulation of F_{tran} depends on the architecture of the conditioning circuit. There are two expressions that tie together the voltage V_0, the charge Q_0 and the energy W_0 of the transducer: $Q_0 = C_{max} V_0$ and $W_0 = Q_0^2/(2C_{max})$, where C_{max} is a local maximum of the variable transducer capacitance $C_{tran}(x)$ and is a dynamic quantity which can vary from one period to another. Therefore, by choosing to fix the value of one of the three quantities V_0, W_0 or Q_0 we will obtain one of three different expressions for the force F_{tran} from the general expression (9.14). The system can display different dynamics depending on which quantity is selected to be fixed. For our analysis, energy W_0 is the quantity that has been fixed since this is the most common case described in [1]. Thus, for a GCT-CQ eVEH, that fixes energy at every period, $V_{tran} = \sqrt{2W_0C_{max}}/2$. This means (9.14) can now be written as:

$$F_{tran}(x, v) = \begin{cases} \frac{W_0}{d(1-x_{max}/d)} & v < 0 \\ 0 & v > 0 \end{cases} \tag{9.15}$$

In this case, the force (9.15) depends upon a local maximum of the resonator displacement x_{max} and therefore is fundamentally nonlinear. This is irrespective of whether a linear or a nonlinear resonator is being utilised. The value of F_{tran} is set at the local maximum of the capacitance C_{tran} and this value is then held until another local maximum is detected. We call this property the 'hold-on effect' and it causes more nonlinear phenomena in the system.

9.2.4 The Complete Normalised Mathematical Model

Now that we have an expression (9.15) for the transducer force F_{tran}, we can now use (9.1) to numerically model a GCT-CQ eVEH in its dimensional form. To make our analysis easier by reducing the number of variables in the system, and to make the application of the semi-analytical methods studied in the latter half of this chapter more straightforward, we can normalise (9.1):

$$y'' + 2\beta y' + y + \sum_{n=2}^{N} \kappa_n y^n = \alpha \cos(\Omega\tau) + f_{tran}(y, y') \tag{9.16}$$

where y is the normalised displacement and the prime denotes the derivative with respect to normalised time τ. Normalising (9.15), the dimensionless transducer force f_{tran} can be represented as:

$$f_{tran}(y, y') = \begin{cases} \frac{\nu_W}{1-y_{max}} & y' < 0 \\ 0 & y' > 0 \end{cases} \tag{9.17}$$

where we introduced the normalised coefficient $\nu_W = W_0/(d^2 m\omega_0^2)$. This normalised coefficient ν_W defines the magnitude of electrical damping in the system. All of the normalised parameters of the system are defined in Table 9.1. Equations (9.1) and (9.16), both represent mathematical models of an eVEH. Depending on the form of the oscillator, whether it is linear or nonlinear, there may be none or many $\kappa_n (n \geq 2)$ terms in the equation. As seen in Sect. 9.2.3, the form that the transducer force F_{tran} takes depends upon which energy conversion cycle is being implemented. A mathematical model like the one in (9.16) is a very important tool for studying the behaviour of a system. It allows one, through computer simulation, to determine very accurately the local behaviour of a system at various different control parameter values. With regards to an eVEH, (9.1) or (9.16) can be used to understand fundamental properties of the system, e.g. how the displacement, velocity and capacitance of the transducer vary with time. Sample steady-state results from simulations of a GCT-CQ eVEH, with a linear resonator, found using (9.1) can be seen in Fig. 9.5.

Such a mathematical model, although a very accurate representation of the dynamics of an eVEH, has its limitations. First, despite the obvious improvements in computing power over the past few decades, such numerical simulations can still be very computationally intensive and as result, particularly if the system is stiff or depends upon frequent event detections, they can be very time consuming too. Conversely, if the system can be described analytically, then its dynamics, in a steady-state period-1 regime at least, can be determined very quickly.

Secondly, a stand-alone numerical simulation tells us nothing about the stability of the system. Once again, if the system can be described analytically (or semi-analytically), there are techniques like the ones outlined in Sect. 9.3 that allow us to quantify how stable a system is at particular values of its control parameters. Nevertheless, to apply these analytical or semi-analytical techniques, a good mathematical model of a system is required.

9.3 Semi-analytical Methods

Conversely to the mathematical model presented in the previous section, in this section, some semi-analytical techniques that can be used to determine the behaviour of an eVEH are discussed. There is no general closed form solution of nonlinear equations like (9.16) but there are various approaches one can take such as:

- Taylor series expansion [3]
- Harmonic Balance Method [4, 5]
- Multiple Scales Method [6, 7]
- Mechanical Impedance Method [2]
- Projection method [3]
- Averaging/Van der Pol method [6]

Semi-analytical methods are very useful for eVEHs. Firstly, they can be used to simulate the behaviour of the system, both very quickly and accurately, and are not

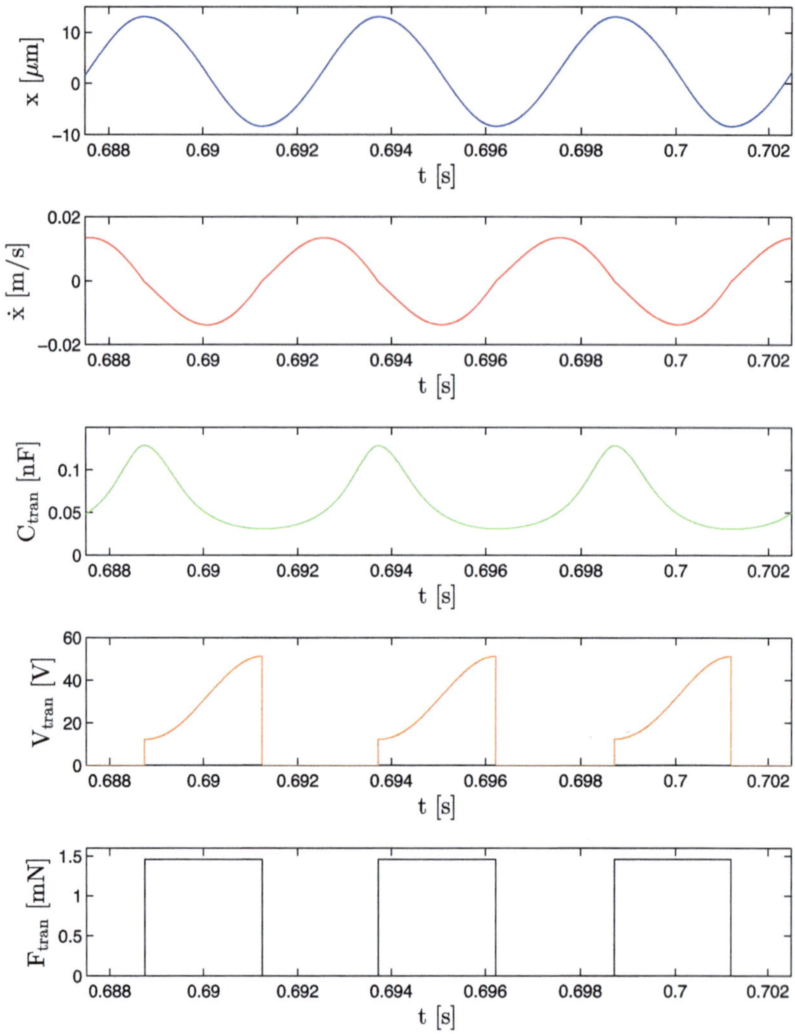

Fig. 9.5 Here we can see an example of the normal operating cycles of a GCT-CQ eVEH in dimensional form. These results were found using a linear resonator with $W_0 = 10$ nJ, $A_{\text{ext}} = 5\ \text{ms}^{-2}$ and $f_{\text{ext}} = 189$ Hz. Resonator displacement and velocity along with transducer capacitance, voltage and force can be seen

computationally intensive. They give an approximate solution to the mathematical model of the system but this approximate solution is generally very accurate provided the system being studied is operating in a steady-state, period-1 regime. "Steady-state, period-1" means that the response of the system is periodic for some period T. For eVEHs, this response is assumed to be sinusoidal due to the inclusion of an assumed high-Q resonator in the model. When performing numerical integration of an ODE,

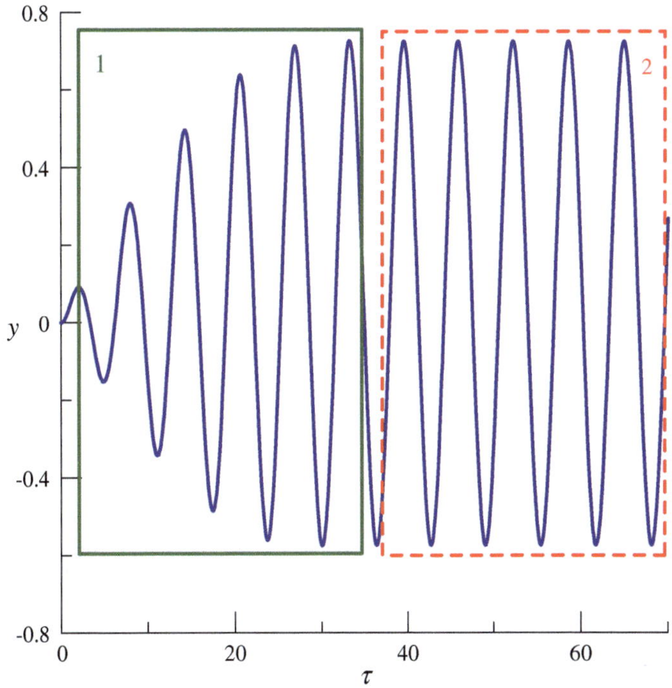

Fig. 9.6 Normalised displacement of the resonator found through numerical integration. In the *solid box* (marked '1'), the transient region of the response can be seen. After a number of cycles, the response eventually settles a steady-state, period-1 regime as can be seen in the *broken box* (marked '2')

for example (9.16), which describes the behaviour of an eVEH, depending on initial conditions, the response will have some form of transient behaviour before it settles to steady-state behaviour. Both the transient element and the steady-state period-1 region of a response found through numerical integration of (9.16) can be seen in Fig. 9.6. All semi-analytical methods covered in this chapter are applied to eVEHs assumed to be operating with steady-state period-1 behaviour.

A steady-state theory for eVEHs with high-Q linear resonators and resonators with squared and cubic nonlinear terms first appeared in [8]. Such a steady-state theory is essential since numerical analysis of these eVEH systems can be quite time consuming and the desired region of operation of the systems is one of "quasi-harmonic" oscillations. This steady-state analysis allows for certain practical properties of the circuit, e.g. converted power, to be examined with relative ease [9]. The following subsections will give the reader a brief introduction in applying the harmonic balance method, the mechanical impedance method and the multiple scales method to a GCT-CQ eVEH configuration.

9.3.1 Harmonic Balance Method

The harmonic balance method is a semi-analytical technique for the steady-state analysis of nonlinear differential equations [4, 5]. The application of this method to an eVEH operated with an area overlap transducer can be seen in [5]. It has advantages over other steady-state analysis techniques for eVEHs since it is better suited for larger parameter ranges (e.g. a large transducer displacement range).

The steady-state response of an eVEH, in either the mechanical or electrical domain, can be found using the harmonic balance method. One assumes that a solution of the system can be represented by a linear combination of sinusoids. After substituting this assumed solution into a system of equations describing an eVEH (e.g. equations of motion or Kirchoff's current law), one then finds the steady-state solution by "balancing" the various harmonics, hence the name "harmonic balance" method is descriptive of the method.

First Harmonic Method

If the external acceleration is periodic, it can be assumed that all mechanical quantities of the harvester are periodic as well. Although this is not exactly true on large timescales for complex conditioning circuits [10], where the electric parameters of the conditioning circuit change slowly in time, at a timescale of several vibration periods, F_{tran} can still be considered periodic, though it is not necessarily sinusoidal. For example, a quasi-linear piezoelectric transducer associated with a sinusoidally vibrating resonator generates a sinusoidal force if connected with a linear resistor, but the force is non-sinusoidal if the conditioning circuit includes a diode bridge and a reservoir capacitor in parallel with the load resistance [11]. It can be seen from Fig. 9.5 that the transducer force for a GCT-CQ eVEH is also non-sinusoidal.

The narrowband hypothesis stated in Sect. 9.3 implies that if a resonator is excited with a non-sinusoidal periodic external force and if this force has the first (fundamental) harmonic inside the passband of the resonator, the upper harmonics of the force are attenuated by the resonator frequency response, and the oscillations of the resonator can be considered as sinusoidal at the fundamental frequency of the external excitation. In this case, the higher harmonics of the force can be neglected. This assumption is the basis of the first harmonic method, which is a simplified version of the harmonic balance method [12]. This assumption is explained graphically in Fig. 9.7. The hormonic balance analysis outlined in this chapter is only valid for the cases where this assumption is true, i.e. where the energy injected in the mechanical system by the fundamental harmonic of the nonlinear force is much greater than the energy injected by the higher harmonics.

Since it can be assumed that the displacement of an eVEH resonator moves sinusoidally, when implementing the harmonic balance method, a solution of the normalised displacement y from (9.16) can be written as

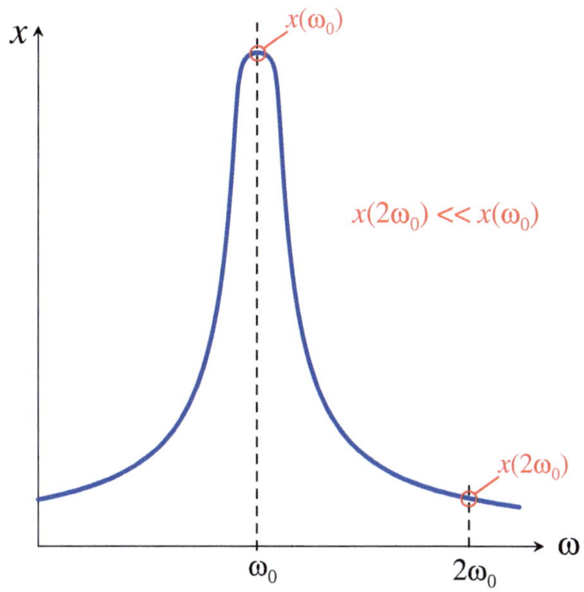

Fig. 9.7 The High-Q assumption which forms the basis of the presented analytical methods. This is the typical response of an GCT-CQ eVEH operated using a linear resonator. Clearly the amplitude of oscillations decreases dramatically when the frequency of operation deviates even slightly from the resonant frequency ω_0. It is this characteristic that allows one to neglect any harmonics of the force higher than the fundamental

$$y = Y_0 \cos(\Omega\tau + \varphi_0) \tag{9.18}$$

Y_0 corresponds to the normalised amplitude of oscillations and φ_0 the phase. The index '0' is used to emphasize that this amplitude and phase are steady-state characteristics. Following on from (9.18), clearly

$$\begin{aligned} y' &= -\Omega Y_0 \sin(\Omega\tau + \varphi_0) \\ y'' &= -\Omega^2 Y_0 \cos(\Omega\tau + \varphi_0) \end{aligned} \tag{9.19}$$

By substituting (9.18) and (9.19) into (9.16), and assuming the use of a linear resonator for ease of analysis, i.e. ($\kappa_n = 0, \ \forall \text{ n} > 2$), one is left with an equation in terms of Y_0, $\cos(\Omega\tau + \varphi_0)$ and $\sin(\Omega\tau + \varphi_0)$.

$$\begin{aligned} &-\Omega^2 Y_0 \cos(\Omega\tau + \varphi_0) - 2\beta\Omega Y_0 \sin(\Omega\tau + \varphi_0) + Y_0 \cos(\Omega\tau + \varphi_0) \\ &= \alpha \cos(\Omega\tau) + f_{\text{tran}}(y, y') \end{aligned} \tag{9.20}$$

It is clear to see from Fig. 9.5 that a GCT-CQ eVEH has a non-sinusoidal f_{tran}. Noting that $y = Y_0 \cos(\Omega\tau + \varphi_0)$ can be written as $y = Y_0 \cos\theta$, then f_{tran} can be described using a Fourier series as follows

$$f_{\text{tran}}(Y_0 \cos\theta, -Y_0 \sin\theta) = \frac{a_0^{\text{tran}}(Y_0)}{2} + \sum_{n=1}^{N} \left(a_{\text{n}}^{\text{tran}}(Y_0) \cos n\theta + b_{\text{n}}^{\text{tran}}(Y_0) \sin n\theta\right) \tag{9.21}$$

where

$$a_n^{\text{tran}}(Y_0) = \frac{1}{\pi}\int_0^{2\pi} f_{\text{tran}}(Y_0\cos\theta, -Y_0\sin\theta)\cos(n\theta)d\theta$$
$$b_n^{\text{tran}}(Y_0) = \frac{1}{\pi}\int_0^{2\pi} f_{\text{tran}}(Y_0\cos\theta, -Y_0\sin\theta)\sin(n\theta)d\theta \tag{9.22}$$

So by taking this Fourier series expansion of f_{tran}, limited to the first harmonic (i.e. the first harmonic method), and remembering that $\theta = \Omega\tau + \varphi_0$, (9.20) can now be written as

$$\begin{aligned}&-\Omega^2 Y_0\cos(\Omega\tau+\varphi_0) - 2\beta\Omega Y_0\sin(\Omega\tau+\varphi_0) + Y_0\cos(\Omega\tau+\varphi_0)\\ &= \alpha\cos(\Omega\tau) + a_1^{\text{tran}}(Y_0)\cos(\Omega\tau+\varphi_0) + b_1^{\text{tran}}(Y_0)\sin(\Omega\tau+\varphi_0)\end{aligned} \tag{9.23}$$

Making use of well known trigonometric identities $\cos(A+B) = \cos(A)\cos(B) - \sin(A)\sin(B)$ and $\sin(A+B) = \sin(A)\cos(B) + \cos(A)\sin(B)$ two equations can be found from "balancing" the harmonics in (9.23), i.e. by equating the $\cos(\Omega\tau)$ terms and also by equating the $\sin(\Omega\tau)$ terms. This leaves two equations in terms of the two unknown quantities of our assumed solution (9.18), Y_0 and φ_0

$$\begin{aligned}\alpha &= P(Y_0)\cos\varphi_0 - Q(Y_0)\sin\varphi_0\\ 0 &= Q(Y_0)\cos\varphi_0 + P(Y_0)\sin\varphi_0\end{aligned} \tag{9.24}$$

where $P(Y_0) = \left(Y_0(1-\Omega^2) - a_1^{\text{tran}}(Y_0)\right)$ and $Q(Y_0) = \left(2\beta\Omega Y_0 + b_1^{\text{tran}}(Y_0)\right)$. Squaring both equations in (9.24), adding them together and then using another trigonometric identity, $\sin^2\varphi_0 + \cos^2\varphi_0 = 1$, one is left with one equation in Y_0

$$P(Y_0)^2 + Q(Y_0)^2 - \alpha^2 = 0 \tag{9.25}$$

In its most general form (9.25) can be written as

$$\left(Y_0(1-\Omega^2) - a_1^{\text{tran}}(Y_0)\right)^2 + \left(2\beta\Omega Y_0 + b_1^{\text{tran}}(Y_0)\right)^2 = \alpha^2 \tag{9.26}$$

Since the transducer force f_{tran} in a GCT-CQ eVEH is a piecewise function, one must be most careful when calculating the Fourier coefficients a_n and b_n. It is known from (9.15) that when $y' > 0$, $f_{\text{tran}} = 0$ and, considering the assumed solution for the normalised displacement, $y = Y_0\cos\theta$, that when $y' < 0$, f_{tran} has a non-zero value of $\nu_W/(1-Y_0)$. Thus, since steady-state, sinusoidal behaviour with a period 2π is assumed, it can be said that for half of the period $[0 \to \pi]$, $f_{\text{tran}} = \nu_W/(1-Y_0)$ and for the other half of the period $[\pi \to 2\pi]$, $f_{\text{tran}} = 0$. Taking only the first harmonic of the Fourier Series we find that the Fourier series coefficients of f_{tran} for a GCT operating in CQ mode are

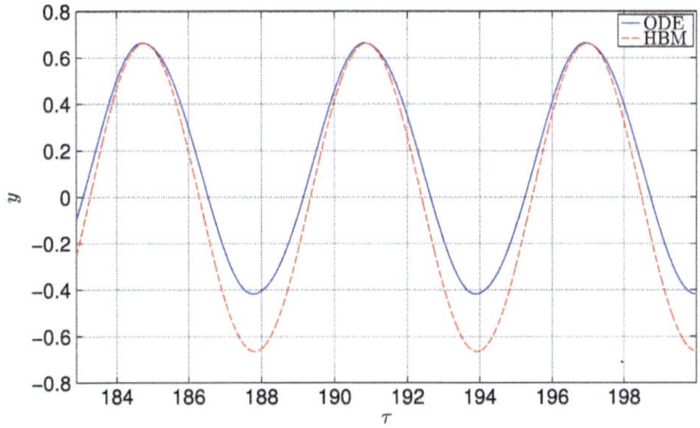

Fig. 9.8 The results of the first harmonic method (*broken red line*) compared with a numerical simulation (*solid blue line*). A severe descrepancy can be seen between the nuumerical and first harmonic method results at the troughs of the oscillations. This descrepancy is a result of neglecting the DC offset in the first harmonic solution. Here $Y_0 = 0.66329$ and $\varphi_0 = -1.81555$. These simultaions were done with $A_{ext} = 5\ \text{ms}^{-2}$, $W_0 = 10$ nJ, $d = 20\ \mu$m, $\sigma = 0.03$, $y(0) = y'(0) = 0$

$$\begin{aligned} a_1^{\text{tran}}(Y_0) &= \frac{1}{\pi}\int_0^{\pi} f_{\text{tran}}\cos(\theta)d\theta + \frac{1}{\pi}\int_{\pi}^{2\pi}(0)\cos(\theta)d\theta = 0 \\ b_1^{\text{tran}}(Y_0) &= \frac{1}{\pi}\int_0^{\pi} f_{\text{tran}}\sin(\theta)d\theta + \frac{1}{\pi}\int_{\pi}^{2\pi}(0)\sin(\theta)d\theta = \frac{2\nu_W}{\pi(1-Y_0)} \end{aligned} \tag{9.27}$$

Substituting the Fourier series coefficients (9.27) into (9.26) results in the following equation in Y_0 specifically for a GCT-CQ eVEH:

$$(Y_0(1-\Omega^2))^2 + \left(2\left(\beta\Omega Y_0 + \frac{\nu_W}{\pi(1-Y_0)}\right)\right)^2 = \alpha^2 \tag{9.28}$$

Solving for Y_0 using (9.28) one is left with multiple solutions but in the case of an eVEH with a linear resonator, a GCT and operating in the CQ mode, there can only be one realistic solution for Y_0. In some other cases, for example, using an AOT or nonlinear resonators, multistability appears in the system and there can be more than one realistic solution for Y_0. After a realistic value for Y_0 has been found from (9.28), a corresponding value for φ_0 can then be found from either of the equations in (9.24).

To investigate the accuracy of the harmonic balance method, a comparison can be made between numerical simulations of (9.16) and the analytical result obtained using (9.18), (9.24) and (9.28). In Fig. 9.8 such a comparison can be seen. A quick look at this figure will immediately suggest to the reader that there is something incorrect with the analytical solution.

It is clear to see from Fig. 9.8 and from (9.18) that the analytical solution oscillates around zero. It is also clear to see from the numerical results in Fig. 9.8 that the transducer does not actually oscillate around zero but rather some non-zero point.

As a result, we must take into account the effect of the zeroth harmonic of the Fourier series expansion of the transducer force f_{tran}. Note that for some transducer geometries and operation modes the effect of the zeroth harmonic of f_{tran} is negligible and can be neglected but in the case of a GCT operating in the CQ mode this is not the case.

The Effect of the Zeroth Harmonic

To extend our harmonic balance analysis to include the effect of the zeroth harmonic of f_{tran} we must now assume that the solution of the system takes the form

$$y = y_{\text{av},0} + Y_0 \cos(\Omega\tau + \varphi_0) \tag{9.29}$$

where $y_{\text{av},0}$ represents the average shift caused by the zeroth harmonic of f_{tran}. Noting now that $y_{\max} = y_{\text{av},0} + Y_0$, the Fourier series coefficients of f_{tran} for a GCT-CQ eVEH are now functions of not only Y_0 but the average shift $y_{\text{av},0}$ too.

$$\begin{aligned} a_0^{\text{tran}}(y_{\text{av},0}, Y_0) &= \frac{\nu_W}{1 - y_{\text{av},0} - Y_0} \\ a_1^{\text{tran}}(y_{\text{av},0}, Y_0) &= 0 \\ b_1^{\text{tran}}(y_{\text{av},0}, Y_0) &= \frac{2\nu_W}{\pi(1 - y_{\text{av},0} - Y_0)} \end{aligned} \tag{9.30}$$

Taking the new assumed solution (9.29) and the updated Fourier series coefficients (9.30), and following the procedure outlined previously, (9.16) can now be written as

$$\begin{aligned} &-\Omega^2 Y_0 \cos(\Omega\tau + \varphi_0) - 2\beta\Omega Y_0 \sin(\Omega\tau + \varphi_0) + y_{\text{av},0} + Y_0 \cos(\Omega\tau + \varphi_0) = \alpha \cos(\Omega\tau) + \\ &\frac{a_0^{\text{tran}}(y_{\text{av},0}, Y_0)}{2} + a_1^{\text{tran}}(y_{\text{av},0}, Y_0) \cos(\Omega\tau + \varphi_0) + b_1^{\text{tran}}(y_{\text{av},0}, Y_0) \sin(\Omega\tau + \varphi_0) \end{aligned} \tag{9.31}$$

By balancing the harmonics of (9.31), and by incorporating the same trigonometric identities used previously, all that remains are three equations with three unknown quantities $y_{\text{av},0}$, Y_0 and φ_0.

$$\begin{aligned} \alpha &= P(y_{\text{av},0}, Y_0) \cos\varphi_0 - Q(y_{\text{av},0}, Y_0) \sin\varphi_0 \\ 0 &= Q(y_{\text{av},0}, Y_0) \cos\varphi_0 + P(y_{\text{av},0}, Y_0) \sin\varphi_0 \\ y_{\text{av},0} &= \frac{a_0^{\text{tran}}(y_{\text{av},0}, Y_0)}{2} \end{aligned} \tag{9.32}$$

where, $P(y_{\text{av},0}, Y_0) = \big(Y_0(1 - \Omega^2) - a_1^{\text{tran}}(y_{\text{av},0}, Y_0)\big)$ and $Q(y_{\text{av},0}, Y_0) = (2\beta\Omega Y_0 + b_1^{\text{tran}}(y_{\text{av},0}, Y_0))$. Using the system of equations (9.32) to solve for $y_{\text{av},0}$, Y_0 and φ_0, one can then use (9.29) to gain an insight into the behaviour of an eVEH. In Fig. 9.9 a comparison can be seen between results found from a numerical simulation of (9.16) and analytical results found using the updated harmonic balance method. These simulations were done for the same parameter values as Fig. 9.8 but it is clear to see that in this instance there is nearly perfect correspondence between the numerical and analytical results.

9.3.2 The Multiple Scales Method (MSM)

The only semi-analytical method studied in detail thus far, the harmonic balance method, can be described as a non-perturbation technique. On the other hand, the multiple scales method (MSM), can be described as a perturbation technique. A perturbation technique is a semi-analytical method for finding approximate solutions to problems that cannot be solved exactly. Using a perturbation method to solve a problem leads to an approximate solution in the form of a power series in terms of some small perturbation parameter, called ε here. ε is a small positive dimensionless parameter that is artificially introduced to establish the different orders of magnitude. The results obtained are independent of this parameter, and ε is ultimately absorbed back into the solution [4]. The multiple scales method (MSM) is a perturbation technique used to find approximate analytical solutions to weakly nonlinear oscillators. It was applied to free and forced nonlinear oscillators in Sects. 3.2.1 and 3.2.2 respectively. The MSM introduces different time scales $T_k = \varepsilon^k \tau$ where ε is the small arbitrary perturbation parameter, alluded to earlier, that appears in the method and k = 0, 1,

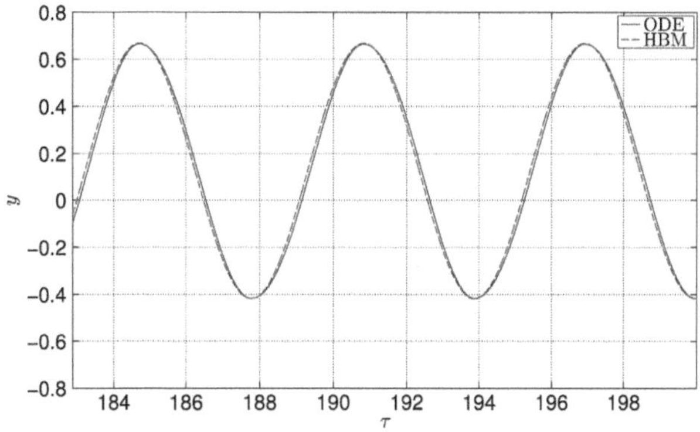

Fig. 9.9 The results of the harmonic balance method including the effect of the DC offset (*broken red line*) compared with a numerical simulation (*solid blue line*). This shows great correspondence between the numerical and semi-analytical methods and the descrepancy seen when using the first harmonic method, caused by the absence of the DC component, no longer exists. Here $y_{av,0} = 0.125603$, $Y_0 = 0.542797$ and $\varphi_0 = -1.77042$. These simultaions were done with $A_{ext} = 5\text{ms}^{-2}$, $W_0 = 10$ nJ, $d = 20\ \mu\text{m}$, $\sigma = 0.03$, $y(0) = y'(0) = 0$

2, 3 …. Applying the MSM to Eq. (9.16), one determines the behaviour of y, not as a function of τ but rather as a function of these different scales $T_0, T_1, T_2, T_3 \ldots$. This method has been applied to eVEHs with linear oscillators in [7] and to eVEHs with nonlinear oscillators in [8, 13] and has proven to be very accurate and useful in describing the behaviour of these eVEHs. In the following section, the application of the MSM to a GCT-CQ eVEH with a nonlinear oscillator is described in detail.

Applying the MSM to an eVEH

Following on from the application of the MSM to free and forced nonlinear oscillators in Sects. 3.2.1 and 3.2.2, the MSM is now applied to a GCT-CQ eVEH operated with a nonlinear oscillator.

In order to apply the MSM, the terms in (9.16) are arranged in appropriate order of the small arbitrary parameter ε that appears in the method.

$$y'' + 2\varepsilon\tilde{\beta}y' + y + \varepsilon\sum_{\mathrm{n}=2}^{\mathrm{N}} \tilde{f}_{\mathrm{NL,n}}(y) = \varepsilon\tilde{f}_{\mathrm{tran}}(y, y') + \varepsilon\tilde{\alpha}\cos(\tau + \varepsilon\tilde{\sigma}\tau) \tag{9.33}$$

where the tilde over the parameters denotes the original parameters divided by ε and $\tilde{f}_{\mathrm{NL,n}}(y) = \tilde{\kappa}_{\mathrm{n}} y^{\mathrm{n}}$ is a nonlinear force.

For displacement y, a standard perturbation technique is to assume that

$$y = y_0(T_0, T_1) + \varepsilon y_1(T_0, T_1), \qquad \frac{d}{d\tau} = D_0 + \varepsilon D_1, \qquad \frac{d^2}{d\tau^2} = D_0^2 + 2\varepsilon D_0 D_1 \tag{9.34}$$

where $T_{\mathrm{k}} = \varepsilon^{\mathrm{k}}\tau$ are time scales introduced by the MSM and where $D_{\mathrm{k}} = \partial/\partial T_{\mathrm{k}}$.

Now, (9.33) can be written as

$$\begin{aligned} &(D_0^2 + 2\varepsilon D_0 D_1)(y_0 + \varepsilon y_1) + 2\varepsilon\tilde{\beta}(D_0 + \varepsilon D_1)(y_0 + \varepsilon y_1) \\ &+ y_0 + \varepsilon y_1 + \varepsilon\sum_{\mathrm{n}=2}^{\mathrm{N}} \tilde{f}_{\mathrm{NL,n}}[y_0 + \varepsilon y_1] = \\ &\varepsilon\tilde{f}_{\mathrm{tran}}[y_0 + \varepsilon y_1,\ (D_0 + \varepsilon D_1)(y_0 + \varepsilon y_1)] + \varepsilon\tilde{\beta}\cos(T_0 + \tilde{\sigma}T_1) \end{aligned} \tag{9.35}$$

Collecting orders 0 and 1 of the parameter ε, and neglecting order 2 and higher, two equations are obtained:

$$D_0^2 y_0 + y_0 = 0 \tag{9.36a}$$

$$\begin{aligned} D_0^2 y_1 + y_1 &= -2D_0 D_1 y_0 - 2\tilde{\beta} D_0 y_0 - \sum_{\mathrm{n}=2}^{\mathrm{N}} \tilde{f}_{\mathrm{NL,n}} \\ &+ \tilde{f}_{\mathrm{tran}}(y_0, D_0 y_0) + \tilde{\alpha}\cos(T_0 + \tilde{\sigma}T_1) \end{aligned} \tag{9.36b}$$

Note that in (9.35) the terms with ε in the arguments of $\tilde{f}_{\text{tran}}$ and $\sum_{n=2}^{N} \tilde{f}_{\text{NL,n}}$ give second order terms in their respective expansions over the powers of ε, hence they have been neglected in (9.36b). The general solution of the expression in (9.36a) can be written as

$$y_0 = A(T_1)e^{iT_0} + c.c. = Y(T_1)\cos(\tau + \varphi(T_1)) \tag{9.37}$$

where the slow complex amplitude $A = (Y/2)\exp(j\varphi)$ is expressed through the amplitude and phase, Y and φ, both of which are real. Note also that $c.c.$ stands for the complex conjugate.

In (9.36b), the forces $\tilde{f}_{\text{tran}}$ and $\sum_{n=2}^{N} \tilde{f}_{\text{NL,n}}$ are periodic functions of T_0 with period 2π. Similarly to the harmonic balance method outlined in Sect. 9.3.1, a Fourier series limited to the first harmonic (since the system is high-Q resonant) is used to represent these forces. The complex representation of the Fourier series seen in (9.38) is used:

$$\begin{aligned} \tilde{f}(Y) &= \frac{\tilde{a}_0(Y)}{2} + [\tilde{c}_1(Y)e^{j(T_0+\varphi)}] + c.c. \\ \tilde{c}_1(Y) &= \frac{\tilde{a}_1(Y) - j\tilde{b}_1(Y)}{2} \end{aligned} \tag{9.38}$$

The Fourier series coefficients are calculated in the same way as the coefficients in (9.27). Substituting the solution for y_0 from (9.37) into (9.36b) we collect the terms that contain e^{jT_0} since these are the terms that lead to resonance of the undamped system. Thus, Eq. (9.36b) yields one equation for y_1 and another for the slow complex amplitude $A(T_1)$:

$$D_0^2 y_1 + y_1 = \tilde{f}_0^{\text{tran}}(Y) - \sum_{n=2}^{N} \tilde{f}_0^{\text{NL,n}}(Y) \tag{9.39a}$$

$$-2\dot{A}j - 2\tilde{\beta}Aj - \sum_{n=2}^{N} \tilde{c}_1^{\text{NL,n}}(Y)e^{j\varphi} + \tilde{c}_1^{\text{tran}}(Y)e^{j\varphi} + \frac{\tilde{\alpha}}{2}e^{j(\tilde{\sigma}T_1+\theta_0)} + c.c. = 0 \tag{9.39b}$$

Here in equations (9.39a) and (9.39b), the superscripts ^{tran} and $^{\text{NL,n}}$ are used to label the Fourier coefficients belonging to the transducer force and the nth mechanical nonlinearity force respectively.

From expression (9.39a), it follows that $y_1 = \tilde{f}_0^{\text{tran}}(Y) - \sum_{n=2}^{N} \tilde{f}_0^{\text{NL,n}}(Y)$ and thus, εy_1 represents the average shift of the mobile mass displacement due to the transducer force and the force of the mechanical nonlinearities. This is denoted as:

$$y_{\text{av}} = \varepsilon y_1 = f_0^{\text{tran}}(Y) - \sum_{n=2}^{N} f_0^{\text{NL,n}}(Y) \tag{9.40}$$

Therefore, the total solution will take the form

$$y(\tau) = y_0 + \varepsilon y_1 = y_{av} + Y\cos(\tau + \varphi) \tag{9.41}$$

Similar to the first harmonic method outlined in Sect. 9.3.1, the theory described above works very well when the average shift y_{av} is relatively small in comparison to the amplitude of oscillations Y. However, when using a gap closing transducer, the constant shift in oscillations y_{av} can be non-negligible, as seen in Fig. 9.8 when comparing the first harmonic method with a numerical simulation. So just as was done in the harmonic balance analysis, when analysing a GCT-CQ eVEH using the MSM, the effect of average shift, y_{av}, must be taken into account. Otherwise, the forces f_{tran} and $f_{NL,n}$ will be underestimated. By incorporating the fact that these forces depend on y_{av} as well as Y, then we must note that the Fourier coefficients of these forces are also functions of both y_{av} and Y.

Taking this into account and noting that $\dot{A} = \frac{1}{2}\left(\dot{Y} + jY\dot{\varphi}\right)e^{j\varphi}$, one obtains equations for the slow amplitude Y and the phase $\psi = \tilde{\sigma}T_1 - \varphi$, by dividing (9.39b) into its real and imaginary parts

$$\begin{aligned} \dot{\psi}a &= \tilde{\sigma}Y + \frac{\tilde{a}_1^{tran}(y_{av}, Y)}{2} - \sum_{n=2}^{N} \frac{\tilde{a}_1^{NL,n}(y_{av}, Y)}{2} + \frac{\tilde{\alpha}}{2}\cos\psi \\ \dot{Y} &= -\tilde{\beta}Y - \frac{\tilde{b}_1^{tran}(y_{av}, Y)}{2} + \sum_{n=2}^{N} \frac{\tilde{b}_1^{NL,n}(y_{av}, Y)}{2} + \frac{\tilde{\alpha}}{2}\sin\psi \end{aligned} \tag{9.42}$$

where $\dot{\psi} = \tilde{\sigma} - \dot{\varphi}$. It is important to note that this system of differential equations (9.42) provides information about the transient dynamics of the system, and allows one to explore the dynamics around multiple stable points and identify different possible stable modes. This ability to study multistability in the system is very important when it comes to seeing the effect of mechanical nonlinearities on the effective bandwidth.

The steady-state solution Y_0 and ψ_0 can be found from the condition $\dot{Y} = 0$ and $\dot{\psi} = 0$. For the phase ψ_0 one obtains a set of equations (assuming only non-zero harmonic coefficients)

$$\begin{aligned} \frac{\tilde{\alpha}}{2}\cos\psi_0 &= -Y_0\tilde{\sigma} - \frac{\tilde{a}_1^{tran}(y_{av,0}, Y_0)}{2} + \sum_{n=2}^{N} \frac{\tilde{a}_1^{NL,n}(y_{av,0}, Y_0)}{2} \\ \frac{\tilde{\alpha}}{2}\sin\psi_0 &= \tilde{\beta}Y_0 + \frac{\tilde{b}_1^{tran}(y_{av,0}, Y_0)}{2} - \sum_{n=2}^{N} \frac{\tilde{b}_1^{NL,n}(y_{av,0}, Y_0)}{2} \end{aligned} \tag{9.43}$$

The equation for the amplitude Y_0 can now be found from (9.43)

$$\frac{\tilde{\alpha}^2}{4} = \left(Y_0 \tilde{\sigma} + \frac{\tilde{a}_1^{\text{tran}}(y_{\text{av},0}, Y_0)}{2} - \sum_{n=2}^{N} \frac{\tilde{a}_1^{\text{NL,n}}(y_{\text{av},0}, Y_0)}{2} \right)^2 + \left(\tilde{\beta} Y_0 + \frac{\tilde{b}_1^{\text{tran}}(y_{\text{av},0}, Y_0)}{2} - \sum_{n=2}^{N} \frac{\tilde{b}_1^{\text{NL,n}}(y_{\text{av},0}, Y_0)}{2} \right)^2 \quad (9.44)$$

where the index '0' is used to emphasize that y_{av}, Y and ψ are steady-state characteristics.

Note, by multiplying both sides of (9.44) by ε^2, the equation can be written in the same form but with the original values of the parameters.

It must be noted that Eq. (9.44) is a very general equation describing the system. It is not a transcendental equation and thus it can be reduced to a polynomial which can be solved analytically to give a closed form solution or it can be reduced to a nonlinear equation and solved numerically.

The Fourier coefficients for f_{tran} were already calculated in Sect. 9.3.1 and can be seen in (9.30). Similarly, using (9.22), the Fourier coefficients for the nonlinear spring forces, $f_{\text{NL,n}} = \kappa_n y^n$, can also be calculated. Unlike the transducer force, these forces are not piecewise. It is noted that in this chapter the nonlinear oscillators that are studied only have κ_2 and κ_3 terms and so below, only the respective Fourier coefficients for these terms are shown. If one needed to calculate Fourier coefficients for more κ_n terms then it can be done in the same way using (9.22).

For κ_2 the Fourier coefficients are:

$$a_0^{\text{NL,2}}(y_{\text{av}}, Y) = \kappa_2(Y^2 + 2y_{\text{av}}^2), \qquad a_1^{\text{NL,2}}(y_{\text{av}}, Y) = 2\kappa_2 y_{\text{av}} Y, \qquad b_1^{\text{NL,2}}(y_{\text{av}}, Y) = 0$$

and for κ_3:

$$a_0^{\text{NL,3}}(y_{\text{av}}, Y) = \kappa_3(3Y^2 y_{\text{av}} + 2y_{\text{av}}^3), \quad a_1^{\text{NL,3}}(y_{\text{av}}, Y) = \frac{3}{4}\kappa_3(Y^3 + 4Y y_{\text{av}}^2), \quad b_1^{\text{NL,3}}(y_{\text{av}}, Y) = 0$$

Interestingly, since a_1^{tran} and $b_1^{\text{NL,n}}$ always equal 0 for a GCT-CQ eVEH we can simplify (9.43) to

$$\frac{\tilde{\alpha}}{2} \cos \psi_0 = -Y_0 \tilde{\sigma} + \sum_{n=2}^{N} \frac{\tilde{a}_1^{\text{NL,n}}(y_{\text{av},0}, Y_0)}{2}$$
$$\frac{\tilde{\alpha}}{2} \sin \psi_0 = \tilde{\beta} Y_0 + \frac{\tilde{b}_1^{\text{tran}}(y_{\text{av},0}, Y_0)}{2} \quad (9.45)$$

and so (9.44) simplifies to

$$\frac{\tilde{\alpha}^2}{4} = \left(Y_0 \tilde{\sigma} - \sum_{n=2}^{N} \frac{\tilde{a}_1^{\text{NL,n}}(y_{\text{av},0}, Y_0)}{2} \right)^2 + \left(\tilde{\beta} Y_0 + \frac{\tilde{b}_1^{\text{tran}}(y_{\text{av},0}, Y_0)}{2} \right)^2 \quad (9.46)$$

Thus, expressions (9.41), (9.45) and (9.46) define the steady-state response of the nonlinear oscillator (9.33) to both the external driving and the nonlinear force f_{tran}. The steady-state solution therefore is

$$y(\tau) = y_{\text{av},0} + Y_0 \cos(\Omega\tau - \psi_0) \tag{9.47}$$

9.3.3 Mechanical Impedance Method

So far, the mathematical model presented in Sect. 9.2, the harmonic balance analysis presented in Sect. 9.3.1 and the multiple scales method presented in Sect. 9.3.2, though they are all coupled electromechanical models, focused on the mechanical displacement of the resonator. It is possible though to study this electromechanical system from an electrical viewpoint. One method of doing this is via the mechanical impedance method. Note, that this method is presented using dimensional parameters rather than the normalised parameters used throughout Sects. 9.3.1 and 9.3.2.

Figure 9.10 shows the transducer as a nonlinear operator which receives a sinusoidal signal at the input (the transducers mobile mechanical terminal displacement) and outputs a non-sinusoidal periodic force. Since the resonator in the system is assumed to be high-Q, and similar to the HB and MSM analyses, only the fundamental harmonic of the system quantities are considered here. This means that the transducer must be represented by a model which reproduces a summarised contribution of the nonlinear properties of the transducer. This model is only characterized by a ratio between the amplitudes of the input and output quantities and by the phase shift, i.e. by a complex transmission coefficient which is redefined for each amplitude value of the input signal. Therefore, the equivalent model is not linear, but it can be viewed as being linear when the input amplitude is constant, i.e. when the eVEH is operating in a steady-state, period-1 regime (see Fig. 9.6). Thus, in the type of analysis outlined here, the nonlinear element of the network in Fig. 9.11 is replaced by a dipole which takes a sinusoidal current (velocity) at its input and outputs a sinusoidal voltage (transduce's force): Such an element can be characterized by a mechanical impedance, and the whole network can be analysed as linear. Table 9.2 contains the equivalent mechanical and electrical quantities present in the model.

The mechanical impedance method for analysis of a vibration energy harvester was proposed in [2]. It is a variation on the first harmonic method presented in Sect. 9.3.1. Similarly to the first harmonic method, when using the mechanical

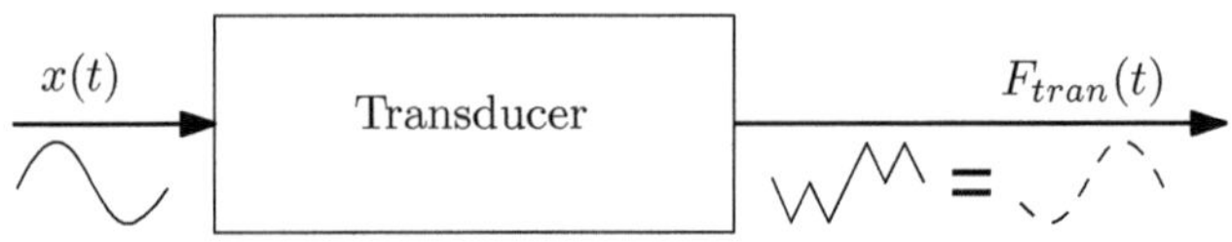

Fig. 9.10 Representation of a nonlinear electromechanical transducer in the first harmonic method

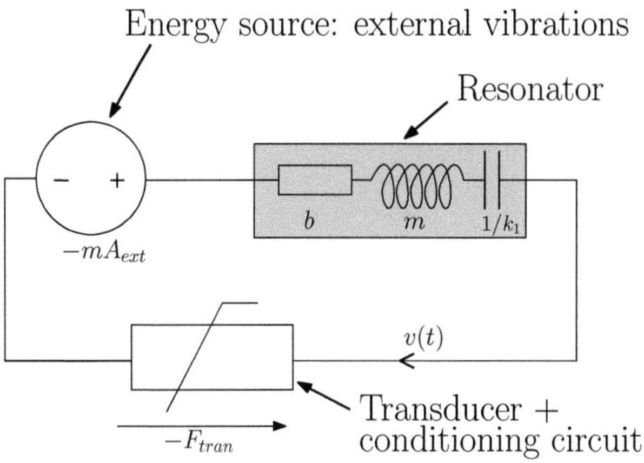

Fig. 9.11 Equivalent electrical representation of the harvester in the mechanical domain

impedance method, it is assumed that the mobile mass of an eVEH moves sinusoidally, i.e. it is assumed that the displacement, and velocity of the resonator in their dimensional forms can be expressed as:

$$
\begin{aligned}
x(t) &= x_0 \cos(\omega_{\text{ext}} t) \\
v(t) &= v_0 \sin(\omega_{\text{ext}} t)
\end{aligned} \tag{9.48}
$$

where x_0 and $v_0 = -\omega_{\text{ext}} x_0$ are the amplitudes of dimensional displacement and velocity oscillations respectively. When applying the mechanical impedance method, all forces applied to the mass are represented by mechanical impedances Ψ equal to the ratio between minus the first (fundamental) harmonic of the force expressed in its phasor (complex amplitude) form ($\bar{F}$) and the phasor of the velocity of the mass motion ($\bar{v}_0$), where a bar, $\bar{}$, is used to represent a phasor. Therefore, the impedance and the power of mechanical energy conversion, related to a force with impedance Ψ, are given by:

$$
\begin{aligned}
\Psi &= -\bar{F}/\bar{v}_0 \\
P &= |\bar{v}_0| \text{Re}\{\Psi\}
\end{aligned} \tag{9.49}
$$

The power P is positive if energy leaves the mechanical domain. The imaginary part of the impedance provides important information about the resonance frequency shift due to the nonlinearity of the resonator/transducer. As explained in [2], the mechanical impedance of a transducer, Ψ_{tran}, depends on the amplitude of the displacement x_0. Knowing the $\Psi_{\text{tran}}(x_0)$ function, it is possible to find the amplitude of mobile mass vibration in the coupled electromechanical mode. It is important to understand that although the transducer impedance relates two mechanical quantities, it is obtained by analysing the electrical side of the system.

Table 9.2 Equivalent mechanical and electrical quantities

Mechanical quantity	Electrical quantity
Force (F)	Minus voltage $(-V)$
Displacement (x)	Charge (q)
Velocity (v)	Current (i)
Mass (m)	Inductance (L)
Spring stiffness (k_1)	Reciprocal of capacitance $(1/C)$
Damping factor (b)	Resistance (R)

Mechanical Impedance of a Nonlinear Transducer in an eVEH

Usually, an impedance is defined for linear electrical systems submitted to sinusoidal excitations. As can be seen from (9.49), the mechanical impedance is defined equivalently to its electrical counterpart as minus the ratio of the complex amplitudes of the sinusoidal force and velocity of the point of force application. The complex amplitude of the fundamental harmonic of F_{tran} is given by

$$\bar{F}_{\text{tran}} = \frac{\omega_{\text{ext}}}{\pi} \int_0^T F_{\text{tran}}(t) e^{-j\omega_{\text{ext}} t} dt \tag{9.50}$$

The expression in (9.50) can be used to find the complex amplitude of any quantity in the equivalent circuit model by replacing F_{tran} by the relevant quantity. Because of the nonlinear relationship between the force and the resonator displacement, the mechanical transducer's impedance depends on the mobile terminal displacement amplitude x_0. The transducer's force depends on the transducer's electrical state defined by the conditioning circuit. Hence, the definition of the mechanical impedance of a transducer can only be done for the whole system composed of the transducer and of the conditioning circuit operating in some specific mode. Calculation of the transducer's mechanical impedance is quite complex even for simple harvesting systems and, in many cases, can only be obtained numerically. The following general procedure is proposed, which is applicable to VEHs with any nature of electromechanical transducer. First, since the energy harvester is assumed to be operating in a steady-state period-1 regime, then the value of the normalised velocity amplitude for which the transducer's impedance is calculated is assumed to be fixed at v_0 (from (9.48)). Then, under this hypothesis of the mobile mass displacement, the electrical quantities defining the transducer's force are calculated using the model of the transducer and of the conditioning circuit. From these quantities, the force generated by the transducer on one mobile mass vibration period is calculated. The mechanical impedance of the transducer is then calculated using (9.49), with $\bar{F}_{\text{tran}}$. The calculation of F_{tran} is the most complex step of this procedure. It depends on the entire system, i.e. the transducer+conditioning circuit as seen in Sect. 9.2.3. For eVEHs, a detailed description of the transducer's impedance calculation can be seen in [2]. It is clear to see that the second Newtonian law for the mobile mass displacement of the resonator is equivalent to the mesh equation for the network in Fig. 9.11:

$$-m\bar{A}_{\text{ext}} = (\Psi_{\text{tran}}(x_0) + \Psi_r)\,\bar{v}_0 \tag{9.51}$$

where $\bar{v}_0 = j\omega_{\text{ext}}x_0$ and Ψ_r is the mechanical impedance of the resonator given by

$$\Psi_r = b + \frac{k_1}{j\omega_{\text{ext}}} + j\omega_{\text{ext}}m \tag{9.52}$$

and $\bar{A}_{\text{ext}}$ is the complex amplitude of the external vibration's acceleration. Since $\bar{v}_0$ and Ψ_{tran} both depend on x_0, (9.51) is a nonlinear equation. Its solution can be found by simply equating the absolute values of both sides of it:

$$mA_{\text{ext}} = |(\Psi_{\text{tran}}(x_0) + \Psi_r)|\omega_{\text{ext}}x_0 \tag{9.53}$$

The amplitude of the resonator displacement, x_0, can now be found from (9.53).

9.3.4 Comparison of the Semi-analytical Methods

Having covered three different semi-analytical techniques for the study of eVEHs, it is now time to compare the relative merits and disadvantages of the three methods. The harmonic balance method is a non-perturbation technique (i.e. it does not need the small perturbation parameter ε outlined in Sect. 9.3.2) and gives very accurate results even when limited to the fundamental harmonic (and DC offset). When adding higher harmonics, the method can become very complicated, although this is not a concern when analysing VEHs with high-Q resonators. The multiple scales method, though initially very daunting to look at for some engineers, is, once understood, very well suited to the addition of "hardening" mechanical nonlinearities and higher harmonics. It is a perturbation-based technique though and that requires an understanding of the perturbation parameter ε. The mechanical impedance method, similar to the harmonic balance method, is also a non-perturbation technique and is a very accessible technique for users versed in circuit analysis. Unfortunately though, it is not suitable when analysing an eVEH operated with a nonlinear mechanical resonator and the calculation of the transducer mechanical impedance Ψ_t can be very complicated for particular transducer geometries and conditioning circuitry combinations. It also cannot account for the constant displacement shift present in the dynamics of an GCT-CQ eVEH.

An interesting comparison can be made between the general equations (9.26) and (9.44) found from both the harmonic balance and multiple scales methods respectively. In its most general form, for an eVEH operated with a linear resonator, (9.44) can be written as

$$\left(2Y_0\sigma + a_1^{\text{tran}}(y_{\text{av},0}, Y_0)\right)^2 + \left(2\beta Y_0 + b_1^{\text{tran}}(Y_0)\right)^2 = \alpha^2 \tag{9.54}$$

The similarities between these two general solutions, (9.26) and (9.54) are immediately clear but so too are the slight differences between them. It must be noted that some information about the effect of the small frequency mismatch σ on the dynamics of the system is lost when using the MSM. This loss of information should be negligible since σ is generally quite small due to the high-Q nature of the VEHs under analysis here (see Fig. 9.7). Both of these general equations, (9.26) and (9.54) are very general equations, applicable to any eVEH.

9.4 Visualisation of Results

In the previous sections of this chapter, some of the possible semi-analytical methods that can be used to analyse the steady-state behaviour of an eVEH have been outlined. But why do we want to study this behaviour? As previously mentioned, eVEHs can display many nonlinear phenomena in their dynamics and semi-analytical techniques allow one to study the dynamics both quickly and accurately. A simple way to show the nonlinear effects in a GCT-CQ is by applying a slowly growing ramp of acceleration amplitude A_{ext} to the system and investigating the system's response to this changing acceleration amplitude using the original, normalised, mathematical model (9.16). This process can be considered to be quasistatic since the ramp of A_{ext} changes very slowly. It is clear from Fig. 9.12 that the system's dynamics undergo various changes as a result of varying A_{ext}. Figure 9.12 can be seen as an analogue of a bifurcation diagram with A_{ext} being the bifurcation parameter. Bifurcation diagrams are explained in more detail in the following subsection. By examining Fig. 9.12, the following qualitative changes (bifurcations) can be seen, as a result of changing A_{ext}: (i) the appearance of steady-state harmonic oscillations (fragment 1 in the figure); (ii) period-doubling bifurcation (fragment 2) and (iii) transition to chaos (fragment 3).

9.4.1 Bifurcation Diagrams

It is desirable when studying a nonlinear dynamical system to understand what different types of behaviour can occur as a function of the system's control parameters. It is beneficial to be able to qualify how the dynamics of a system change and evolve as control parameters are varied. Obviously, for an eVEH there are an infinite number of control parameter values, but by taking discrete steps of one of these values over a specified range and monitoring the system dynamics at each one of these steps, conclusions can be drawn on the global evolution of the behaviour of an eVEH. A bifurcation diagram is a useful way of presenting this evolution. An analogue of a bifurcation diagram can be seen in Fig. 9.12.

To construct a one-dimensional bifurcation diagram, a control parameter is chosen to be varied across a particular range. This chosen parameter can be referred to as the *bifurcation parameter*. All other parameters of the system remain fixed. Some

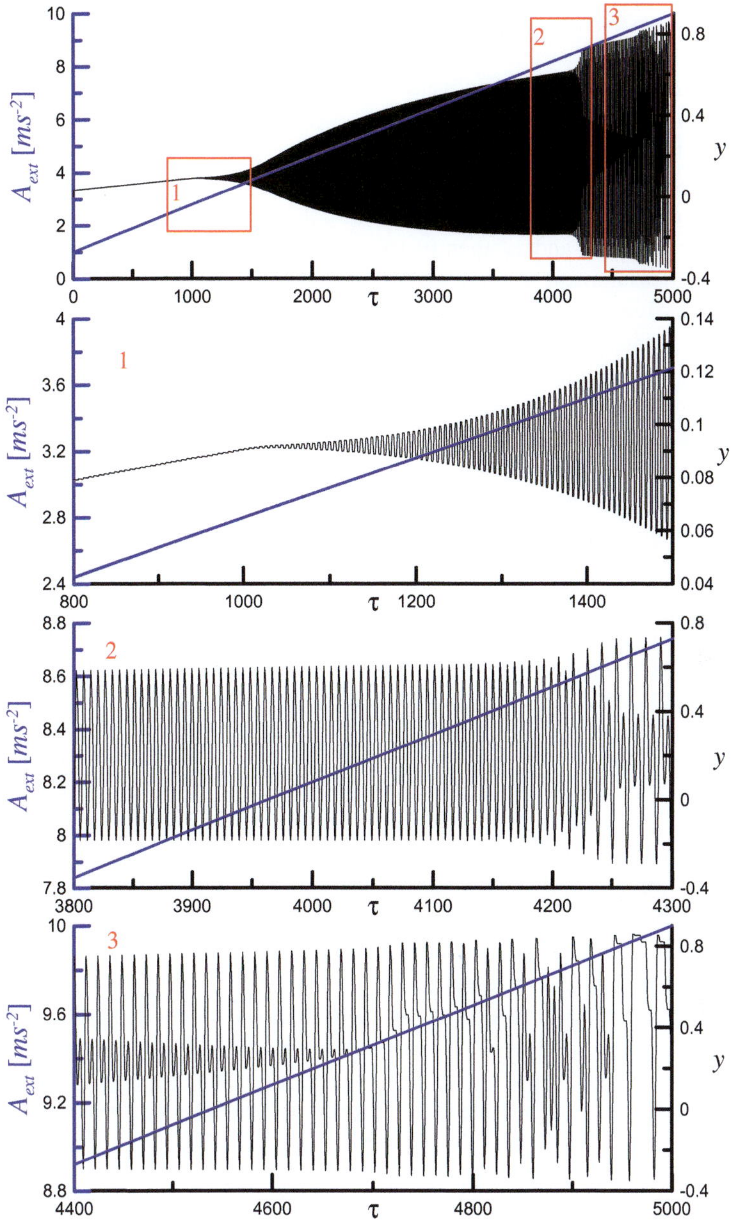

Fig. 9.12 The response of a GCT-CQ eVEH, where $W_0 = 20$ nJ, to a growing external acceleration amplitude. In the *top* subfigure, a slowly growing ramp of A_{ext} can be seen along with the normalised resonator displacemnt y. The different changes that the system's dynamics undergo have been highlighted and magnified. In fragment 1, the appearance of steady-state harmonic oscillations can be seen. In fragment 2, period-doubling bifurcation; and in fragment 3, the transition to chaos. This figure is analagous to a bifurcation diagram

output of the system is recorded for an initial value of the bifurcation parameter. The bifurcation parameter is then changed, and the same output is recorded again. This process is repeated until the final value of the bifurcation parameter is found. Every recording or measure of the output quantity in question is completely independent of every other one. For example, in the case of an eVEH modelled by (9.16), A_{ext} can be chosen as the bifurcation parameter. Choosing the resonator displacement y to be the measured output quantity, and fixing the initial conditions of displacement and velocity to be y_0 and y_0', then (9.16) is evaluated for a particular value of A_{ext}. A discrete measurement of y is then recorded, after y has reached steady-state, i.e. the transient dynamics are neglected. This procedure is then repeated for different values of A_{ext} but with every other quantity fixed and with the initial conditions reset to y_0 and y_0' on each iteration. It is the fact that the output quantity does not reach steady-state before it is recorded, which makes the trace in Fig. 9.12 analogous to a bifurcation diagram rather than a true bifurcation diagram.

How the discrete measurement of the output quantity is undertaken in a true bifurcation diagram can differ depending on the system in question or what information is being sought. A rule or a mapping is outlined and when the conditions for this mapping are met, a value of the output quantity is recorded. These mappings can be, for example, based on characteristic events that occur in a system's dynamics, when a system quantity reaches an extremum, or periodic. A very common periodic mapping used in the construction of bifurcation diagrams is a Poincaré section.

Poincaré Sections

A Poincaré section is a periodic mapping that can be used to construct bifurcation diagrams for nonlinear oscillators. It is named after French mathematician Jules Henri Poincaré. To construct a Poincaré section for a three-dimensional system like an eVEH, a two-dimensional surface called a Poincaré plane is chosen. When a trajectory evolving in three-dimensional space intersects this plane, then a measurement of the quantity of interest is recorded.

Take for example a three-dimensional autonomous system that consists of three variables X, Y and Z. If the XY plane is chosen as the Poincaré plane, the value of Z is recorded every time a trajectory crosses the plane. Depending again on the information desired from such a measurement, this recording may be limited to every occasion the trajectory crosses the plane from $+Z$ to $-Z$ or vice versa.

In the case of a non-autonomous system like the eVEH system studied here, which has a natural period associated with it, i.e. the normalised period $T_0 = 2\pi/\Omega$ of the external mechanical vibrations, then the plane could be a surface corresponding to a definite (but perhaps arbitrarily chosen) phase of the external vibrations. This is sometimes also called a "stroboscopic mapping" [14]. The principle of the mapping is presented visually in Fig. 9.13.

In the case of eVEHs, if the system is displaying steady-state period-1 behaviour for a particular value of A_{ext} and a stroboscopic mapping is performed, then for whichever quantity is being recorded (either y or y'), each recording should give the

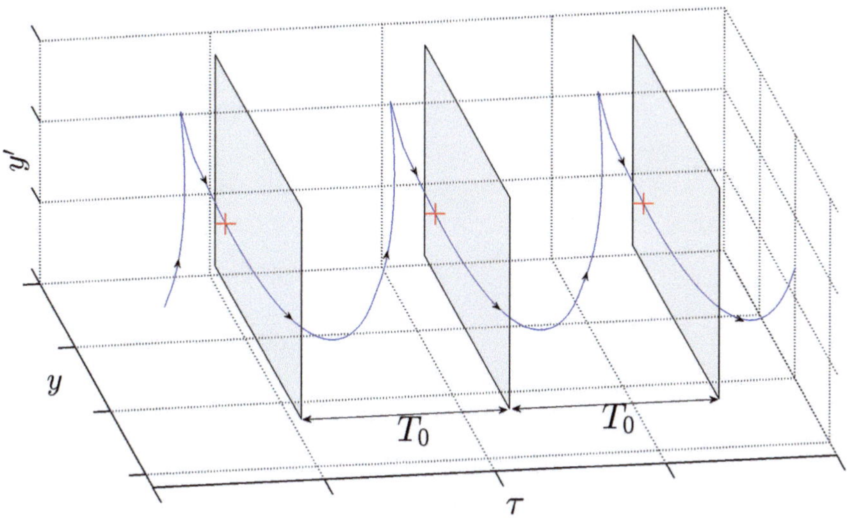

Fig. 9.13 Stroboscopic mapping. A Poincaré plane (the grey rectangle) is constructed once per period and the values of *y* and *y'* where the trajectory intersects the Poincaré plane are recorded. These intersections are marked by the red crosses. Since the system is operating in a period-1 regime in this figure, the intersection occurs at the same values of *y* and *y'* every period.

same value. This is the nature of the forced oscillatory behaviour of the system and an example of this can be seen in Fig. 9.14. If for another value of A_{ext} the system displays period-2 behaviour, then the stroboscopic mapping will record two different values, alternating on each period. If for yet another value of A_{ext} the system displays period-4 behaviour, then the stroboscopic mapping will record four different values that repeat every four periods/recordings. Thus, if the system displays period-N behaviour, the stroboscopic mapping will record N different values that repeat every N periods/recordings. If the system displays chaotic behaviour then the stroboscopic mapping will record many different values that show no discernible repetition. All of these recordings assume that the transient of the trajectory has been neglected (see Fig. 9.14).

EVEH Bifurcation Diagrams

To construct a bifurcation diagram for an eVEH, the stroboscopic mapping must first be defined. Recalling that the period of the external vibrations is $T_0 = 2\pi/\Omega$, its phase can be introduced as $\theta = 2\pi\tau/T_0 \,(\text{mod}\, 2\pi) \in \mathbb{S}$. Starting from a point in time τ_0 after transient behaviour has died down, the Poincaré section can be constructed by stroboscopically monitoring the state variables y and y' in the time interval of length T_0. Therefore the mapping can be written as

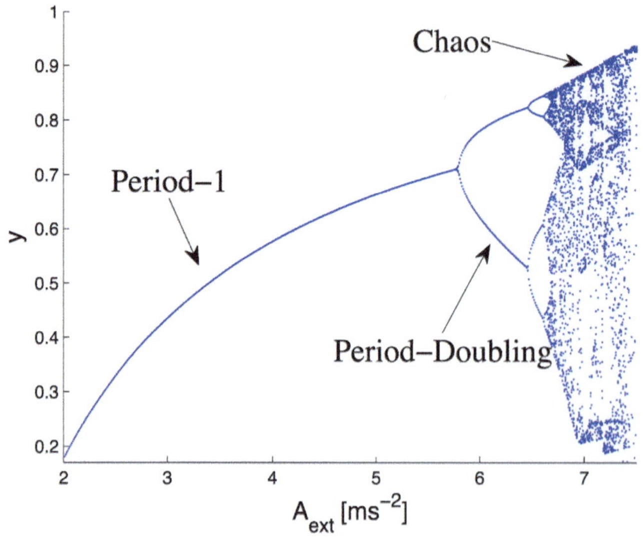

Fig. 9.14 Bifurcation diagram for a GCT-CQ eVEH operated with a linear resonator for $W_0 = 10$ nJ. In this instance the bifurcation parameter is A_{ext} and y is the parameter being plotted. The values of y are found through numerical integration with the same set of initial conditions $y = y' = 0$ for every value of A_{ext}. A bifurcation diagram like this gives a lot of information about the global dynamics of a system. Different regions of periodic behaviour can be seen along with the system's transitioning to chaos. Windows of periodicity can even be seen within the chaos at higher values of A_{ext}

$$\mathcal{P} = \left\{ (y, y', \theta) \in \mathbb{R} \times \mathbb{R} \times \mathbb{S} \mid \theta = \theta_0 = \frac{2\pi\, \tau_0}{T_0} \right\} \tag{9.55}$$

The diagram constructed using (9.55), the bifurcation diagram for the system, although very time-consuming to produce, gives a lot of very useful information about the global dynamics of the system, regions of steady-state behaviour, bifurcation points and tells us how the system transitions to chaos. An example of such a bifurcation diagram for an eVEH can be seen in Fig. 9.14. In this instance, the acceleration of the external vibrations, A_{ext}, is the chosen bifurcation parameter and it is the evolution of the normalised displacement y that is being observed as A_{ext} is being varied. It is clear to see the benefits of numerically constructed bifurcation diagrams from Fig. 9.14. A quick glance at the figure allows us to determine the global dynamical behaviour of the system, how it transitions to chaos and approximate bifurcation points.

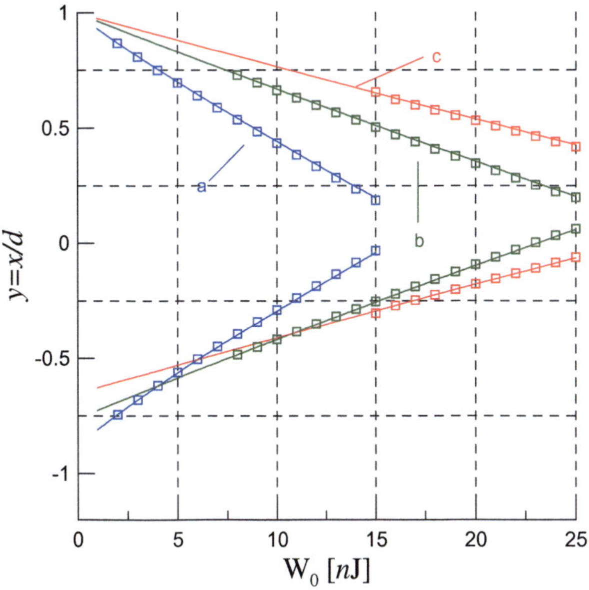

Fig. 9.15 Envelope plot—Comparing the envelope of oscillations (the y_{max} and y_{min} values) calculated using the steady-state theory (the *solid lines*) against numerically obtained values (the *squares*) for three different values of A_{ext}: line a (*blue*)—$A_{ext} = 3$ ms^{-2}, line b (*green*)—$A_{ext} = 5$ ms^{-2} and line c (*red*)—$A_{ext} = 7$ ms^{-2}

9.4.2 *Results of the MSM*

The MSM steady-state theory will now be validated by comparing results found from (9.41), (9.45) and (9.46) with results found through numerical simulations of (9.16). A GCT-CQ eVEH with a linear oscillator, first presented in [7, 15], will be examined and some differences that appear in nonlinear oscillator cases are reviewed. Finally, the effect of electrical damping on the system is examined.

The application of the developed theory to the linear and nonlinear spring cases of a GCT-CQ eVEH have been well examined in [7, 13, 15]. Noting that $y_{max} = y_{av,0} + Y_0$ and $y_{min} = y_{av,0} - Y_0$, one can use (9.40) and (9.46) to find the maximum and minimum values of the steady-state displacement of the resonator. We call y_{max} and y_{min}, for a particular combination of A_{ext} and W_0, the "envelope" of the steady-state oscillations. In Fig. 9.15, one can see a plot comparing analytically calculated values of the envelope with numerically calculated ones for three different values of A_{ext} and varying W_0. The analytical values are represented by the solid lines and the numerical values are represented using the squares. It is clear that for the linear oscillator case there is great correspondence between the numerical simulations and the theory developed in Sect. 9.3.2. The same is also true of the nonlinear oscillator cases, as can be seen in [13].

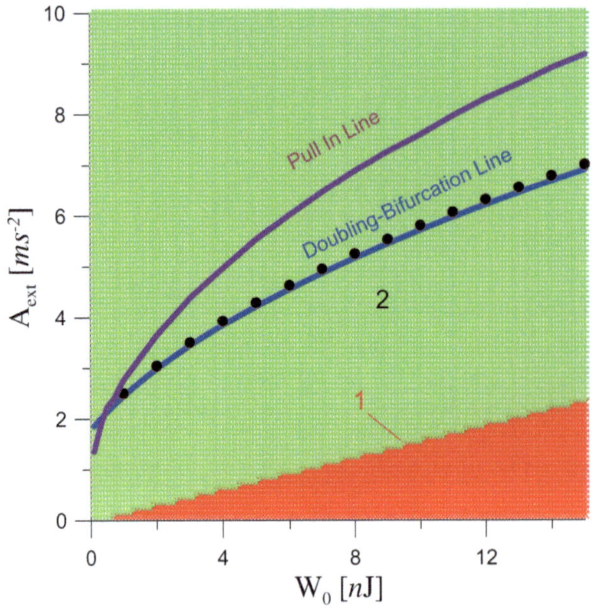

Fig. 9.16 W_0-A_{ext} plane of parameters for linear case—Region 1 (*red*) corresponds to the area where no oscillations occur, i.e. no stable roots were found. Region 2 (*green*—between *red* region and *blue/purple lines*) corresponds to areas where single stable roots were found. The doubling bifurcation line is also presented: *blue line* (analytical) and *black dots* (numerical)

Using equations (9.40) and (9.46), and by choosing A_{ext} and W_0 as the control parameters to be varied, a plane of parameters can be created for the system. Such a plane, like the one in Fig. 9.16 for a linear oscillator, displays the number of realistic roots for the various combinations of these aforementioned control parameters (A_{ext} and W_0). The red region (region '1') corresponds to W_0-A_{ext} combinations where no oscillations occur, i.e. there are no stable roots found from (9.46). If one tries to operate the system under the conditions covered by region '1', no oscillations will occur and thus the system will not work. The green region (marked '2' on the figure) shows where one stable (and realistic) root was found for the parameter pairing. This region has an upper bound formed by the combination of a pull-in line (at very low values of W_0) and a doubling bifurcation line. Pull-in is a characteristic phenomenon of electrostatic MEMS devices where the capacitor plates "snap" together. This happens in the eVEH studied here when $y = 1$. The purple pull-in line on display in the plane of parameters in Fig. 9.16 was found numerically. The doubling bifurcation line marks the onset of period-doubling oscillations in the system. The blue line in the plane represents the doubling bifurcation points that were calculated analytically using the stability theory developed in Sect. 9.5 whereas the black dots represent the doubling bifurcation points that were found numerically; there is great correspondence between these two methods. At low values of W_0, pull-in actually occurs before a doubling bifurcation happens. For any points on the plane above either the pull-in or doubling bifurcation line the steady-state theory no longer holds. Numerical analysis in the region bounded by both the doubling bifurcation and pull-in lines produces period doubling bifurcations, sliding and chaos which are characteristic phenomena of GCT-CQ eVEHs.

In Fig. 9.17, we can see a plane of parameters, found once again using the MSM, but this time for a nonlinear oscillator rather than a linear one. This nonlinear oscillator, with $\kappa_2 = 0$, $\kappa_3 = 0.3$ and κ_n (n > 3) = 0, can be described as a "hardening" spring since it causes the system to display well known hardening characteristics. The plane of parameters for this hardening case once again has red (marked as '1') and green (marked as '2') regions similar to those found in the linear plane of parameters representing the areas where no roots and one stable root are found respectively. One will also notice though that there now appears a small purple region (marked as '3') that was not on display in the linear plane of parameters. This region corresponds to control parameter value combinations that cause multistability, i.e. the existence of two or more stable solutions in the system. This is the effect of the nonlinear oscillator. The pull-in line (calculated numerically) can also be seen along with the doubling bifurcation line (calculated both analytically and numerically). Comparing these pull-in and bifurcation lines with the ones found in the linear plane of parameters, it is clear to see that they occur at higher values of A_{ext} for the nonlinear case than they do in the linear case. The nonlinearities have a stabilising effect on the system allowing it to remain operating in the steady-state region at higher values of A_{ext}. This also means that the amplitude of the steady-state displacement can be larger which is an important factor in the value of the maximum convertible power of the system [13].

Using the MSM, it is also possible to obtain an analytical frequency response of an eVEH. Figure 9.18a, shows the evolution of y_{max} versus σ (the normalised frequency mismatch) for the linear system. Note the similarity with the high-Q representation in Fig. 9.7. This is an expected frequency response for a high-Q system that clearly displays no multistability in its plane of parameters (Fig. 9.16). The frequency sweep for the hardening case, as seen in Fig. 9.18b, is a bifurcation diagram that allows us to trace the amplitude of coexisting stable oscillations as a function of the normalised frequency mismatch σ. At certain values of $\sigma > 0$ three roots were found using (9.46). This multistability is caused by the nonlinear resonator, though not all of the solutions are stable. In the figure, the blue points represent stable solutions and the red points, unstable solutions. In this frequency sweep one can observe the well known hysteresis frequency response typical of nonlinear driven oscillators. The effective bandwidth has been increased in comparison to the linear response which is the desired effect of using a nonlinear resonator in an eVEH.

Surprisingly though, if one examines the plane of parameters for the nonlinear case in Fig. 9.17, we do not see a very large area of multistabilty. This purple area (labelled '3') is relatively small considering how large the value of the nonlinearities being used are. If the nonlinear case being studied here was used in the standard Duffing equation one would expect to see a much larger area of multistability than is on display in Fig. 9.17. The only difference between (9.16) and the Duffing equation

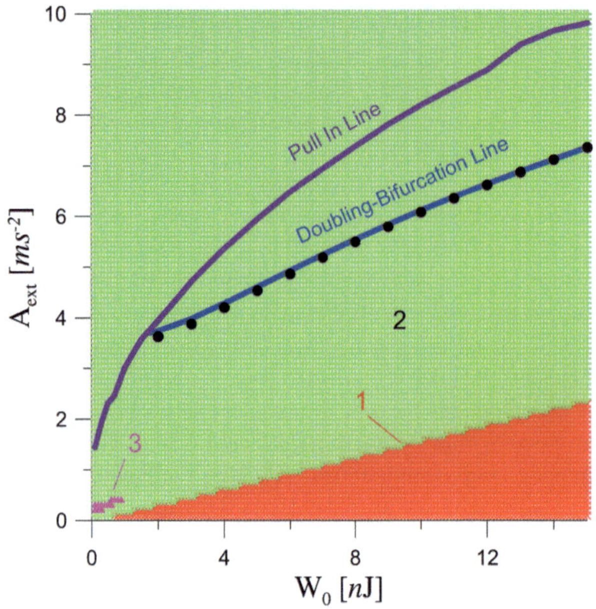

Fig. 9.17 W_0-A_{ext} plane of parameters for hardening case—Region 1 (*red*) corresponds to the area where no oscillations occur, i.e. no stable roots were found. Region 2 (*green*—between red region and *blue/purple lines*) corresponds to areas where single stable roots were found. Region 3 (*purple*—seen in the two nonlinear cases) are areas where the system displays multistability, i.e. more than one stable root was found. The doubling bifurcation line is also presented: *blue line* (analytical) and *black dots* (numerical)

is the addition of the transducer force f_{tran}. Thus, if f_{tran} was set to equal 0, (9.16) reduces to the Duffing equation.

So by examining (9.17), it can be noted that f_{tran} is directly proportional to the fixed energy of the system W_0. Thus, if W_0 is increased then the transducer force increases. Increasing W_0 in the system results in a "flattening" of the frequency response of the system. This can be clearly seen in Fig. 9.19 where $A_{ext} = 0.4\ \mathrm{ms}^{-2}$ and a softening spring with coefficients $\kappa_2 = 0.3$, $\kappa_3 = -0.1$ and κ_n $(n > 3) = 0$ have been used. Four frequency responses are shown in the figure for four different values of W_0. One can clearly observe that as energy is increased, the frequency response is flattened. Thus, the force f_{tran} which is proportional to W_0, is analogous to electrical damping in the system, and so ν_W can be seen as the dimensionless coefficient of electrical damping.

This electrical damping compensates the hysteresis effect which is caused by the nonlinear oscillator, and so, this widening of the frequency response is only observed at low values of W_0. As has been noted seen in [9, 13], the optimum region for power conversion is close to the doubling bifurcation line and it is clear to see from the plane of parameters plot in Fig. 9.17 that multistability never appears near this line.

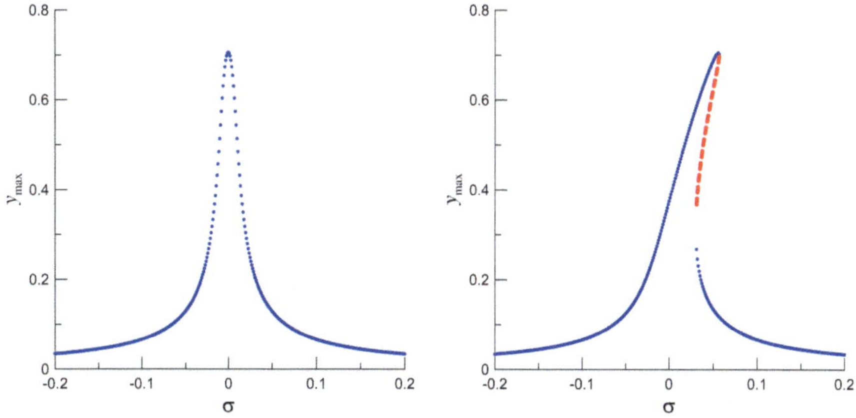

(a) Linear case (b) Hardening case

Fig. 9.18 Frequency response of a GCT-CQ eVEH found using multiple scales method for **a** a linear resonator and **b** a nonlinear resonator. It is clear that there is no multistability in the linear case but that multistability can be found in the hardening (nonlinear) case for some positive values of σ

As a result GCT-CQ eVEHs cannot fully exploit the benefits of adding nonlinearities to the system.

9.5 Stability Analysis

It was first noted in [16] that eVEHs implementing the constant charge energy conversion cycle display irregular and chaotic behaviour. Considering this, it is important that we know when these regimes occur so, when using the steady-state theories developed in Sect. 9.3, that they are only used for parameter values located within the steady-state region (marked '2' on the planes of parameters plots in Figs. 9.16 and 9.17). This requires some form of stability analysis to determine an upper bound on the steady-state region.

In smooth systems, the stability analysis of periodic orbits is based upon adding a small perturbation to the orbit and if this perturbed solution converges back to the original orbit then the orbit is said to be stable (See Fig. 9.20). This is typically carried out by employing fundamental and monodromy matrices and Floquet multipliers that are the eigenvalues of the monodromy matrix [4, 17]. Briefly, the monodromy matrix shows the local rate of convergence/divergence of small perturbations around the analysed periodic orbit. For the nonlinear dynamical system in the form $\dot{\mathbf{x}} = \mathbf{F}(\mathbf{x}, t)$, $\mathbf{x} \in \mathbb{R}^n$ that has a periodic solution $\mathbf{x_0}$, the perturbed solution is $\mathbf{x} = \mathbf{x}_0 + \tilde{\mathbf{x}}$. This small perturbation $\tilde{\mathbf{x}}$ is described by $\dot{\tilde{\mathbf{x}}} = \frac{\partial \mathbf{F}}{\partial \mathbf{x}} \tilde{\mathbf{x}}$ where $\frac{\partial \mathbf{F}}{\partial \mathbf{x}}$ is the Jacobian matrix. This n dimensional linear system (with time-varying coefficients) has n independent solutions $\{\tilde{\mathbf{x}}_i\}$ that are called the fundamental set of solutions. These can be expressed

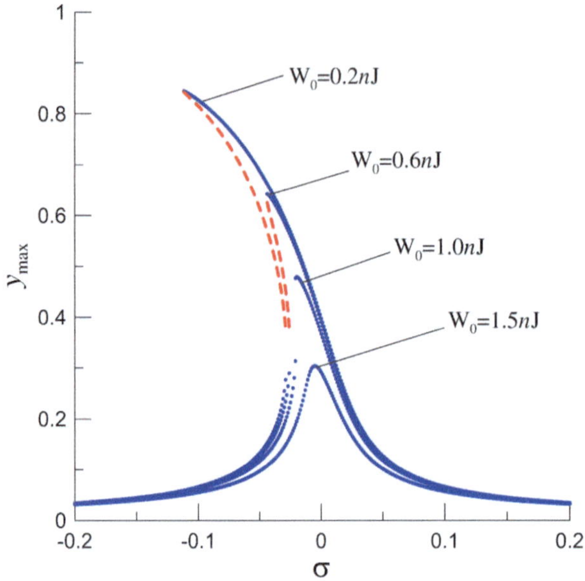

Fig. 9.19 The effect of electrical damping in the system. Four frequency responses are shown for four different values of W_0: 0.2, 0.6, 1 and 1.5 nJ. As W_0 is increased the response of the system is 'flattened' due to electrical damping. This electrical damping (which is proportional to W_0) limits the appearance of multistability in the system. In the calculations for this plot $A_{ext} = 0.4\ \mathrm{ms}^{-2}$ and a softening oscillator with oefficients $\kappa_2 = 0.3$, $\kappa_3 = -0.1$ and κ_n (n > 3) = 0 was used

in the form of a square matrix that is called the fundamental solution matrix: $\Phi(t) = [\tilde{\mathbf{x}}_1, \tilde{\mathbf{x}}_2, \ldots, \tilde{\mathbf{x}}_n,]$. The fundamental solution matrix calculated over the period $T = 2\pi/\Omega$ of the periodic orbit is called the monodromy matrix and its eigenvalues are called the Floquet multipliers [4].

By contrast though, it is clear by examination of (9.16) and (9.17) that an eVEH operating in constant charge mode is not a smooth system but rather a piecewise smooth system since it is governed by different vector fields in domains of state space separated by a switching surface Σ given by the expression $y' = 0$. This makes a GCT-CQ eVEH a Filippov type system [18]. As a result, when performing stability analysis on the system, it is more convenient to present the system in its standard dynamical form. By introducing the variable vector $\mathbf{x} = (x_1, x_2, x_3)$ where $x_1 = y$, $x_2 = y'$, $x_3 = \tau$ and the vector representing the parameters $\mathbf{P} = (\beta, \alpha, \Omega, \nu_W, \mathrm{K})$, (where $\mathrm{K} = (\kappa_2, \kappa_3, \ldots \kappa_N)$ is a vector of mechanical nonlinearities depending on resonator architecture) we can write the second-order differential equation in a standard form:

$$\dot{\mathbf{x}} = \begin{cases} \mathbf{F}_1(\mathbf{x}, \mathbf{P}), & H(\mathbf{x}) > 0 \\ \mathbf{F}_2(\mathbf{x}, \mathbf{P}), & H(\mathbf{x}) < 0 \end{cases} \tag{9.56}$$

where $\mathbf{F}_1$ and $\mathbf{F}_2$ are the vector functions

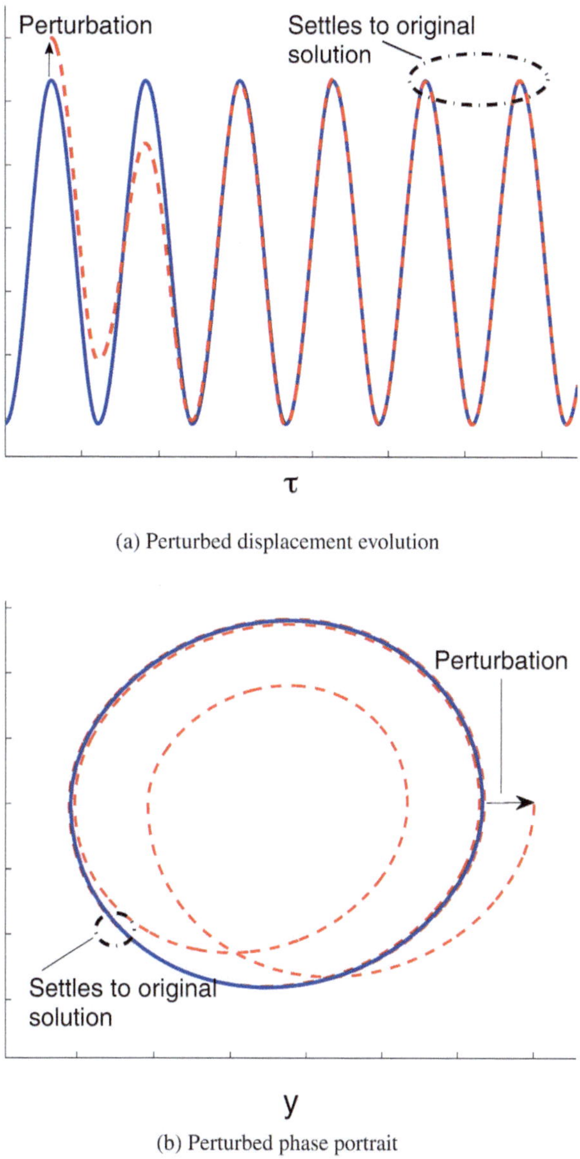

Fig. 9.20 The evolution of a perturbed stable orbit. If a trajectory is dynamically stable this means that even if a trajectory is perturbed, it will settle back to the original unperturbed orbit. This can be seen in the above figures. The *solid line* represents the original unperturbed orbit, whereas the broken line represents the perturbed solution. In (**a**), it is clear to see that the perturbed solution returns to the original solution after a few cycles of dimensionless time. The equivalent phase portrait for the perturbed and unperturbed solutions can be seen in (**b**)

$$\mathbf{F}_1(\mathbf{x},\mathbf{P}) = \begin{cases} x_2 \\ -2\beta x_2 - x_1 - \sum_{n=2}^{N} \kappa_n x_1^n + \alpha\cos\Omega x_3 + f_{tran} \\ 1 \end{cases} \tag{9.57}$$

and

$$\mathbf{F}_2(\mathbf{x},\mathbf{P}) = \begin{cases} x_2 \\ -2\beta x_2 - x_1 - \sum_{n=2}^{N} \kappa_n x_1^n + \alpha\cos\Omega x_3 \\ 1 \end{cases} \tag{9.58}$$

The scalar function $H(\mathbf{x})$ given by the switching condition

$$H(\mathbf{x}) = -x_2 = 0 \tag{9.59}$$

defines the switching surface Σ (or the switching manifold) in the state space. Now we can write that the switching surface Σ is

$$\Sigma = \left\{\mathbf{x} \in R^3 \mid H(\mathbf{x}) = 0\right\} \tag{9.60}$$

System (9.56) belongs to the class of Filippov systems in which the two smooth functions that define the behaviour of the system in the two sub-spaces are not equal to each other on the switching manifold, i.e. $\mathbf{F}_1 \neq \mathbf{F}_2$ on Σ. In order to avoid ambiguity in the definition of solutions, Filippov suggested that the function $\mathbf{F}_s$ that defines the system behaviour exactly on the switching surface when $H(\mathbf{x}) = 0$ must be a set valued function [18], whose limits are the values of $\mathbf{F}_1$ and $\mathbf{F}_2$ immediately before and after the switching event. Therefore, formally, Eq. (9.56) must be supplied with an additional expression: $\dot{\mathbf{x}} = \mathbf{F}_s(\mathbf{x},\mathbf{P})$ if $H(\mathbf{x}) = 0$. Here $\mathbf{F}_s$ is the closed convex set containing both $\mathbf{F}_1$ and $\mathbf{F}_2$. The function $\mathbf{F}_s$ governing the sliding flow can be constructed using either Filippov's convex method [18] or Utkin's equivalent control method [19]. Given the piecewise smooth nature of the system, the approach used for stability analysis in smooth systems must be modified for an eVEH since the original and perturbed orbits may not cross the switching manifold at the same time. This possibility must be accounted for when determining the monodromy matrix of a discontinuous system like the one presented here [17, 18].

Figure 9.21 shows an example of the evolution of a trajectory of the studied system on the (y, y') phase plane. It is clear to see that the switching manifold Σ is represented in this figure by the x axis, i.e. the switching condition $y' = 0$ and the times τ_0 and τ_1 mark the points when the orbit crosses the manifold at $y = y_{max}$ and $y = y_{min}$ respectively.

The approach described in [18] considers small perturbations to the initial conditions of an orbit and examines how these perturbations evolve over a period T. To account for the discontinuity in the system, a map is required that will describe the evolution of the fundamental matrix at the switching manifold and thus relate the perturbation vectors before and after the switching event. This map is called a

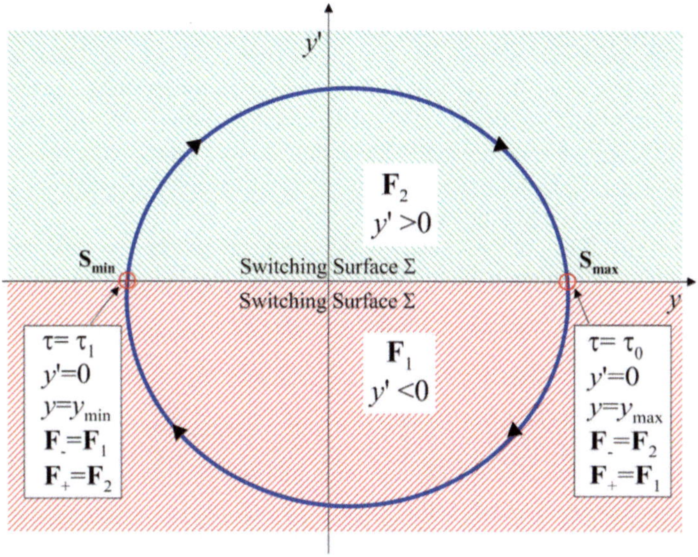

Fig. 9.21 A representation of a trajectory on the (y,y') phase plane. The trajectory moves through two different vector fields $\mathbf{F}_1$ and $\mathbf{F}_2$ which are separated by the switching manifold Σ (given by the expression $y' = 0$). $\mathbf{F}_1$ and $\mathbf{F}_2$ are described by expressions (9.57) and (9.58) respectively. The time instants τ_0 and τ_1 correspond to the moments when the trajectory crosses the switching manifold (y-axis)

saltation matrix **S** [20] and is defined as

$$\Phi(\tau_0^+) = \mathbf{S}\Phi(\tau_0^-). \tag{9.61}$$

where

$$\mathbf{S} = \mathbf{I} + \frac{(\mathbf{F}_+ - \mathbf{F}_-)\mathbf{n}^{\mathrm{T}}}{\mathbf{n}^{\mathrm{T}}\mathbf{F}_-} \tag{9.62}$$

and I is the identity matrix. Note that the trajectory spends an infinitesimally small time at the switching manifold Σ. The vector **n** is a normal vector directed as shown:

$$\mathbf{n}^{\mathrm{T}} = [\nabla H(\mathbf{x})]^{\mathrm{T}} = [\partial H(\mathbf{x})/\partial x_1, \ldots, \partial H(\mathbf{x})/\partial x_{\mathrm{n}}] \tag{9.63}$$

The function $\mathbf{F}_-$ denotes the governing vector field just before (symbolically denoted as τ^-) a trajectory crosses the switching manifold and $\mathbf{F}_+$ denotes the governing vector field just after (symbolically denoted as τ^+) a trajectory crosses the switching manifold.

Therefore, if one starts the analysis of the orbit from Fig. 9.21 at the moment τ_0^+, the monodromy matrix for one cycle of oscillations takes the form:

$$\mathbf{M} = \Phi(\tau_0, \tau_0 + 2\pi/\Omega) = \mathbf{S}_{\max}\Phi(\tau_1^+, \tau_0^- + 2\pi/\Omega)\mathbf{S}_{\min}\Phi(\tau_0^+, \tau_1^-) \tag{9.64}$$

where $\mathbf{S}_{\max} = \mathbf{S}(\tau = \tau_0 + T)$ and $\mathbf{S}_{\min} = \mathbf{S}(\tau = \tau_1)$ are the saltation matrices (9.62). Since the discontinuity boundary does not depend on time in this instance, x_3 can be neglected from the stability analysis and this reduces the systems of equations in (9.57) and (9.58). This in turns simplifies the stability analysis and the calculation of the saltation and monodromy matrices.

The normal vector (directed towards the subspace governed by the vector field $\mathbf{F}_1$ is $\mathbf{n}^{\mathrm{T}} = [0, -1]$. At the time instant $\tau = t_0$ according to Fig. 9.21, $\mathbf{F}_- = \mathbf{F}_2$ and $\mathbf{F}_+ = \mathbf{F}_1$.
Therefore

$$\mathbf{S}_{\max} = \mathbf{I} + \left.\frac{(\mathbf{F}_1 - \mathbf{F}_2)\mathbf{n}^{\mathrm{T}}}{\mathbf{n}^{\mathrm{T}}\mathbf{F}_2}\right|_{\tau=\tau_0+T} = \begin{pmatrix} 1 & 0 \\ 0 & 1 - \frac{\nu_W}{(1-x_{1,\max})(x_{1,\max} - \alpha\cos(\Omega(\tau_0+T)) + \sum_{\mathrm{n}=2}^{\mathrm{N}} \kappa_{\mathrm{n}} x_{1,\max}^{\mathrm{n}})} \end{pmatrix} \tag{9.65}$$

Note that x_1 is replaced by $x_{1,\max}$ and $x_2 = 0$ since $\tau_0 + T$ is the moment when the velocity is zero and the displacement passes its maximum.

At the time instant $\tau = \tau_1$, according to Fig. 9.21, $\mathbf{F}_- = \mathbf{F}_1$ and $\mathbf{F}_+ = \mathbf{F}_2$.
Therefore

$$\mathbf{S}_{\min} = \mathbf{I} + \left.\frac{(\mathbf{F}_2 - \mathbf{F}_1)\mathbf{n}^{\mathrm{T}}}{\mathbf{n}^{\mathrm{T}}\mathbf{F}_1}\right|_{\tau=\tau_1} = \begin{pmatrix} 1 & 0 \\ 0 & 1 + \frac{\nu_W}{(1-x_{1,\max})(x_{1,\min} + \sum_{\mathrm{n}=2}^{\mathrm{N}} \kappa_{\mathrm{n}} x_{1,\min}^{\mathrm{n}} - \alpha\cos(\Omega\tau_0) - \frac{\nu_W}{(1-x_{1,\max})})} \end{pmatrix} \tag{9.66}$$

In this case, x_1 is replaced by $x_{1,\min}$ and $x_2 = 0$ since τ_1 is the moment when the velocity is zero and the displacement passes its minimum. However, the term $f_{\mathrm{tran}} = \nu_W/(1 - x_{1,\max})$ still has the same value as it had at $\tau = \tau_0$ since the value of f_{tran} is fixed for the whole period when displacement is at its maximum, i.e. this is the hold-on term.

Finally, the variation equation that is used to find the fundamental matrix Φ_i at $\tau_1 < \tau < \tau_0 + 2\pi/\Omega$ is obtained simply from the perturbation equation

$$\dot{\tilde{\mathbf{x}}} = D(\mathbf{F_2})\tilde{\mathbf{x}} \quad \Rightarrow \quad \begin{pmatrix} \dot{x}_1 & \dot{x}_2 \\ \dot{\upsilon}_1 & \dot{\upsilon}_2 \end{pmatrix} = \begin{pmatrix} \frac{\partial \mathbf{F}_{2_1}}{\partial x_1} & \frac{\partial \mathbf{F}_{2_1}}{\partial x_2} \\ \frac{\partial \mathbf{F}_{2_2}}{\partial x_1} & \frac{\partial \mathbf{F}_{2_2}}{\partial x_2} \end{pmatrix} \begin{pmatrix} x_1 & x_2 \\ \upsilon_1 & \upsilon_2 \end{pmatrix} \tag{9.67}$$

and, therefore, for the fundamental system of solutions one obtains

$$\begin{pmatrix} \dot{x}_1 & \dot{x}_2 \\ \dot{\upsilon}_1 & \dot{\upsilon}_2 \end{pmatrix} = \begin{pmatrix} 0 & 1 \\ -(1 + \sum_{\mathrm{n}=2}^{\mathrm{N}} n\kappa_{\mathrm{n}} x_1^{\mathrm{n}-1}) & -2\beta \end{pmatrix} \begin{pmatrix} x_1 & x_2 \\ \upsilon_1 & \upsilon_2 \end{pmatrix} \tag{9.68}$$

where the following initial conditions are used:

$$\begin{pmatrix} x_1 & x_2 \\ v_1 & v_2 \end{pmatrix} = \mathbf{I} \tag{9.69}$$

where **I** is the identity matrix as in the formulas above. Solving the system of equations (9.68) on the interval $(\tau_1, \tau_0 + 2\pi/\Omega)$ one obtains $\Phi(\tau_1^+, \tau_0^- + 2\pi/\Omega)$.

The perturbation equation for $\tau_0 < \tau < \tau_1$, is $\dot{\tilde{\mathbf{x}}} = D(\mathbf{F_1})\tilde{\mathbf{x}}$. In order to obtain this, $\mathbf{F}_1$ is presented as $\mathbf{F}_1 = \mathbf{F}_1^* + \mathbf{F}_t$. Here $\mathbf{F}_1^*$ is a differentiable part of $\mathbf{F}_1$ with respect to **x**, and $\mathbf{F}_t = (0, f_{\text{tran}})^T$ can be seen as an operator since it contains $x_{1,\max}$ and is not differentiable with respect to **x**. The perturbation of $\mathbf{F}_1^*$ can be easily found and happens to give the same result as the one for $\mathbf{F}_2$ (see expression (9.68)). Assuming the original orbit being analysed is $\mathbf{x}_0(\tau_0)$, then the perturbed trajectory will be $\mathbf{x}(t_0) = \mathbf{x}_0 + \tilde{\mathbf{x}}$. Thus, the perturbation equation for $\tau_0 < \tau < \tau_1$ is

$$\begin{pmatrix} \dot{x}_1 & \dot{x}_2 \\ \dot{v}_1 & \dot{v}_2 \end{pmatrix} = \begin{pmatrix} 0 & 1 \\ -(1 + \sum\limits_{n=2}^{N} n\kappa_n x_1^{n-1}) & -2\beta \end{pmatrix} \begin{pmatrix} x_1 & x_2 \\ v_1 & v_2 \end{pmatrix} + \begin{pmatrix} 0 & 0 \\ \frac{v_W}{[1-x_{1,0}(\tau_0)]^2} & 0 \end{pmatrix} \begin{pmatrix} x_1 & x_2 \\ v_1 & v_2 \end{pmatrix}\Bigg|_{\tau=\tau_0} \tag{9.70}$$

It must be noted that (9.70) is linear and contains time dependent periodic coefficients with $T = 2\pi/\Omega$.

Using (9.65), (9.66), (9.68) and (9.70), the mondoromy matrix in (9.64) can be formed. Upon obtaining the monodromy matrix, M, its eigenvalues can be found. These eigenvalues are called Floquet multipliers [4] and they allow the stability of the orbit to be determined. If even one of these Floquet multipliers exists outside the unit circle for a set of parameter values then that particular orbit is unstable.

Floquet multipliers can also give us information about what type of bifurcation occurs when an orbit leaves the steady-state region. If a Floquet multiplier leaves the unit circle through -1, the original fixed point or orbit undergoes a doubling bifurcation; if it leaves through +1, a transcritical, symmetry breaking or cyclic fold bifurcation occurs, and, finally, if two complex conjugate multipliers leave the circle, a Hopf bifurcation takes place in the system.

By performing this stability analysis on orbits in a GCT-CQ eVEH one will note that the Floquet multipliers in these results leave the unit circle through -1 and so the steady-state region of oscillations is limited by a doubling bifurcation. The monodromy matrix (9.64) was used to calculate the blue doubling bifurcation lines that appear in the planes of parameters plots in Figs. 9.16 and 9.17 and thus any roots that appear above this doubling bifurcation boundary in these plots are no longer steady-state. The stability theory developed here generally shows great correspondence with numerical simulations of when the doubling bifurcation occurs as can be seen by the black dots in the plane of parameters plots in Figs. 9.16 and 9.17.

References

1. Meninger, S., Mur-Miranda, J., Amirtharajah, R., Chandrakasan, A., & Lang, J. (2001). Vibration-to-electric energy conversion. *IEEE Transactions on Very Large Scale Integration (VLSI) Systems*, *9*(1), 64–76.
2. Galayko, D., & Basset, P. (2011). A general analytical tool for the design of vibration energy harvesters (VEHs) based on the mechanical impedance concept. *IEEE Transactions on Circuits and Systems I*, *99*, 299–311.
3. Juillard, J. (2014). A comparative study of reduced-order modeling techniques for nonlinear mems beams. In *DTIP 2014*, pp. 261–265.
4. Nayfeh, A. H., & Balachandran, B. (2008). *Applied nonlinear dynamics* (Vol. 24). Wiley-VCH.
5. Nguyen, C. H., & Halvorsen, E. (2014). Harmonic-balance analysis of nonlinear energy harvester models In *IEEE International Symposium on Circuits and Systems (ISCAS)*, pp. 2608–2611.
6. Nayfeh, A. (1993). *Introduction to perturbation techniques*. Wiley.
7. Blokhina, E., Galayko, D., Basset, P., & Feely, O. (2013). Steady-state oscillations in resonant electrostatic vibration energy harvesters. *IEEE Transactions on Circuits and Systems I*, *60*, 875–884.
8. Blokhina, E., Fournier-Prunaret, D., Harte, P., Galayko, D., & Feely, O. (2013). Combined mechanical and circuit nonlinearities in electrostatic vibration energy harvesters. In *Proceedings of the IEEE International Symposium on Circuits and Systems 2013, Beijing, China, 19–23 May 2013*, 2013.
9. Blokhina, E., Galayko, D., Harte, P., Basset, P., & Feely, O. (2012). Limit on converted power in resonant electrostatic vibration energy harvesters. *Applied Physics Letters*, *101*, 173904.
10. Yen, B. C., & Lang, J. H. (2006). A variable-capacitance vibration-to-electric energy harvester. *IEEE Transactions on Circuits and Systems I*, *53*, 288–295.
11. Shu, Y., & Lien, I. (2006). Efficiency of energy conversion for a piezoelectric power harvesting system. *Journal of Micromechanics and Microengineering*, *16*, 2429.
12. Levitan, E. (1960). Forced oscillation of a spring-mass system having combined coulomb and viscous damping. *The Journal of the Acoustical Society of America*, *32*, 1265.
13. Harte, P., Blokhina, E., Feely, O., Fournier-Prunaret, D., & Galayko, D. (2014). Electrostatic vibration energy harvesters with linear and nonlinear resonators. *International Journal of Bifurcation and Chaos*, *24*(11), 1430030.
14. Hilborn, R. C. (2000). *Chaos and nonlinear dynamics: An introduction for scientists and engineers*. Oxford: Oxford University Press. www.summon.com.
15. Blokhina, E., Galayko, D., Wade, R., Basset, P., & Feely, O. (2012). Bifurcations and chaos in electrostatic vibration energy harvesters. In *Proceedings of the IEEE International Symposium on Circuits and Systems 2012, Seoul, Korea, 20–24 May 2012*, 2012, pp. 397–400.
16. Galayko, D., Guillemet, R., Dudka, A., Basset, P. (2011). Comprehensive dynamic and stability analysis of electrostatic vibration energy harvester (E-VEH). In *Proceedings of International Conference on Solid-State Sensors, Actuators and Microsystems (TRANSDUCERS)*, 2011, pp. 2382–2385.
17. Giaouris, D., Banerjee, S., Zahawi, B., & Pickert, V. (2008). Stability analysis of the continuous-conduction-mode buck converter via Filippov's method. *IEEE Transactions on Circuits and Systems I: Regular Papers*, *55*(4), 1084–1096.
18. Filippov, A. (1998). *Differential equations with discontinuous righthand sides: Control systems*. Springer.
19. Utkin, V. (1992). *Sliding modes in control optimization*. Springer.
20. Leine, R., & Nijmeijer, H. (2006). *Dynamics and bifurcations of non-smooth mechanical systems*. Springer.

Chapter 10
Nonlinear Conditioning Circuits for Piezoelectric Energy Harvesters

Adrien Badel and Elie Lefeuvre

10.1 Introduction

Design and analysis of piezoelectric vibration energy harvesters is a complex multi-physics problem related to mechanics, materials science, and electronics. The analysis of works published in the field of piezoelectric energy harvesting over the last decade shows that numerous papers focused on mechanical optimization without taking into account the actual constraints and requirements on the electrical side of the systems (i.e., the electric circuit was often modeled as a simple resistor). Conversely, other works aimed at optimizing systems from the electrical point of view without taking into account the mechanical effects induced by the energy conversion process. Consequently, in both cases the solutions proposed were not truly optimal or remained very far from practical applications.

To highlight the main aspects of this multi-physics problem, this chapter begins with general considerations about harvested power based on the simple and well-known model proposed by William and Yates [1]. Starting from this model, the maximal power and the frequency bandwidth of the system is analyzed, and a figure of merit taking into account both the power and the bandwidth of the energy harvester is proposed.

This model is then refined to include the description of piezoelectric electromechanical coupling, leading to accurate and reliable behavioral representation of most linear inertial piezoelectric vibration energy harvesters.

Using the classical analogies between electrical and mechanical figures, an equivalent electrical circuit representing the whole electromechanical system is described. Such circuit may be very convenient to study the system associated to its electronic

A. Badel (✉)
SYMME, Université Savoie Mont Blanc, 74000 Annecy, France
e-mail: adrien.badel@univ-savoie.fr

E. Lefeuvre
IEF, Université Paris Sud, 91405 Orsay Cedex, France

E. Blokhina et al. (eds.), *Nonlinearity in Energy Harvesting Systems*,
DOI 10.1007/978-3-319-20355-3_10

interface using SPICE-type simulations. In order to get the model parameters representing an actual vibration energy harvester, an identification procedure based on the measurement of the complex admittance is presented.

The performances of the system using optimal impedance matching approach is then analyzed, followed by a brief outline of a possible implementation using a PWM inverter circuit.

The association of the piezoelectric vibration energy harvester with the classical rectifier circuit is then analyzed, showing the need for an additional maximum power point tracking system.

After a short description of the main nonlinear interface circuits developed over the last decade, a detailed analysis of the so-called OSECE interface circuit is exposed.

Finally, this chapter discusses the possibility of tuning the resonant frequency of the piezoelectric energy harvester through its interface circuit. The theoretical analysis of a new interface developed for this purpose and termed FTSECE is presented.

10.2 A First Very Simple Model for Kinetic Energy Harvesters

The power generated by a vibration energy harvester depends on the transducer used for the electromechanical energy conversion and on the way the transducer is implemented in a mechanical structure that is suitable to capture ambient mechanical energy.

In most vibration energy harvesters (VEH) an inertial mass driven by the ambient acceleration is used to transmit mechanical energy to the electromechanical transducer. A simple {mass–spring–damper} system, initially proposed in. [1] can be used to model this behavior. A schematic of such an inertial vibration energy harvester is shown in Fig. 10.1, where D_L is a damper that embodies mechanical and electrical losses and D_H is a damper that corresponds to the electromechanical transducer. The assumption of modeling the transducer by a simple damper is valid in the case of sinusoidal motion of the inertial mass, choosing the value of D_H such as the energy dissipated into the damper during one mechanical period corresponds to the harvested energy.

10.2.1 Consideration on the Harvested Power

The equation governing the motion of the inertial mass is given by (10.2.1), where $\gamma = \ddot{y}$ is the ambient acceleration.

$$M\ddot{x} + (D_H + D_L)\dot{x} + Kx = -M\gamma \tag{10.2.1}$$

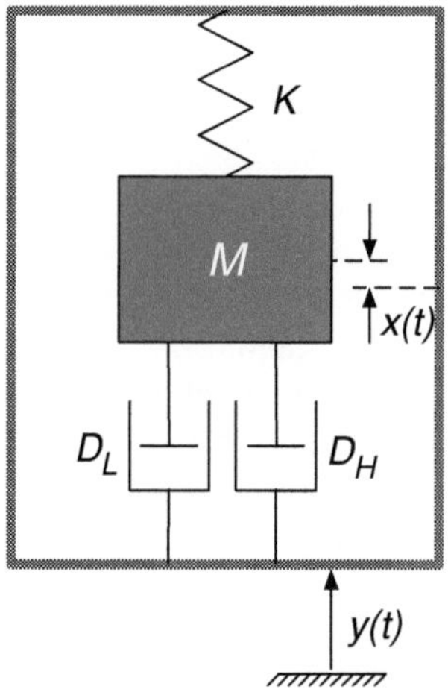

Fig. 10.1 Inertial vibration energy harvester

The relative motion x of the mass with respect to the housing can then be expressed in the frequency domain as

$$\underline{x} = -\frac{\frac{\gamma}{\omega_0^2}}{\left(1-\Omega^2\right) + 2j\Omega\left(\xi_L + \xi_H\right)} \text{ with } \begin{cases} \omega_0 = \sqrt{\frac{K}{M}} \\ \xi_L = \frac{D_L}{2}\sqrt{\frac{1}{KM}} \\ \xi_H = \frac{D_H}{2}\sqrt{\frac{1}{KM}} \\ \Omega = \frac{\omega}{\omega_0} \end{cases} \tag{10.2.2}$$

where ω is the operation angular frequency, ω_0 is the mechanical oscillator natural angular frequency, Ω is the normalized operation frequency and ξ_L and ξ_H are the damping ratios due to intrinsic losses and energy harvesting, respectively.

Normalizing x with respect to γ_M/ω_0^2 (which is the ambient displacement magnitude) leads

$$\underline{x}' = -\frac{1}{\left(1-\Omega^2\right) + 2j\Omega\left(\xi_L + \xi_H\right)} \tag{10.2.3}$$

The harvested power, calculated as the power dissipated in the damper D_H is then

$$P = \frac{D_H\omega^2 x_M^2}{2} = \frac{\xi_H\Omega^2}{\left(1-\Omega^2\right)^2 + 4\Omega^2\left(\xi_L + \xi_H\right)^2}\frac{M\gamma_M^2}{\omega_0} \tag{10.2.4}$$

This power reaches a maximum at the resonant frequency ($\Omega = 1$) and when the damping ratios are equals ($\xi_H = \xi_L$):

$$P_{\lim} = \frac{D_H\omega^2 x_M^2}{2} = \frac{M\gamma_M^2}{16\omega_0\xi_L} \tag{10.2.5}$$

The maximal power that can be harvested is proportional to the inertial mass and to the square of the ambient acceleration magnitude. It is also inversely proportional to the losses (mechanical and electrical) in the harvester and to the natural angular frequency.

This power limit is independent from the transduction principle itself. This means that whatever the transduction mechanism, the maximal power can be harvested provided that the electromechanical coupling is high enough to reach the intrinsic damping ratio [2].

The harvested power can be normalized as shown in (10.2.6). This allows to compare the performances of different energy harvesters independently from their resonant frequencies inertial masses and from the ambient acceleration magnitude. The normalized power cannot exceed $1/(2\xi_L) = Q_M$ which is the quality factor of the mechanical oscillator.

$$P' = P\frac{8\omega_0}{M\gamma_M^2} = \frac{8\xi_H\Omega^2}{\left(1-\Omega^2\right)^2 + 4\Omega^2\left(\xi_L+\xi_H\right)^2} \tag{10.2.6}$$

It is worthy of note that this result is only valid for inertial linear vibration energy harvesters driven around their resonance frequencies by sinusoidal vibrations.

10.2.2 Consideration on the Frequency Bandwidth

When considering an inertial vibration energy harvester, not only the maximal power but also the frequency bandwidth has to be studied. Equation (10.2.5) shows that intrinsic losses represented by ξ_L should be minimized to increase the maximal power. The goal of this subsection is to highlight the effect of ξ_L on the frequency response of the VEH.

We consider here the case in which the damping ratio due to energy harvesting ξ_H is equal to ξ_L in order to maximize the power at the resonance frequency. In a first consideration, ξ_H is kept constant over the frequency. The normalized power is then given by:

$$P' = \frac{8\xi_L\Omega^2}{\left(1-\Omega^2\right)^2 + 16\Omega^2\xi_L^2} \tag{10.2.7}$$

we define the angular frequency bandwidth $\Delta\omega$ as the angular frequency range where the power is at least 50 % of its maximal value ($P'_{\lim} = Q_M$). It is obtained seeking Ω_1 and Ω_2 so that $P'(\Omega_1) = P'(\Omega_2) = P'_{\max}/2$. The normalized bandwidth $\Delta\Omega$ is then given by (10.2.8), and the bandwidth by $\Delta\omega = \Delta\Omega \cdot \omega_0$.

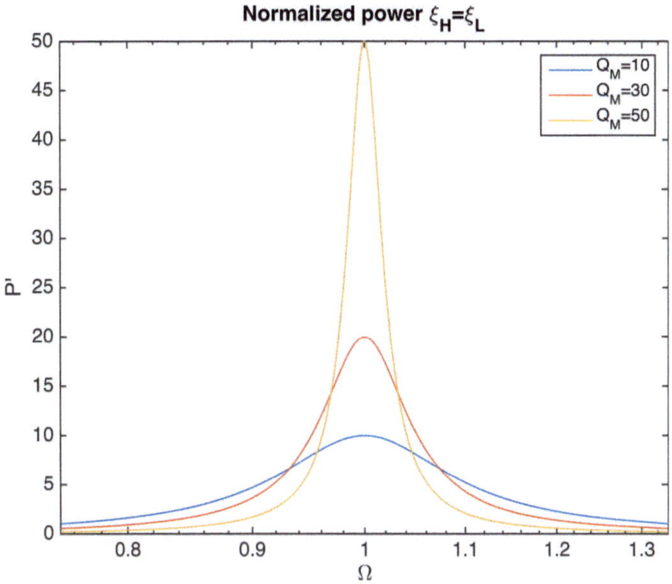

Fig. 10.2 Normalized power versus frequency for different values of $Q_M = 1/(2\xi_L)$ when $\xi_H = \xi_L$

$$\begin{cases} \Omega_1 = \sqrt{8\xi_L^2 + 1 - 4\xi_L\sqrt{4\xi_L^2 + 1}} \\ \Omega_2 = \sqrt{8\xi_L^2 + 1 + 4\xi_L\sqrt{4\xi_L^2 + 1}} \end{cases} \Rightarrow \Delta\Omega = \Omega_2 - \Omega_1 = 4\xi_L \tag{10.2.8}$$

Figure 10.2 shows the normalized power as a function of the frequency, for three different values of the intrinsic losses. It is clearly shown that decreasing the intrinsic losses increases the maximal power but decreases the bandwidth.

It is possible to increase the bandwidth if ξ_H can be adjusted as a function of the frequency. An ideal case is now considered, where ξ_H can be tuned without any limitation. The optimal value of ξ_H that maximizes the harvested power can be obtained as a function of Ω by looking for the roots of the derivative of Eq. (10.2.6) with respect to ξ_H. It is given by:

$$\xi_{Hopt} = \frac{\sqrt{\Omega^4 + 4\Omega^2\xi_L^2 - 2\Omega + 1}}{2\Omega} \tag{10.2.9}$$

And the expression of the normalized power is then

$$P'\left(\xi_{Hopt}\right) = \frac{4\Omega\sqrt{\Omega^4 + 4\Omega^2\xi_L^2 - 2\Omega + 1}}{4\Omega^2\left(\xi_L + \frac{\sqrt{\Omega^4+4\Omega^2\xi_L^2-2\Omega+1}}{2\Omega}\right)^2 + \left(\Omega^2 - 1\right)^2} \tag{10.2.10}$$

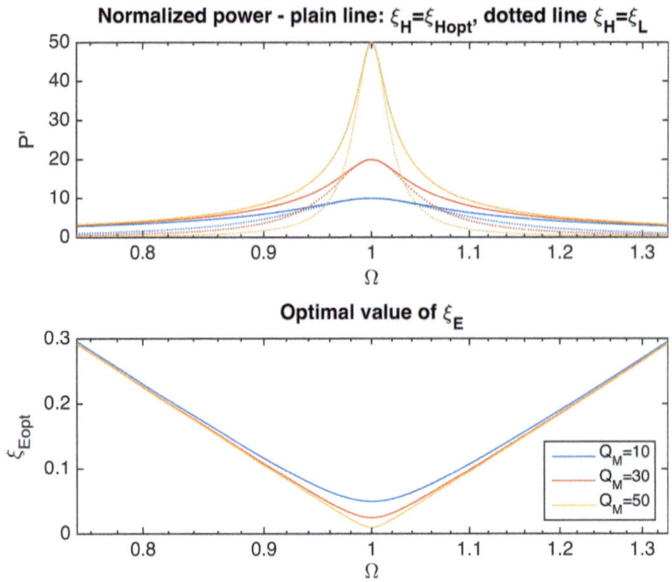

Fig. 10.3 *Upper plot* normalized power versus frequency for different values of $Q_M = 1/(2\xi_L)$ when $\xi_H = \xi_{Hopt}(\Omega)$ (*plain line*) and when $\xi_H = \xi_L$ (*dotted line*), *Lower plot* optimal value of ξ_H

The new expressions of the half-power normalized angular frequencies are given by (10.2.11), together with the normalized bandwidth, which is shown to be $\sqrt{2}$ times larger than the case where ξ_H is kept equal to ξ_L.

$$\begin{cases} \Omega_1 = \sqrt{16\xi_L^2 + 1 - 4\xi_L\sqrt{16\xi_L^2 + 2}} \\ \Omega_2 = \sqrt{16\xi_L^2 + 1 + 4\xi_L\sqrt{16\xi_L^2 + 2}} \end{cases} \Rightarrow \Delta\Omega = \Omega_2 - \Omega_1 = 4\sqrt{2}\xi_L \quad (10.2.11)$$

An interesting result is shown here: whatever the strategy used to tune ξ_H, the bandwidth is proportional to ξ_L, whereas the maximal power is inversely proportional to it.

Figure 10.3 exhibits the normalized power and the optimal value of ξ_H as a function of the frequency, for different values of $Q_M = 1/(2\xi_L)$. It is shown that the bandwidth can be enlarged provided that $\xi_H$can be adequately increased while Ω gets away from 1.

10.2.3 Figure of Merit

An ideal VEH would exhibit a large output power over a wide frequency range. For a given VEH, a relevant figure of merit can then be obtained multiplying its maximal normalized power by its normalized bandwidth.

Using Williams and Yates model, this figure of merit equals a constant, as shown in (10.2.12).

$$FoM = P'_{\max} \cdot \Delta\Omega = \begin{cases} 2 & \text{if } \xi_H = \xi_L \\ 2\sqrt{2} & \text{if } \xi_H = \xi_{Hopt}(\Omega) \end{cases} \tag{10.2.12}$$

From this analysis based on Williams and Yates simple model, some general rules for VEHs can be drawn:

- Increasing the intrinsic losses implies that higher damping induced by energy harvesting is required to reach the maximal power. This means that the electromechanical coupling has to be larger for VEH exhibiting large intrinsic losses.
- Lowering the intrinsic losses of the VEH increases the maximal power but decreases the frequency bandwidth. The *FoM* is, however, unchanged provided that the optimal energy harvesting damping can be reached.
- The performances can be increased by tuning the damping induced by energy harvesting as a function of the frequency. Higher electromechanical coupling is then required out of the resonance frequency.

Figure 10.4 (upper plot) shows the normalized maximal power and bandwidth that can be derived from Williams and Yates model as a function of the damping ratio ξ_{Hmax}/ξ_L, where ξ_{Hmax} is the maximal energy harvesting induced damping. It is shown that for $\xi_{Hmax} < \xi_L$, the normalized power limit (Q_M) cannot be obtained. If $\xi_{Hmax} \geq \xi_L$, the power limit is reached at the resonance frequency and for $\xi_H = \xi_L$.

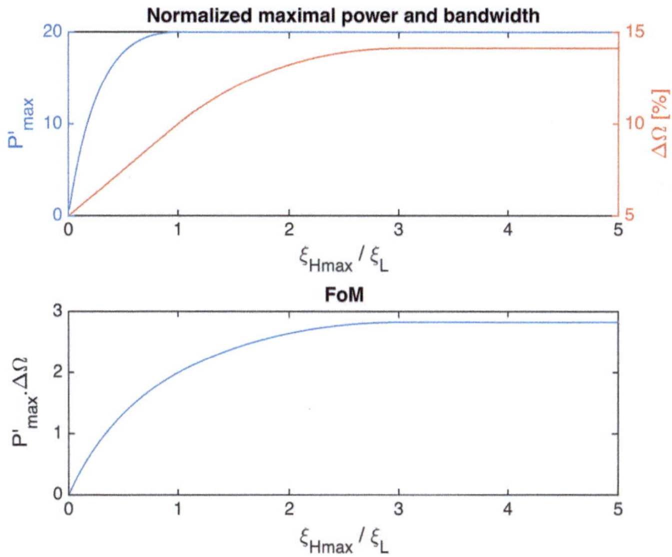

Fig. 10.4 *Upper plot* normalized power and bandwidth as a function of the ratio ξ_{Hmax}/ξ_L (for $\xi_L = 0.025$). *Lower plot FoM* as a function of the ratio ξ_{Hmax}/ξ_L (this plot is independent on ξ_L)

The normalized bandwidth limit ($4\sqrt{2}\xi_L$) is reached for $\xi_{Hmax} \geq 3\xi_L$. In this case, ξ_H has to be set equal to ξ_L for $\Omega = 1$, and has to be increased for Ω larger lower than 1. ξ_H will thus be equal to $3\xi_L$ for $\Omega = \Omega_1$ and $\Omega = \Omega_2$ Eq. (10.2.11).

In conclusion of this subsection, Williams and Yates model is very simple but allows understanding the relationship between the maximal power, the bandwidth and the damping induced by both the energy harvesting process and the intrinsic losses. It however exhibits a major limitation: In this model, the electrical load has no effect on the natural frequency of the VEH; it only affects the damping. Yet, in practical VEH, especially in the case of piezoelectric VEH, the electrical side of energy harvesting (i.e., the interface circuit) also impacts the resonance frequency. This effect will be further evidenced in the next sections of this chapter.

10.3 Modeling and Parameter Identification for Piezoelectric VEH

10.3.1 Model for Piezoelectric Vibration Energy Harvester

Based on electromechanical transduction principles and the electrical and mechanical equations of equilibrium, lumped electromechanical modeling of electromechanical transducers is known as a very effective method [3, 4]. In theory, such models are exact with no restrictions other than linearity, within the limits of the assumptions on boundary conditions and within the frequency range covered by the modeled resonant frequencies.

Most inertial piezoelectric vibration energy harvesters are based on a linear mechanical oscillator (a cantilever beam with a tip mass for instance) including one or several piezoelectric patches.

The very simple PVEH lumped model presented in Fig. 10.5 then provides an accurate and reliable behavioral representation of such PVEH excited around one of their resonant frequencies, as described in [2].

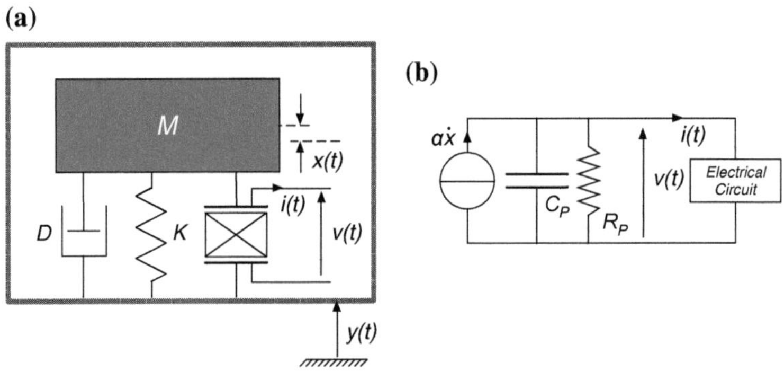

Fig. 10.5 Schematic of a PVEH, **a** mechanical point of view, **b** electrical point of view

A piezoelectric element represents the piezoelectric transducer(s) (electrically connected together in series or parallel if several are used). It is characterized by its electromechanical force factor α, its clamped capacitance C_P and its parasitic resistance R_P; $A_P = 1/R_P$ is the parasitic conductance. M is the equivalent dynamic mass, D is the damping coefficient corresponding to the mechanical losses and K represents the stiffness of the system when the piezoelectric element is short-circuited.

An external vibration $y(t)$ is applied to the system, inducing a relative displacement $x(t)$ between the mass and the housing. As a consequence of the mechanical stress variation, an AC voltage v appears between the piezoelectric electrodes and a current i is generated if an electrical load is connected. Since most applications require a DC voltage, the electrical load usually implements an AC/DC conversion stage.

The governing equations of such an electromechanical system are given by Eq. (10.3.1), where γ is the ambient acceleration.

$$\begin{cases} M\ddot{x} + Kx + D\dot{x} + \alpha V = M\gamma \\ I = \alpha\dot{x} - C_P\dot{V} - A_P V \end{cases} \tag{10.3.1}$$

ω_0, the natural frequency of the short-circuited PVEH is given by:

$$\omega_0 = \sqrt{\frac{K}{M}} \tag{10.3.2}$$

Three dimensionless parameters are used for the characterization of PVEH: the electromechanical coupling coefficient squared k^2, the mechanical losses damping ratio ξ_M and the resistive losses coefficient ξ_E, whose expressions are

$$k^2 = \frac{\alpha^2}{KC_0 + \alpha^2} \xi_M = \frac{D}{2\sqrt{KM}} \xi_E = \frac{1}{2R_P C_P \omega_0} \tag{10.3.3}$$

k^2 describes the effectiveness of quasi-static energy conversion between electrical and mechanical forms. For a PVEH in open-circuit subjected to a quasi-static stress, it is equal to the electrostatic energy divided by the total energy in the system. A modified coupling coefficient k_m^2, defined as (10.3.4), is introduced to simplify theoretical expressions. It is, too, an indicator for the electromechanical coupling, insofar as it represents the quotient between the electrostatic energy and the elastic energy in the generator (also in quasi-static operation). Note that despite k^2 remains always lower than 1 ($k^2 = 1$ meaning that all the input mechanical energy is converted into electrostatic energy), k_m^2 is not limited. For weakly coupled generators, the values of k^2 and k_m^2 are close one to the other.

$$k_m^2 = \frac{\alpha^2}{KC_P} = \frac{k^2}{1 - k^2} \tag{10.3.4}$$

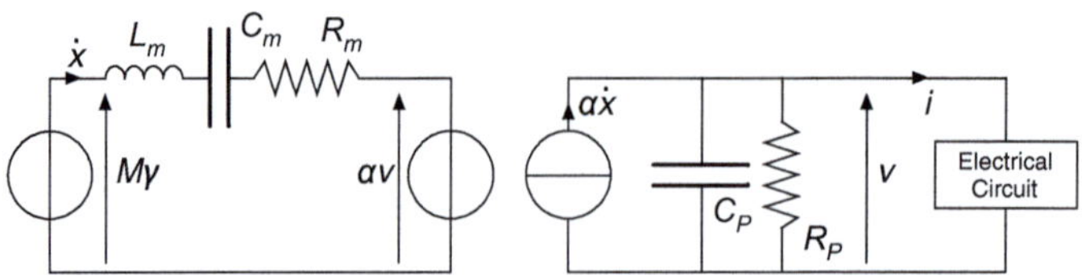

Fig. 10.6 Equivalent electrical network for a PVEH

10.3.2 Electrical Model for SPICE-Type Simulations

Using the classical analogies between electrical and mechanical figures (force and voltage; velocity, and current), the equivalent electrical network shown in Fig. 10.6 can be derived for the modeling of PVEH. The αV voltage source and the $\alpha\dot{x}$ current sources correspond to the electromechanical transduction. The mechanical power absorbed in the voltage source equals the electrical power provided by the current source, reflecting a lossless energy conversion. The mechanical branch includes an electrical oscillator $\{L_m, C_m, R_m\}$ that represents the mechanical resonance. The electrical branch is the same as the one depicted in Fig. 10.5b). The governing equations from this electrical network are identical to Eq. (10.3.1) provided that $L_m = M$, $C_m = K^{-1}$ and $R_m = D$.

This model can be easily implemented in SPICE-based software (SPICE: Simulation Program with Integrated Circuit Emphasis) and can then be used for efficient simulation of PVEH with various electrical interface and ambient acceleration profiles.

As an example, Fig. 10.7 is a screenshot of a schematic implemented in the LTSPICE™ Software (Linear Technologies Corporation), where a classical full bridge rectifier is used as an AC/DC electrical interface between the PVEH and the resistor R_L, which models the input resistance of the circuit to be powered.

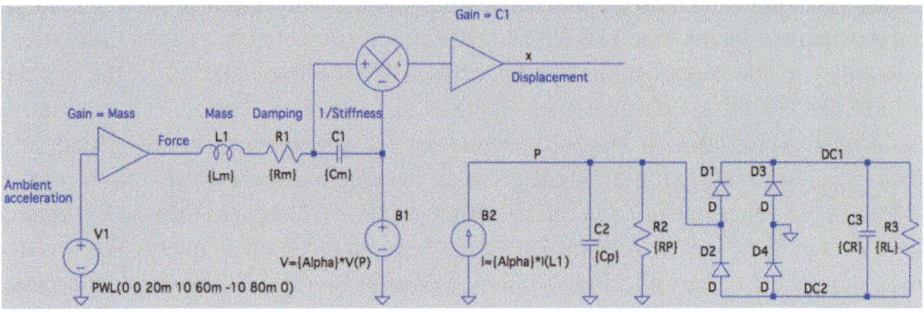

Fig. 10.7 PVEH combined with the classical full bridge rectifier interface (detailed in Sect. 10.5.1)—modeling using LTSPICE™ Software

10.3.3 Model Identification Procedure

The parameters of the model presented in the previous subsection can be derived from a given PVEH structure using analytical or finite element modeling approaches [5, 6], provided that the characteristics and dimensions of the used materials are known with sufficient accuracy. In practice, the mechanical losses related to mechanical assembly of the different parts of a PVEH are not easy to predict (i.e., bonding, clamping, etc.). The piezoelectric characteristics provided by manufacturers may also exhibit important uncertainty and variability. This explains why significant discrepancies are usually observed between theoretical PVEH characteristics and experimental ones.

This subsection details a very convenient procedure to determine experimentally the actual electromechanical parameters of the model based on a simple measurement of the PVEH complex admittance. The measurement of the complex admittance of the PVEH has to be performed around its natural frequency. This can be done using an impedance analyzer. During this measurement, the PVEH should not be excited by ambient acceleration.

Using Eq. (10.3.1) written in the frequency domain, and the dimensionless parameters k_m^2, ξ_E and ξ_M, the expression of the PVEH admittance can be obtained

$$\underline{Y_P} = -\frac{\underline{I}}{\underline{V}} = jC_P\omega\left(1 + \frac{k_m^2}{1-\Omega^2+2j\xi_M\Omega} - \frac{2j\xi_E}{\Omega}\right) \tag{10.3.5}$$

It is shown that the admittance is the one of a C_P capacitor multiplied by a dimensionless factor only function of Ω, k_m^2, k_m^2, ξ_E and ξ_M. Figure 10.8 shows the comparison

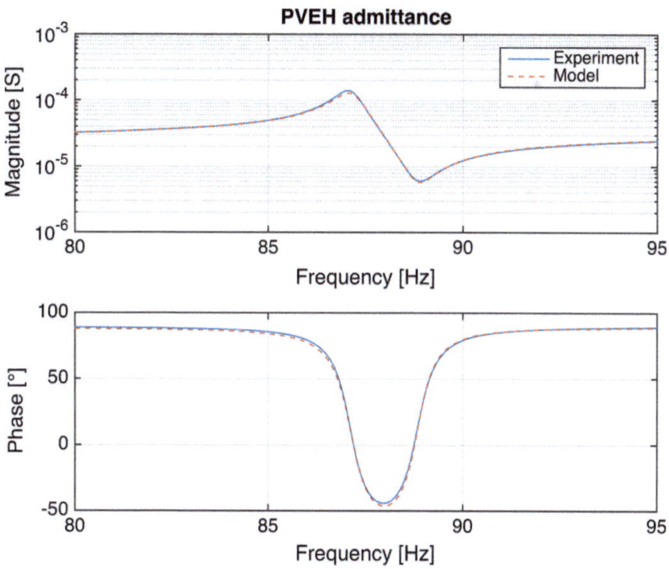

Fig. 10.8 An example of the experimental and modeled admittance of a PVEH

Table 10.1 PVEH identified parameters

C_0	51.7 nF	ξ_E	2.86 10^{-3}
ω_0	547 rad/s	ξ_M	4.14 10^{-3}
k_m^2	3.97 %	M	10.2 g

of the experimental and modeled admittance of a real PVEH. The identification of the parameters has been performed using optimization methods to get the best match between the measurements and the model.

Five parameters can be obtained from the measurement of the admittance: $C_P \omega_0$, k_m^2, ξ_E, and ξ_M. From them, it is not possible to get all the six parameters of the constitutive equations (10.3.1). To get M, K, D, and α, an additional measurement is indeed required. For instance, the ratio in the frequency domain of the open-circuit voltage V_0 to the displacement of the dynamic mass x can be used to determine α, as shown by Eq. (10.3.6). From Eqs (10.3.2) and (10.3.3), M, K, and D can eventually be calculated.

$$\frac{\underline{V_0}}{\underline{x}} = \frac{\alpha}{C_P} \frac{1}{1 - 2j\frac{\xi_E}{\Omega}} \approx \frac{\alpha}{C_P} \tag{10.3.6}$$

Finally, the six independent identified parameters for the considered PVEH model are listed in Table 10.1.

Among other identification procedure which can be found in the literature, a classical one consists in determining the coupling coefficient k^2 from the open-circuit and short-circuit resonance frequencies of the PVEH, and the mechanical damping ratio ξ_M from the -3dB frequency bandwidth of the displacement x of the PVEH with piezoelement short-circuited as electrical boundary condition [7].

The electrical losses coefficient ξ_E is usually much lower than 0.01 and its effect on the admittance and more generally on the PVEH performances is weak in most cases. Consequently, ξ_E is neglected in most of the literature.

10.4 Optimal Impedance Matching

Several works used the well-known optimal impedance matching strategy to maximize the power output of PVEH [5, 8]. In this section the optimal linear electrical load that maximizes the PVEH power generation will be analytically determined. As a first approach, no technical limitation will be taken into account. Results that are not achievable in practice may be obtained, but this study will theoretically define the PVEH performance upper boundary.

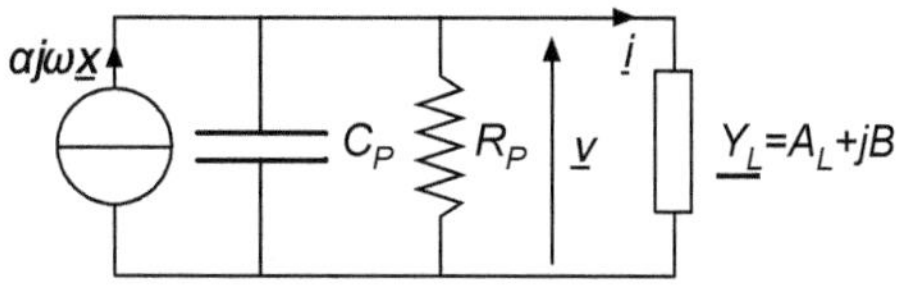

Fig. 10.9 Electrical network using the matching impedance strategy

10.4.1 Theory

An electrical load modeled by the complex admittance $\underline{Y_L} = A_L + jB$ is connected to the PVEH, as shown in Fig. 10.9.

The piezoelectric voltage V can then be expressed in the frequency domain as

$$\underline{V} = \frac{\alpha j \omega \underline{x}}{A_P + A_L + j\left(C_P\omega + B\right)} \tag{10.4.1}$$

which can be rewritten as Eq. (10.4.2), where ν and υ are two dimensionless coefficient, respectively corresponding to the in-phase and the quadrature-phase components of the voltage with respect to the displacement.

$$\underline{V} = \frac{\alpha}{C_P}\left(\nu + j\upsilon\right)\underline{x} \text{ with } \begin{cases} \nu = \frac{C_P\omega(B+C_P\omega)}{(A_P+A_C)^2+(B+C_P\omega)^2} \\ \upsilon = \frac{C_P\omega(A_P+A_C)}{(A_P+A_C)^2+(B+C_P\omega)^2} \end{cases} \tag{10.4.2}$$

Substituting Eq. (10.4.2) in the mechanical constitutive equation (10.3.1) written in the frequency domain leads to the expression of the displacement

$$\underline{x} = \frac{M\underline{\gamma}}{K - M\omega^2 + \frac{\alpha^2}{C_P}\nu + j\left(C\omega + \frac{\alpha^2}{C_P}\upsilon\right)} \tag{10.4.3}$$

Normalizing the displacement with respect to γ_m/ω_0^2 (the ambient displacement magnitude) gives

$$\underline{x}' = \frac{1}{1 - \Omega^2 + k_m^2\nu + j\left(2\xi_m\Omega + k_m^2\upsilon\right)} \tag{10.4.4}$$

This expression shows that the natural frequency of the PVEH can be tuned through ν and that the damping induced by energy harvesting can be tuned through $\upsilon(\upsilon > 0)$. The larger k_m^2, the larger the effect of ν and υ variations.

The harvested energy is considered to be the energy dissipated in the real component of the electrical load (A_L). Of course, this does not correspond to a realistic energy harvesting circuit since the voltage would need to be rectified. A_L can, however, be considered as the equivalent input conductance of the circuit to be supplied. This approach aims at providing an upper limit to the performance of a PVEH.

In these conditions, the harvested power can be expressed as

$$P = \frac{|V|^2}{2} A_L \tag{10.4.5}$$

In the case where ξ_E is neglected ($A_P = 0$), intrinsic losses in the PVEH are modeled by ξ_M only. It follows from Williams and Yates analysis and from Eq. (10.4.4) that the power is maximized when ν is chosen so that the natural frequency of the PVEH matches the operation frequency and υ so that the damping induces by the energy harvesting equals the intrinsic damping, which gives:

$$\begin{cases} \nu = \frac{\Omega^2 - 1}{k_m^2} \\ \upsilon = \frac{2\xi_M \Omega}{k_m^2} \end{cases} \tag{10.4.6}$$

If no technical constraint is taken into account, ν can take any positive or negative value and υ can take any positive value, which means that Eq. (10.4.6) can be verified whatever the value of Ω. In this case, the power limit P_{lim} given by Eq. (10.2.5) is obtained, whatever the operation frequency.

If ξ_E is not neglected ($A_P \neq 0$), the optimal values of A_L and B that maximize the harvested power are obtained as a function of Ω by looking for the roots of the derivatives of the harvested power with respect to A_L and B, respectively. They are given by

$$\begin{cases} A_{Lopt} = C_P \omega \left(\frac{2\xi_E}{\Omega} + \frac{2K_m^2 \xi_m \Omega}{4\Omega^2 \xi_m^2 + (\Omega^2 - 1)^2} \right) \\ B_{opt} = C_P \omega \left(\frac{K_m^2 (\Omega^2 - 1)}{4\Omega^2 \xi_m^2 + (\Omega^2 - 1)^2} - 1 \right) \end{cases} \tag{10.4.7}$$

And the maximal power that can be harvested at a given operation frequency by

$$P_{opt} = P|_{A_L = A_{Lopt} \text{ and } B = B_{opt}} = \frac{M \gamma_M^2}{16\omega_0} \frac{k_m^2 \Omega^2}{k_m^2 \xi_M \Omega^2 + \xi_E \left(4\xi_M^2 \Omega^2 + (\Omega^2 - 1)^2 \right)} \tag{10.4.8}$$

The normalized optimal power is then

$$P'_{opt} = \frac{1}{2\xi_M + \frac{2\xi_E}{k_m^2} \left(4\xi_M^2 + \frac{(\Omega^2 - 1)^2}{\Omega^2} \right)} \tag{10.4.9}$$

As previously mentioned if $\xi_E = 0$, the optimal power reaches $P_{\lim}$ whatever the operation frequency, provided that A_L and B are adequately tuned. If ξ_E is not neglected, it is shown that the optimal power reached a maximum for $\Omega = 1$. Equation (10.4.10) gives the normalized maximal power.

$$P'_{\max} = \frac{1}{2\xi_M \left(1 + 4\frac{\xi_E \xi_M}{k_m^2}\right)} \tag{10.4.10}$$

The PVEH bandwidth can be obtained finding Ω_1 and Ω_2 so that. The normalized bandwidth is then given by (10.4.11), and the bandwidth by $\Delta\omega = \Delta\Omega \cdot \omega_0$.

$$\Delta\Omega = \Omega_2 - \Omega_1 = 2\xi_M \sqrt{1 + \frac{k_m^2}{4\xi_E \xi_M}} \tag{10.4.11}$$

The normalized power and displacement are plotted in Fig. 10.10 as a function of the operating frequency for different values of ξ_E. The corresponding real and complex parts of the optimal electrical load admittance are shown in Fig. 10.11. It is clearly shown that ξ_E drastically affects the bandwidth of the PVEH. From Fig. 10.11, it can be seen that the optimal complex part of the load admittance is independent on ξ_E and that its sign is mainly negative in the frequency range but can also be positive. This concretely means that the optimal load is a resistor in parallel with an inductor in most of the frequency range but that it is sometimes a resistor in parallel with a capacitor (for operation frequency slightly above the natural frequency of the PVEH).

Figure 10.12 exhibits the voltage on the piezoelectric element when the optimal complex electrical load is selected as a function of the operating frequency. It is clearly shown that increasing the bandwidth lead to very high piezoelectric voltage

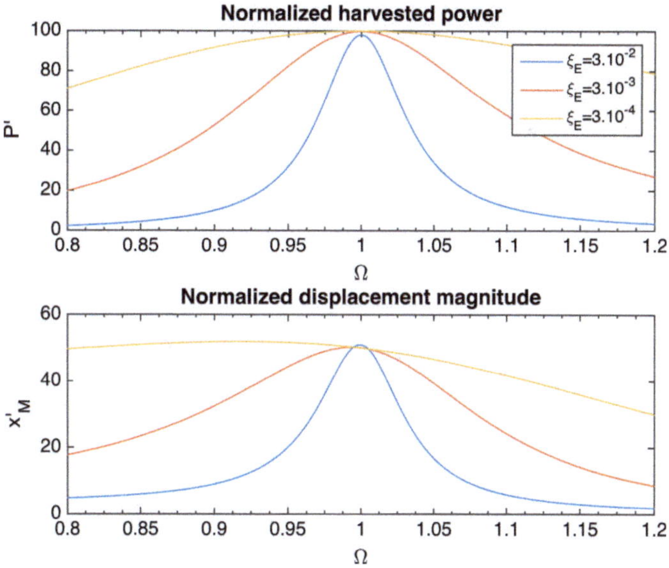

Fig. 10.10 Normalized harvested power (*upper plot*) and displacement magnitude (*lower plot*) as a function of the operation frequency for different values of ξ_E ($k_m^2 = 3\,\%$, $\xi_M = 0.005$)

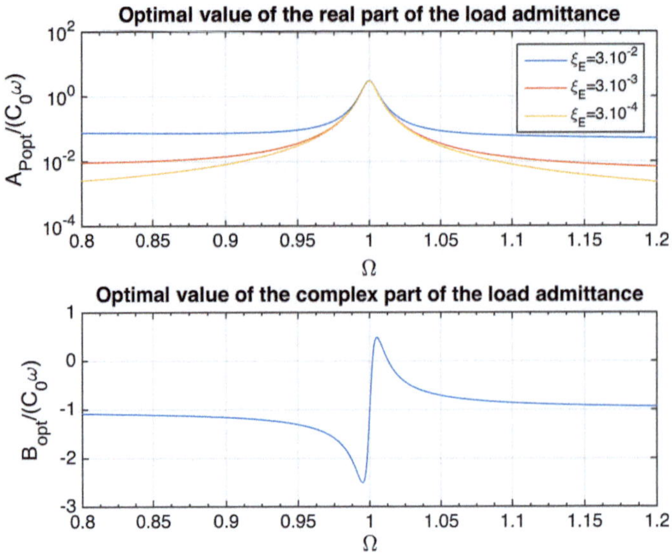

Fig. 10.11 Real (*upper plot*) and complex (*lower plot*) parts of the optimal electrical load admittance as a function of the operation frequency for different values of ξ_E ($k_m^2 = 3\,\%$, $\xi_M = 0.005$)

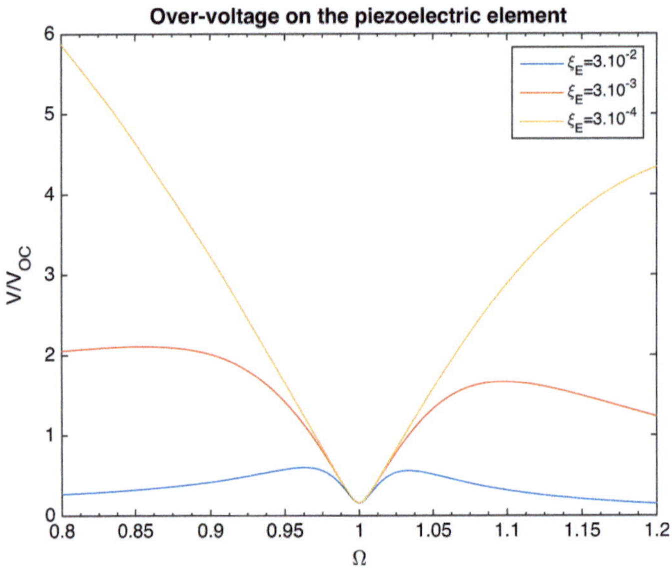

Fig. 10.12 Ratio of the optimal piezoelectric voltage to the maximal open-circuit voltage as a function of the operation frequency for different values of ξ_E ($k_m^2 = 3\,\%$, $\xi_M = 0.005$)

when the operating frequency is shifted away from the resonance frequency (up to six times the maximal open-circuit voltage in the considered cases). This is due to the increase of reactive power alternatively flowing between the capacitance of the piezoelectric element and the complex part of the optimal admittance.

Finally the figure of merit that is equals to the product of the maximal normalized power by the normalized product is given by:

$$FoM = P'_{\max} \cdot \Delta\Omega = \frac{\frac{k_m^2}{4\xi_E\xi_M}}{\sqrt{1+\frac{k_m^2}{4\xi_E\xi_M}}} = \frac{\chi}{\sqrt{1+\chi}} \text{ with } \chi = \frac{k_m^2}{4\xi_E\xi_M} = \frac{\alpha^2 R_P}{D} \tag{10.4.12}$$

where χ is dimensionless parameter that can be seen as an indicator of the performance of the PVEH (it increases with the coupling coefficient and decreases with the electrical and mechanical losses).

The normalized power and bandwidth, as well as the *FoM* are plotted in Fig. 10.13 as a function of χ an be seen that the *FoM* can largely exceed $2\sqrt{2}$, which was the upper limit obtained from the Williams and Yates model. This is because of the frequency tuning mechanism induced by the complex part of the load admittance. For typical PVEH, the order of magnitude for k_m^2 is around 10^{-2}, whereas it is between 10^{-2} and 10^{-3} $\xi_E\ \xi_M$. This means that χ typical values ranges from 10 to 100 about. In this case, the figure of merit can be approximated by $FoM \approx \sqrt{\chi}$.

In practice, this impedance matching strategy can hardly be implemented because it requires the realization of complex electrical load with tuning mechanisms. One

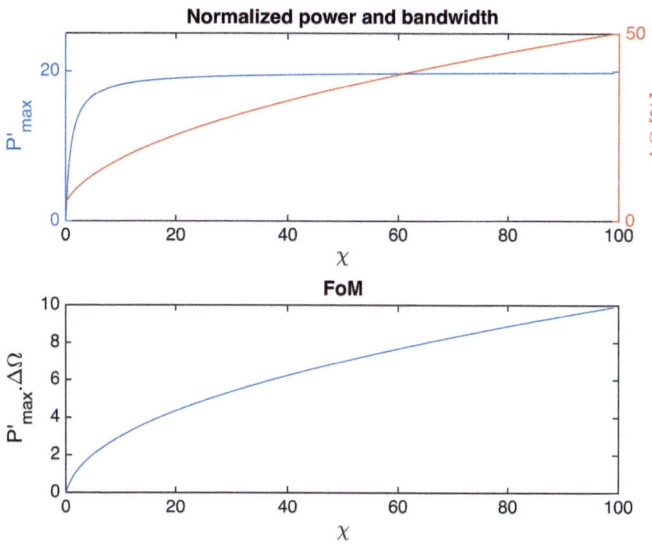

Fig. 10.13 *Upper plot* normalized power and bandwidth as a function of χ (for $\xi_L = 0.025$). *Lower plot FoM* as a function of ξ_{Hmax}/ξ_L (this plot is independent on ξ_L)

could think about using active gyrator circuits to emulate large tunable inductors. Such synthetic inductors based on operational amplifiers, capacitances, and resistances however consume a lot of power and are not a viable option for energy harvesting applications.

Another limitation of this approach can be inferred from Fig. 10.12: the voltage increase due to the reactive part of the complex optimal load may lead to very high piezoelectric voltage for which depoling effect or at least nonlinearities in the piezoelectric coefficients may be observed.

Possible practical solution for implementing this impedance matching strategy will be discussed in the next section

10.4.2 Practical Implementation

Modern power electronics offers many solutions based on switching-mode converters enabling high-efficiency electrical power conversion. Input and output voltages can exhibit various shapes and polarities, and the power transfer can be unidirectional or bi-directional, depending on the considered circuits [9].

In practice, electric loads powered by energy harvesters—such as wireless sensor nodes for instance—do not behave at all like the "optimal matching impedance" defined in the previous subsection. Indeed, such electric loads require DC voltage-regulated power supply, and their consumption may be extremely variable in time. Thus, AC–DC power conversion, energy storage, and voltage regulation are the minimum requirements for the electronic interface, which will be used to transfer the electrical energy produced by the piezoelectric material to the electric load.

Passive AC–DC converters, such as diode rectifiers, are very simple to implement but the shape of their AC input voltage and current is not similar to that of linear impedances. Emulation of linear impedance is however possible using PWM rectifiers. The proposed implementation of the optimal impedance matching strategy is based on the interface circuit represented in Fig. 10.14, which is presented in [10]. In this circuit, the AC input of the PWM rectifier is connected to the piezoelectric element. Through the PWM control of the switches of the active rectifier, it is the-

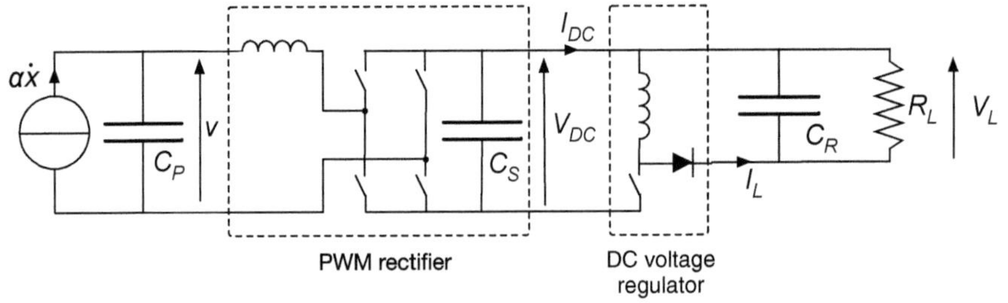

Fig. 10.14 Interface circuit for practical implementation of the optimal impedance matching strategy

oretically possible to emulate any linear or nonlinear load, including the optimal impedance defined previously. The DC output of the active rectifier is connected to and energy storage element (a capacitor or a supercapacitor) whose voltage may vary, depending on the energy stored. The output DC voltage delivered to the electric load can be then regulated using a buck-boost converter. This sub-circuit enables to ensure output voltage regulation whether the output DC voltage V_L is lower or higher than the input DC voltage V_{DC}.

The control strategy of this interface circuit is not very complicated, but it is out of the scope of this chapter and it will not be detailed here. Among important ideas to have in mind, practical implementation of the optimal impedance matching strategy is possible. However, in case of very low harvested power level, typically in the range of a few tenths of microwatts or below, the available power may be too low to implement sophisticated control circuits. Therefore, "less optimal" approaches based on simpler control principles and simpler circuits may turn out to be much more efficient in practice. This is the objective of the various techniques developed in the next sections.

10.5 The Classical Rectifier Followed by a Resistive Load

In this section, the classical rectifier circuit followed by a storage/smoothing capacitor is studied. This circuit is shown in Fig. 10.15, where R_L represents the equivalent input resistance of the device to be supplied. The harvested power is then calculated as the power dissipated into R_L.

For simplicity, it will be assumed that the electrical losses can be neglected $\xi_E =$ 0). This assumption is valid in most of practical cases where R_L is much lower than R_P.

10.5.1 Power and Bandwidth

This circuit has been studied in details by Shu and Lien in [11]. Sinusoidal ambient accelerations around the resonance frequency of the PVEH are considered, and it is assumed that the smoothing capacitor is large enough so that V_{DC} is ripple free. The value of V_{DC} can then be obtained as a function of the magnitude x_M of the dynamic mass relative displacement:

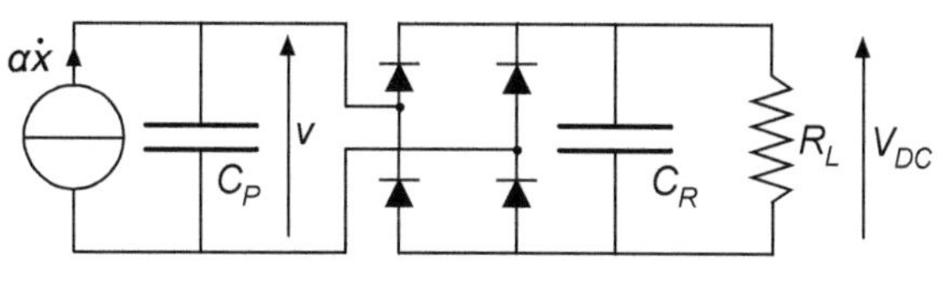

Fig. 10.15 Electrical network using the classical rectifier circuit

$$V_{DC} = \frac{\alpha \omega R_L}{C_P \omega R_L + \frac{\pi}{2}} x_M \tag{10.5.1}$$

Because of the nonlinear behavior of the full-wave rectifier, the piezoelectric voltage V is periodic but not sinusoidal. However, since only its fundamental harmonic frequency is close from the resonance frequency of the VEH, it can be assumed that only this fundamental component V_1, given in the frequency domain by (10.5.2), impacts the displacement of the VEH dynamic mass.

$$\underline{V_1} = \frac{\alpha}{C_P} (\nu + j\upsilon)\, \underline{x} \text{ with } \begin{cases} \nu = \frac{\omega R_L C_P}{\omega R_L C_P + \frac{\pi}{2}} \\ \upsilon = \frac{2\omega R_L C_P}{\left(\omega R_L C_P + \frac{\pi}{2}\right)^2} \end{cases} \tag{10.5.2}$$

The expression in the frequency domain of the displacement and the normalized displacement are then the same as in Sect. 10.4.1, Eqs. (10.4.3) and (10.4.4), respectively, except that the expressions of ν and υ are different.

$$r' = R_L C_P \omega_0 \tag{10.5.3}$$

Let r be the normalized load as given by (3.3), ν and υ can also be expressed as

$$\begin{cases} \nu = \frac{r'\Omega}{r'\Omega + \frac{\pi}{2}} \\ \upsilon = \frac{2r'\Omega}{\left(r'\Omega + \frac{\pi}{2}\right)^2} \end{cases} \tag{10.5.4}$$

It can be seen from (10.5.4) that ν is between 0 (when $r'= 0$) and 1 (when r' tends to infinite). From (10.4.4), this means that the resonance frequency of the PVEH is between ω_0 (short-circuit resonance frequency) and $\omega_0\sqrt{1 + K_m^2}$ (open-circuit resonance frequency).

It can also be seen that υ is between 0 (when $r'= 0$ or when r' tends to infinite) and $1/\pi$ (when $r'\Omega = \pi/2$ From Eq. (10.4.6), this means that the optimal damping cannot be obtained if $k_m^2 < 2\Omega\xi_M\pi$.

The harvested power can be expressed as

$$P = \frac{V_{DC}^2}{R_L} \tag{10.5.5}$$

And the normalized harvested power is finally given by Eq. (10.5.6), where the expression of $\underline{x}'$ is the same as in Eq. (10.4.4)

$$P' = 8k_m^2 \frac{r'\Omega^2}{\left(r'\Omega + \frac{\pi}{2}\right)^2} \left|\underline{x}'\right|^2 \tag{10.5.6}$$

The maximal normalized power and the corresponding optimal normalized load are plotted as a function of the frequency in Fig. 10.16, for different values of k_m^2. When k_m^2 increases, it can be seen that whereas the bandwidth keeps on increasing,

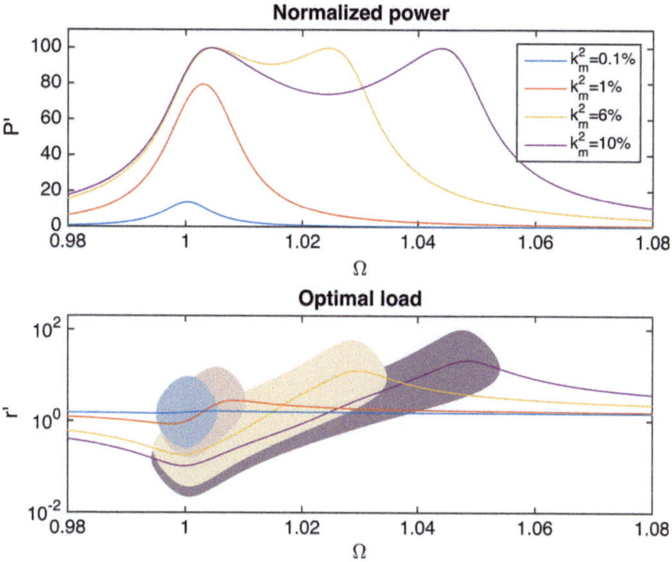

Fig. 10.16 *Upper plot* normalized power as a function of Ω for $\xi_L = 0.005$ ($Q_M = 100$) and different values of k_m^2. *Lower plot* normalized optimal load as a function of Ω (areas corresponds to the load domain for which the power is at least 50 % of the maximal power)

the maximal normalized power first increases and then saturates at Q_M. It is worthy of note that a continuous tuning of the electrical load as a function of the operation frequency is required to get the best performance. The colored areas represent the load domains for which at least 50 % of the maximal power is harvested.

The normalized power, bandwidth, and the figure of merit defined in (10.2.12) are plotted in Fig. 10.17 as a function of k_m^2 for different values of Q_M. The black curve in Figure 10.17a corresponds to $k_m^2 Q_M = \pi$. It is shown that the normalized power equals Q_M for $k_m^2 Q_M > \pi$. Figure 10.17b shows that the bandwidth is quasi-linearly

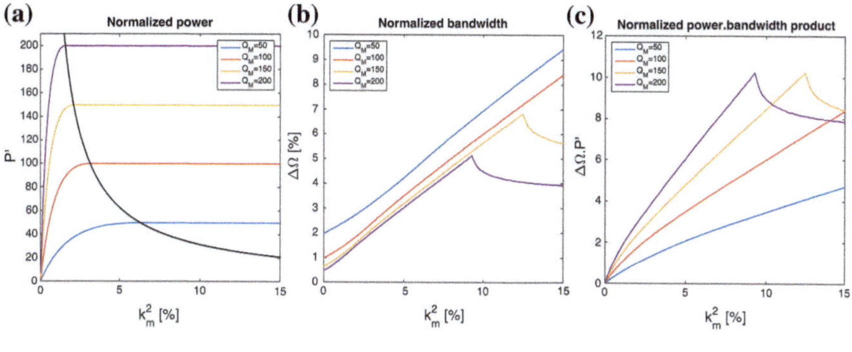

Fig. 10.17 **a** Normalized power versus k_m^2, **b** normalized bandwidth versus k_m^2, **c** FoM versus k_m^2

increasing with k_m^2. For very large values of k_m^2, the bandwidth however decreases because the power in the well between the two maxima (cf. Fig. 10.16 with $k_m^2 > 6\,\%$) becomes less than half of the maximal power.

For typical PVEH, k_m^2 is in the order of a few percent and Q_M is between 10 and a few hundreds. The figure of merit is then usually lower than 5. Figure 10.17c clearly shows that PVEH with high k_m^2 and high Q_M exhibit higher performance.

It is worthy of note that results presented in Fig. 10.17 imply a continuous tuning of the electrical load as a function of the operation frequency, which requires specific power conversion interface and dedicated MPPT control circuit for practical implementation.

10.5.2 Practical Implementation

Several interface circuit based on the classical rectifier circuit principle have been proposed for maximizing the power transfer from the piezoelectric device to the load. In this optimization approach, the main function of the interface circuit is to emulate the optimal load resistance defined in the previous section. Ottman et al. studied one of the first interface circuits [12]. This circuit was based on the buck DC–DC converter represented in Fig. 10.18. Using the appropriate control law of the DC-DC converter, it was shown that the performances of the system could be greatly improved. In order to reduce the power consumption of the control circuit so that it can be self-powered, the authors proposed a new control principle of the buck converter that exhibited similar performances, with much simpler implementation [13].

Following the same approach, Lefeuvre et al. [14] proposed an interface circuit based on a buck-boost DC–DC converter depicted in Fig. 10.19. In discontinuous current mode, this circuit exhibits a constant input resistance for a given duty-cycle control of the electronic switch, making possible the implementation of the interface using only an oscillator with constant duty-cycle as control circuit (IC1 in Fig. 10.19).

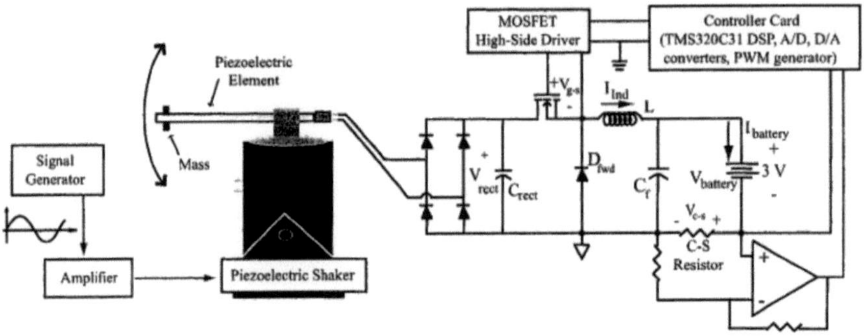

Fig. 10.18 Experimental setup including a buck DC–DC converter (from [12])

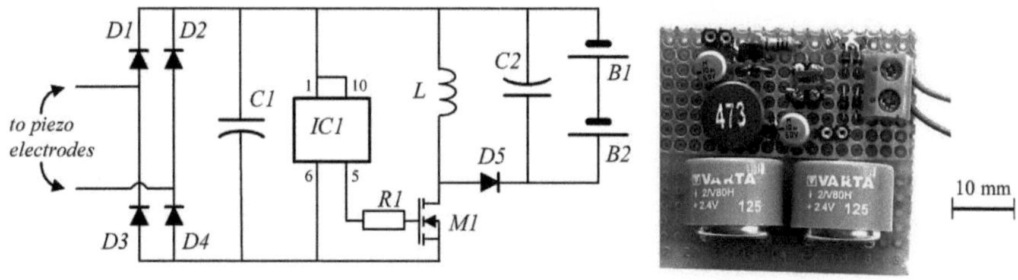

Fig. 10.19 Experimental circuit and setup including a buck-boost DC–DC converter (from [14])

For output powers ranging from 200 μW to 1.5 mW, experimental results showed more than 70 % overall efficiency of this circuit, including the control circuit consumption. In this practical example, the input resistance of the interface circuit was predetermined to be as close as possible to the optimal value at the resonant frequency. Consequently, the robustness was relatively weak with respect to variations of electromechanical characteristics.

To overcome this drawback, maximum power point tracking (MPPT) circuits with ultra-low power consumption were implemented. Yi et al. proposed a low-power interface circuit, based on a switched capacitor DC-DC converter, integrated in a 0.35 μm CMOS process ASIC [15]. In this ASIC, an energy-adaptive MPPT allowed to activate different operation modes according to the available power. Kong et al. presented an interface circuit based on a flyback DC–DC converter [16]. The MPPT algorithm was implemented using a low-power microcontroller unit MSP430 from Texas Instruments. Experimental results indicated that the proposed interface circuit achieved 72 % efficiency around the resonant frequency and around 8.4 mW output power.

In this domain, the current trends are clearly to improve the efficiency of low-power DC–DC converters and to design fast and effective MPPT control circuits with ultra-low power consumption [17, 18].

10.6 Nonlinear Energy Harvesting Circuits

10.6.1 Principle

Nonlinear energy extraction approaches have been developed to optimize the energy extracted from PVEH. From the simple model presented in Sect. 10.3.1, the energy equation (10.6.1) is obtained multiplying both terms of the dynamical mechanical equilibrium equation by the velocity and integrating over the time variable. It shows that the energy provided by the ambient acceleration is divided into kinetic energy, potential elastic energy, mechanical losses and the energy extracted from the piezoelectric element. Nonlinear energy extraction circuits aim at increasing this last term.

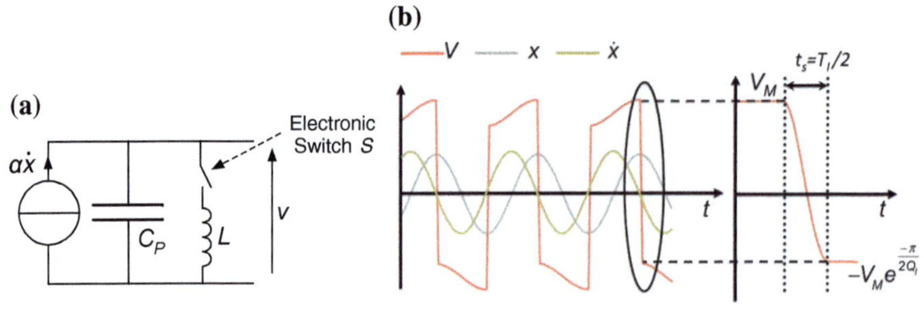

Fig. 10.20 **a** Basic electronic element for nonlinear energy extraction **b** Corresponding typical waveforms

$$\int M\gamma\dot{x}dt = \frac{1}{2}M\dot{x}^2 + \frac{1}{2}Kx^2 + \int D\dot{x}^2dt + \int \alpha V\dot{x}dt \quad (10.6.1)$$

The basic electronic element for nonlinear energy extraction circuit is shown in Fig. 10.20a. It consists of connecting a coil in series with an electronic switch parallel to the piezoelectric element. The switch is almost always open, except when a minimum or maximum of the dynamic mass displacement occurs. At this moment, the switch is closed and C_P in parallel with L forms an electrical oscillating circuit. The switch is kept closed during half of the oscillating period T_I, so that the voltage is reversed. Because there are some electrical losses in the $\{L, C_P\}$ network, characterized by the quality factor Q_I, the absolute value of the voltage after inversion is slightly reduced compared to the one before the switch is closed. Corresponding waveforms for the voltage, displacement, and velocity of the dynamic mass are shown in Fig. 10.20b.

This simple circuit has two effects: First, for constant displacement amplitude, the amplitude of the voltage is largely increased; second voltage and velocity are of the same sign. These two effects clearly induce the increase of the energy extraction term of Eq. (10.6.1).

This very simple circuit alone is, however, not suitable for energy harvesting purpose, since the extracted power is not converted into useful power: It is actually dissipated as heat in the coil and the piezoelectric element. This circuit was in fact initially developed for vibration damping purpose and called SSDI (Synchronized Switch Damping on Inductor) [19].

Adding AC/DC and energy storage stages to this elementary circuit has been the basis of the further developments of nonlinear energy harvesting approaches. For instance, the SSHI (Synchronized Switch Harvesting on Inductor) simply consists in combining this switch-coil circuit to the classical full-wave rectifier approach, connecting them in parallel to the piezoelectric element [7].

10.6.2 Comparison of Various Nonlinear Circuits

Based on the general principle exposed previously, several nonlinear energy harvesting circuits with different features have been proposed over the last decade. This subsection outlines the main properties of each circuit. They have in common to increase the last term of Eq. (10.6.1), that is to say they increase the mechanical damping induced by energy conversion. Low-coupling PVEH particularly benefit from these circuits. Indeed, such increase of the electromechanical energy conversion effectiveness allows getting closer to the optimal damping (see Sect. 10.2.1). This beneficial effect also increases the PVEH performances in case of out-of resonance and pulsed excitation [20]. In the case of strongly coupled PVEH excited at resonant frequency, no gain can be expected if optimal damping is already attained. Specific features of the main nonlinear circuits are detailed hereafter.

The **parallel SSHI** circuit depicted on Fig. 10.21 is one of the first nonlinear PVEH interfaces based on the principle of "synchronized switching" [7, 20, 21]. This circuit is a straightforward association of the classical rectifier circuit of Fig. 10.15 and the circuit of Fig. 10.20a. Compared to the classical rectifier circuit, the parallel SSHI circuit tends to increase the piezoelectric voltage. The optimal load resistance tends also to be higher than with the classical rectifier circuit. This voltage magnification property may be used to get high voltages, or to reduce the energy losses related to the voltage drop of the diodes in the rectifier bridge. This is particularly interesting in the case of low-voltage PVEH microsystems, whose open-circuit voltage is typically lower than 1 V. Shu et al. studied in detail the effect of this circuit on PVEH as a function of the excitation frequency [22].

In case of the **series SSHI** circuit (Fig. 10.22), the coil-switch dipole is connected in series with the PVEH instead of being connected in parallel. This induces slight changes on the piezoelectric voltage waveform, and significant differences on the circuit output voltage and the optimal load resistance. Indeed, in this case the output voltage and the optimal load resistance are smaller than those of the classical rectifier circuit [23]. Therefore, the series SSHI circuit is particularly interesting to get an output voltage lower than the piezoelectric voltage. This circuit was first proposed for this voltage reduction property by Taylor et al. [24].

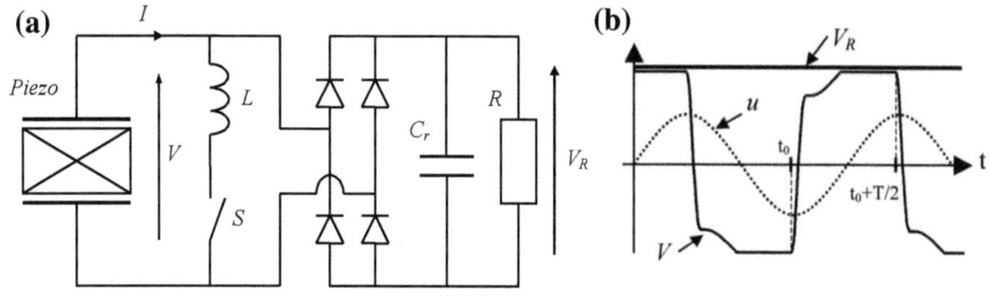

Fig. 10.21 **a** Parallel SSHI circuit schematic and **b** typical waveforms (from [23])

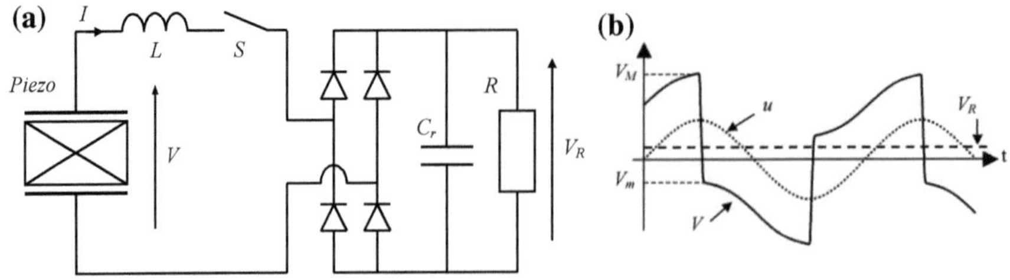

Fig. 10.22 **a** Series SSHI circuit schematic and **b** typical waveforms (from [23])

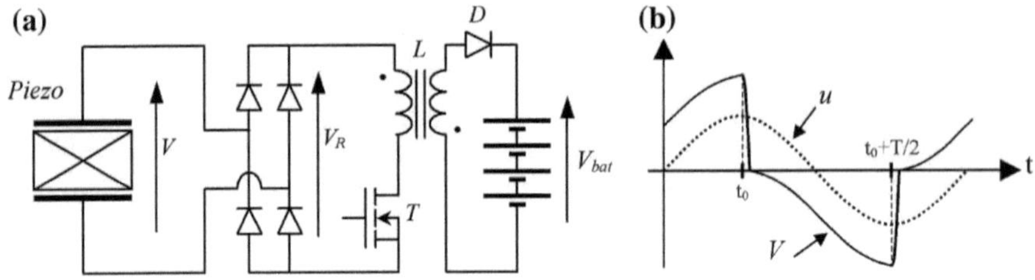

Fig. 10.23 **a** SECE circuit schematic and **b** typical waveforms (from [25])

The principle of the **SECE** circuit consists in extracting promptly and entirely the electric energy converted by the piezoelectric element on each extremum of the piezoelectric voltage. In the SECE circuit represented on Fig. 10.23, the energy transfer is achieved by a flyback-type DC–DC power converter [25]. Alternatively, the flyback circuit can be replaced by a buck-boost DC–DC converter [26]. In theory, the harvested power is independent of the load. This unique property enables to harvest the maximum power without MPPT system.

Several works focused on efficiency improvement of the previous circuits for low-voltage applications. A way for reducing the voltage drops consisted in reducing the number of diodes used for voltage rectification. In this domain, Makihara et al. [27] proposed a half-bridge circuit for the Parallel SSHI circuit. Lallart et al. proposed another half-bridge circuit for low-voltage implementation of the series SSHI interface [28].

The **SSHI-MR circuit** proposed by Garbuio et al. [29] (Fig. 10.24) brought an ultimate reduction of the losses due to the threshold voltage of the diodes. The typical waveforms of this circuit are very similar to those of the series SSHI circuit. The main difference comes from the magnetic transformer, which replaces the coil of the series SSHI circuit. The voltage gain of the transformer in association with the single-diode rectifier enables operation toward ultra-low-voltage PVEH: experimental results showed effective energy harvesting from piezoelectric voltages as low as 30 mV.

The **single-supply pre-biasing** circuit presented by Elliott et al. [30] has piezoelectric waveforms identical to that of the parallel SSHI circuit. The use of MOSFET

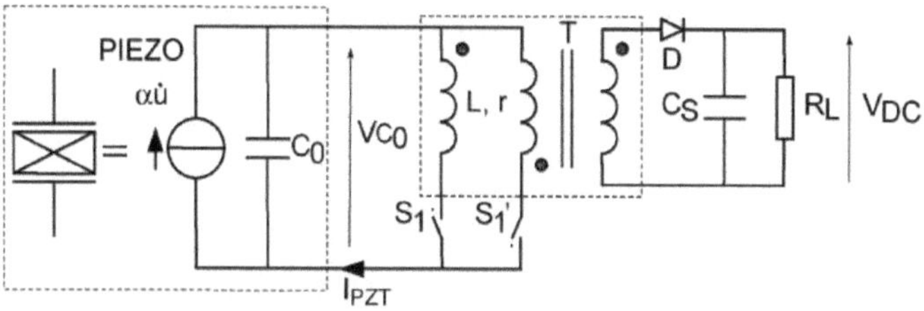

Fig. 10.24 SSHI-MR circuit schematic (from [29])

electronic switches instead of diodes enables efficient operation even with ultra-low PVEH voltages.

The **DSSH circuit** proposed by Lallart et al. is an association of the Series SSHI circuit and the buck-boost DC–DC converter of the SECE circuit [31] (Fig. 10.25). The intermediate energy storage capacitor C_{int} brings an additional degree of freedom to control the energy conversion. This intermediate energy tank is used here for optimizing the trade-off between energy harvesting and mechanical damping. In addition, the buck–boost DC–DC converter makes the harvested power optimal whatever the load characteristics (i.e., no influence of the load equivalent resistance). The DSSH circuit implementation is a little bit more complicated than that of the SECE or the SSHI circuits. Despite the cumulated losses of the two conversion stages, experimental results exhibited much better performances than that of the standard and SECE techniques in the case of PVEH with small k^2Q_m. The so-called **ESSH** circuit proposed by Shen et al. [32] can be considered as an improvement of the DSSH circuit, which allows a finer control of the mechanical damping, induced by energy conversion.

Based on this general nonlinear approach, several other interface circuits have been proposed. Wu et al. proposed the so-called **SSDCI** circuit, based on a circuit similar to the series SSHI, but with a modified switch control [33]. The principle of the method consists of transferring the electrostatic energy available on the piezoelectric element to a storage capacitor through an inductance.

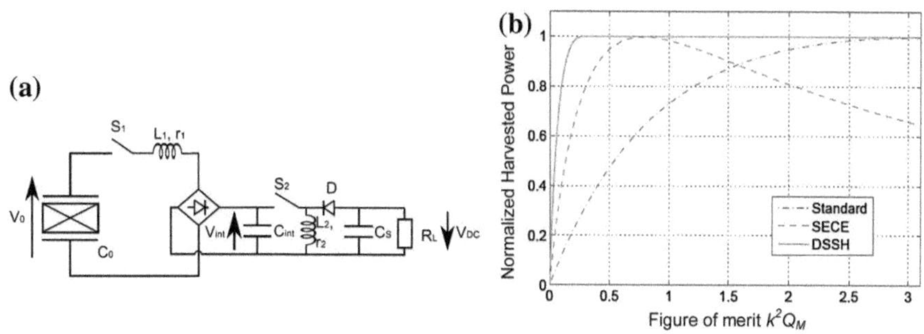

Fig. 10.25 **a** DSSH circuit schematic, and **b** power versus k^2Q_m at resonant frequency (from [31])

Dicken et al. proposed the so-called **Pre-Biasing** circuit [34], which enables to pre-bias the piezoelectric element with the appropriate voltage to finely optimize the harvested power. However, the complexity of the circuit, which includes numerous switches and two different power supplies, may be an obstacle to low-power standalone implementation.

Energy conversion cycles can be actively controlled using PWM inverters [10]. This technique was named "**active energy harvesting**" by Liu et al. [35], and further analyzed for blood-pressure energy harvesting [36]. Such active principle theoretically enables to give any shape to the piezoelectric voltage waveform, including for instance the "optimal impedance emulation" described in Sect. 3.1, yielding outstanding power level in theory. However, power consumption of the PWM control and energy losses of the circuit may limit the actual performances.

In summary, this subsection presented an overview of the main nonlinear circuits proposed until now. The reader will find more detailed analysis of these circuits in the original articles given in the References list. The next subsection presents a detailed analysis of the so-called OSECE circuit [37], which is one of the last proposed nonlinear circuits. It can be seen as a variant of the SSHI-MR circuit, but with simplified control principle, making much easier low-power standalone implementation.

10.6.3 The OSECE Approach

The OSECE (Optimized Synchronous Electrical Charge Extraction) has been developed as an improvement of the SECE approach. It allows to keep its main feature, namely the low dependency of the performance on the electrical load, while simplifying the electronic switches driving signals and enhancing the energy conversion [37].

The OSECE interface is shown in Fig. 10.26. A transformer with two primary and one secondary windings divides this circuit into two parts: the left part is very similar to the SSDI circuit [19], including the switch control signal; the right part is similar to the secondary of a typical flyback DC/DC converter (smoothing capacitor C_r plus equivalent load resistor R_L).

The switch S_1 and S_2 are complementary driven: S_1 is closed and S_2 opened after the displacement (voltage) reaches a maximum and S_1 is opened and S_2 closed after

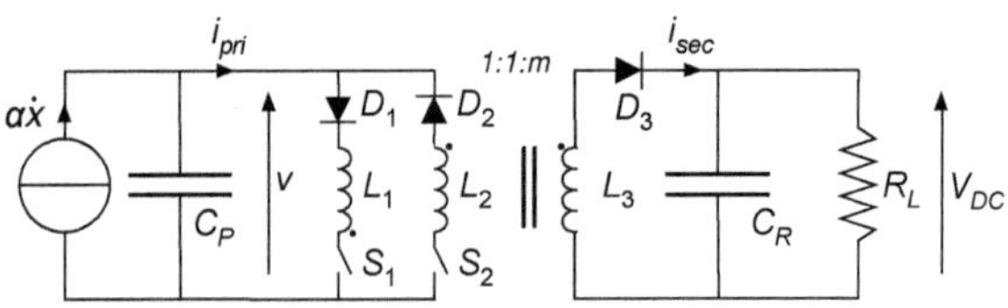

Fig. 10.26 Electrical network using the OSECE strategy

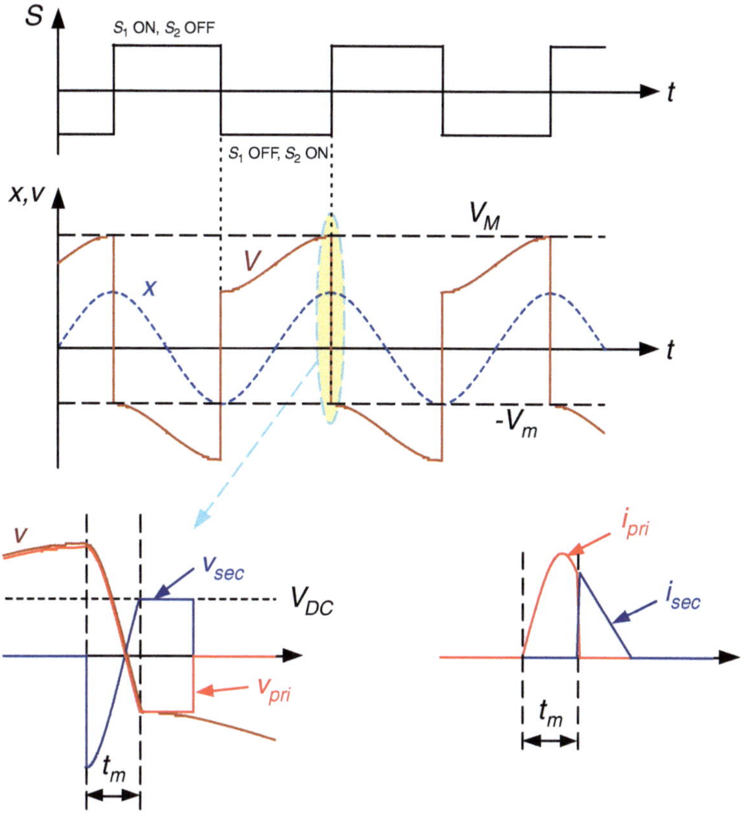

Fig. 10.27 Typical waveforms for the OSECE approach (from [38])

the displacement (voltage) reaches a minimum. This switching strategy allows the voltage to be partly inverted two times a period of vibration.

The voltage is only partly inverted because the inversion phase is stopped as soon as the voltage on the secondary reaches V_{DC} (at this moment, the diode in series with the closed switch becomes reversed biased). A detailed description and the modeling of this interface circuit can be found in [37].

Figure 10.27 shows the typical OSECE waveforms for the displacement, the piezoelectric voltage and the switch control signal for several periods. Details of voltages and currents close to the energy extraction moments are also shown. V_M and V_m are the piezoelectric voltage values just before and just after the energy extraction phase whose duration is t_m.

For the calculation of the performance of the OSECE approach, several assumptions are made: the magnetic circuit is linear; the coupling between the primary and secondary windings is ideal; the on-state voltage induced by the switches and the diodes are neglected; the output voltage V_{DC} is ripple-free; and sinusoidal ambient accelerations around the resonance frequency of the PVEH are considered.

The value of V_{DC} can be obtained as a function of the magnitude x_M of the dynamic mass relative displacement:

$$V_{DC} = -2m\frac{\alpha}{C_P}\frac{\cos(\theta)\,e^{\frac{-\theta}{2Q_I}}}{1+\cos(\theta)\,e^{\frac{-\theta}{2Q_I}}}x_M \ (V_{DC} \geq 0) \tag{10.6.2}$$

where Q_I is the quality factor of the primary $\{L_1, C_P\}$ oscillating circuit, m is the turns ratio of the transformer, and θ is the phase angle corresponding to the duration of the inversion phase, given by:

$$\theta = \pi - \arctan\left(m\sqrt{\frac{2\pi}{R_L C_P \omega}}\right) = \pi - \arctan\left(m\sqrt{\frac{2\pi}{r'\Omega}}\right) \tag{10.6.3}$$

As for the classical rectifier approach, only the fundamental component V_1 of the piezoelectric voltage, given by (10.6.4), is considered to impact the motion of the dynamic mass.

$$\underline{V}_1 = \frac{\alpha}{C_P}(\nu + j\upsilon)\,\underline{x} \text{ with } \begin{cases} \nu = 1 \\ \upsilon = \frac{4}{\pi}\frac{1-\cos(\theta)e^{\frac{-\theta}{2Q_I}}}{1+\cos(\theta)e^{\frac{-\theta}{2Q_I}}} \end{cases} \tag{10.6.4}$$

The expression in the frequency domain of the displacement and the normalized displacement are still the same as in Sect. 10.4.1, Eqs. (10.4.3) and (10.4.4), respectively.

Since $\nu = 1$, the natural angular frequency of the PVEH is the open-circuit angular resonance frequency $\omega_0\sqrt{1+k_m^2}$. It can also be seen that υ is always larger that $4/\upsilon$ (since θ is between $\pi/2$ and π) and depends on θ and Q_I. Compared to the classical rectifier case where υ is lower than $1/\pi$ it is clear that the OSECE approach induces more damping. As for the classical approach, the harvested power can be expressed as (10.5.5), and the normalized harvested power is finally given by Eq. (10.6.5); where the expression of $\underline{x}'$ is the same as in Eq. (10.4.4)

$$P' = \frac{16}{\pi}k_m^2\frac{\Omega\sin(\theta)\,e^{-\frac{\theta}{Q_I}}}{\left(1+\cos(\theta)\,e^{-\frac{\theta}{2Q_I}}\right)^2}\left|\underline{x}'\right|^2 \tag{10.6.5}$$

The maximal normalized power and the corresponding optimal normalized load are plotted as a function of the frequency in Fig. 10.28, for different values of k_m^2. When k_m^2 increases, it can be seen that whereas the bandwidth keeps on increasing, the maximal normalized power first increases and then decreases. This decrease is due to the too large damping effect induced by the OSECE approach in the case of highly coupled PVEH. Consequently comparing the power using the OSECE approach and the classical rectifier approach, it is found that OSECE leads to better performance for structures with low coupling coefficients or structures driven out of their resonance frequency.

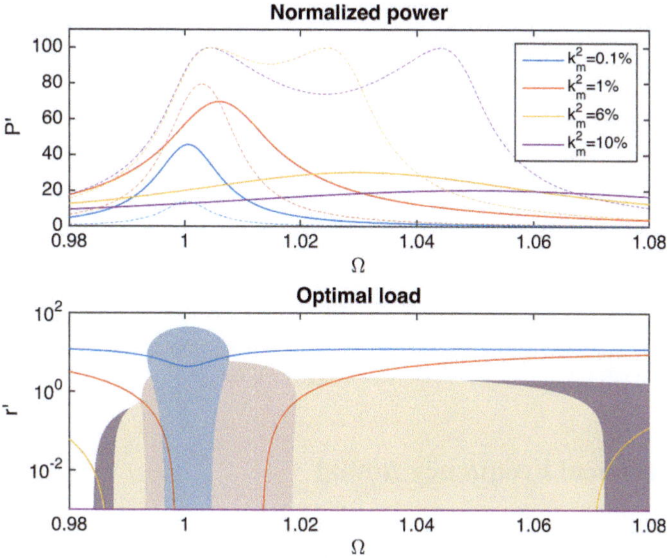

Fig. 10.28 *Upper plot* normalized power as a function of Ω for $\xi_L = 0.005$ ($Q_M = 100$) and different values of k_m^2 (*plain line* OSECE, dashed line: simple rectifier). *Lower plot* normalized optimal load as a function of Ω (areas corresponds to the load domain for which the power is at least 50 % of the maximal power)

The colored areas in the lower plot represent the load domains for which at least 50 % of the maximal power is harvested. It is shown that the dependency on the load is much lower than when using the classical rectifier approach.

The normalized power, bandwidth, and the figure of merit defined in (1.12) are plotted in Sect. 10.7 as a function of k_m^2 for different values of Q_M. Results using the OSECE circuit are plotted as plain lines, and can be compared with results from the classical approach, which are plotted as dashed lines. Figure 10.29a shows that the power using the OSECE approach is larger when $k_m^2 Q_M$ is lower than approximately 0.7 ($k_m^2 Q_M = 0.7$ corresponds to the black curve). This particular value depends on Q_I,which was set to 5 in this calculation. Figure 10.29b shows that the bandwidth using the OSECE approach is always larger, confirming the interest of this approach for PVEHs exited out of their resonances. Finally, Fig. 10.29c shows that the figure of merit is higher using the OSECE approach when $k_m^2 Q_M$ is lower than approximately 2.8 ($k_m^2 Q_M = 2.8$ corresponds to the black line).

It is worthy of note that the performances plotted in Fig. 10.29 are given for the optimal electrical load. Practical implementations of the classical approach may lead to significantly poorer results than theoretical ones because of the strong dependency of the harvested power on the load, which imply the mandatory use of a complex MPPT (Maximum Power Point Tracking) strategy. On the other hand, the OSECE approach is much more tolerant to the load variations, which makes its practical performance closer from theory [38].

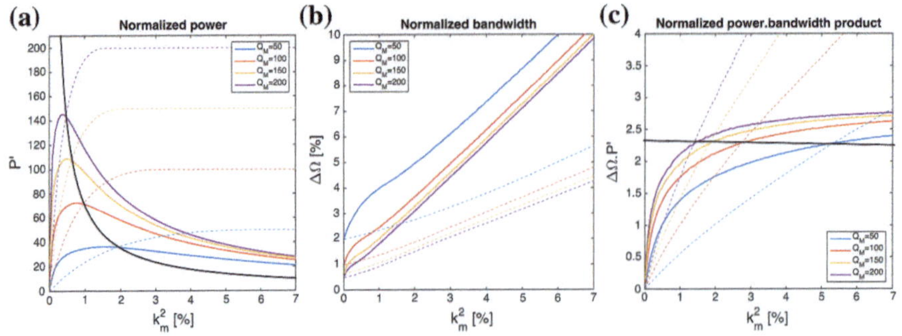

Fig. 10.29 **a** Normalized power versus k_m^2, **b** normalized bandwidth versus k_m^2, c) FoM versus k_m^2 *Plain lines* OSECE (Q_I=5, m=1), *dotted lines* simple rectifier

10.7 Electrical Frequency tuning

10.7.1 The Need for Wideband PEH

Most of PEHs reported in the literature exhibit squared coupling coefficient k_m^2 lower than 5 %. As shown in the previous sections, their bandwidth can be increased using nonlinear energy extraction circuits, but are still limited to a few percent of their resonance frequency. Such PEH can thus efficiently operate in a narrow frequency band tuned to match the excitation frequency.

However, environmental excitations have broadband or time-dependent characteristics in which the energy is distributed over a spread spectrum of frequencies. Two strategies have been investigated: Developing nonlinear wideband oscillators and developing linear oscillator with resonance frequency tuning mechanisms. Nonlinear oscillators (hardening, softening or bistable) are well suited for broadband vibration spectrum [39], whereas resonance frequency tuning is more appropriated for narrow band but time-dependent vibrations (which can be found on a motor whose rotation speed is varying). Nonlinear oscillators approaches are beyond the scope of this chapter since they are not related to the energy extraction circuit but to the architecture of the PEH. Resonance frequency tuning is usually done through an additional mechanical component that passively or actively changes the stiffness or inertia of a linear mechanical oscillator [40]. These mechanical approaches will also not be detailed here.

This section reports a theoretical nonlinear energy extraction approach to tune the resonant frequency of linear inertial PEH through the control strategy of the associated electronic interface circuit. When associated with piezoelectric devices exhibiting high electromechanical coupling, it enables to vary the resonant frequency in large proportions without any additional component.

Seddik et al. previously proposed a principle of control through the electronic interface circuit [41]. It consisted in connecting shunt capacitors to the piezoelectric device. Consequently, the electromechanical structure stiffness was varied, and the resonant frequency was changed accordingly. They showed that it was possible to significantly vary the resonant frequency of highly coupled piezoelectric energy harvesters. However, because the shunt capacitance was varied step by step, the resonant frequency could not be continuously tuned. With the proposed technique, it is expected that a continuous tuning is achievable on a noticeably larger frequency range.

10.7.2 Frequency Tuning SECE (FTSECE)

The FTSECE approach is also derived from the previously developed SECE (Synchronized Electrical Charge Extraction) approach. It is called FTSECE for Frequency Tuning SECE [42]. It consists in letting the PEH in open-circuit condition most of the time, and to extract the generated electrical charges two times a period of vibration. In contrast to the SECE technique, the FTSECE circuit does not extract energy at the piezoelectric voltage extremum, but with a phase shift ϕ. Moreover, whereas the piezoelectric voltage is null after each energy extraction phase in the SECE approach (all the electrical charges are extracted), it can be tuned with the FTSECE technique.

The FTSECE approach has not been practically demonstrated yet, but a possible electronic interface circuit for its realization is schematically depicted in Fig. 10.30. This circuit includes an electronic switch S, whose control is synchronized with the piezoelectric voltage. Two parameters of S are varied through the control circuit: the on-state time duration (t_{ON}) and the lag time duration (t_{lag}) between the voltage extrema and the instant where S is turned on. The phase shift $\phi = \Omega t_{lag}$ can be positive (S is closed after the voltage extremum) or negative (S is closed before the voltage extremum). If V_M is the piezoelectric voltage at the moment where S is turned on, varying t_{ON} will allow to set the voltage to βV_M after the energy extraction phase. If Q_I is the quality factor of the $\{L, C_P\}$ circuit, βbe tuned between 1 ($t_{ON} = 0$) and $-e^{-\pi/(2Q_I)}$ ($t_{ON} \simeq \pi(LC_P)^{1/2}$)

The load resistance value has no influence on the harvested power provided that the DC load voltage is larger than the piezoelectric voltage amplitude. In this way, energy transfers from the inductor to the rectifier only occur right after S is turned off. Varying t_{ON} enables to tune the amount of electrical energy extracted from the piezoelectric elements, and, if needed, to reverse the polarity of the piezoelectric voltage. Varying t_{lag} enables to modify the electromechanical structure stiffness

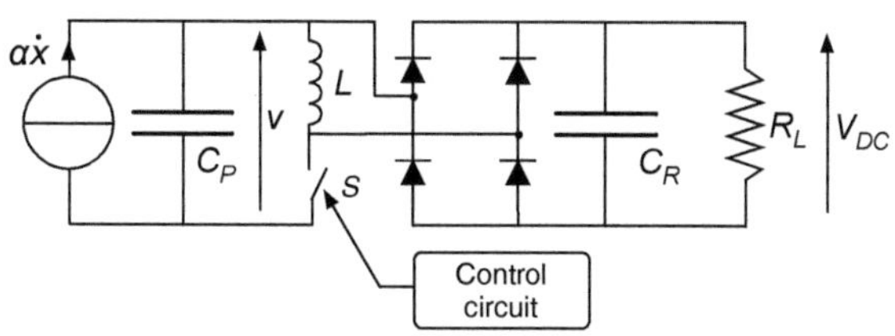

Fig. 10.30 Schematic representation of the FTSECE electronic interface circuit

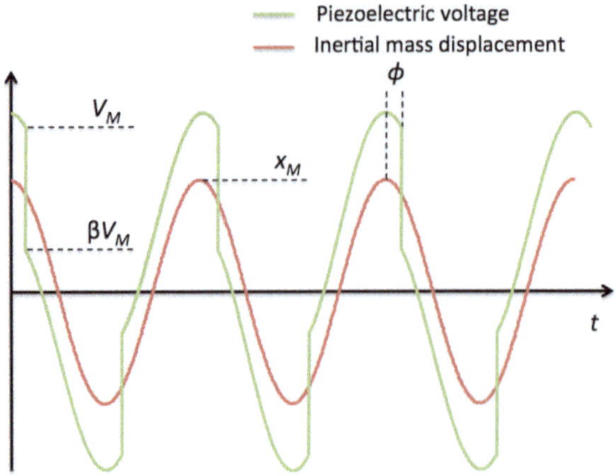

Fig. 10.31 Typical waveforms for the FTSECE approach

by tuning the phase shift between the piezoelectric voltage and the strain in the piezoelectric element.

Typical waveforms for the FTSECE approach are given in Fig. 10.31. Using the same procedure as in previous sections, the fundamental component V_1 of the piezoelectric voltage is calculated:

$$\underline{V}_1 = \frac{\alpha}{C_P}\left(\nu + j\upsilon\right)\underline{x} \text{ with } \begin{cases} \nu = 1 + \frac{2}{\pi}\frac{1-\beta}{1+\beta}\sin\left(2\phi\right) \\ \upsilon = \frac{4}{\pi}\frac{1-\beta}{1+\beta}\cos^2\left(\phi\right) \end{cases} \tag{10.7.1}$$

Equation (10.7.1) confirm that ϕ and β (t_{lag} and t_{ON}) affect both the resonance frequency of the system and its damping. Frequency tuning as well as optimization of the energy transfer is then achievable using the FTSECE approach.

The power extracted from the PEH is calculated from the energy extracted two times a period of the ambient vibration

$$P_{EX} = \frac{\omega}{2\pi} C_P V_M^2 \left(1 - \beta^2\right) \tag{10.7.2}$$

The expression of V_M is given by (10.7.3) where x_M is the magnitude of the dynamic mass displacement.

$$V_M = \frac{\alpha}{C_P} x_M \cos\left(\phi\right) \frac{2}{1+\beta} \tag{10.7.3}$$

Taking into account the losses in the inductor L the normalized harvested power can then be approximated by Eq. (10.7.4), where the expression of $\underline{x}'$ is given by Eq. (10.4.4)

$$P' = \frac{16}{\pi} k_m^2 \Omega \cos^2\left(\phi\right) \frac{\left(1-\beta\right)\left(e^{-\pi/(2Q_I)} + \beta\right)}{\left(1+\beta\right)^2} \left|\underline{x}'\right|^2 \tag{10.7.4}$$

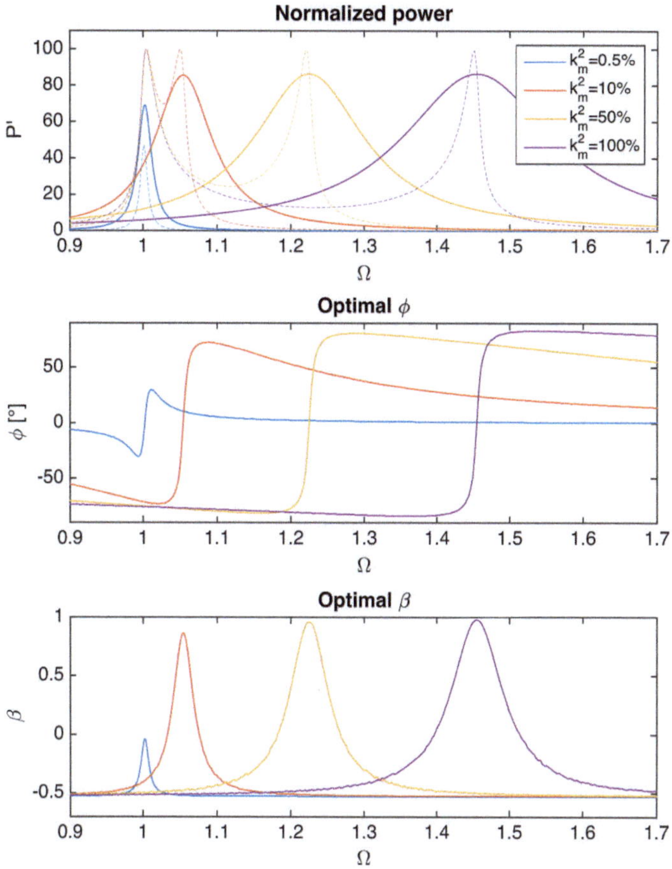

Fig. 10.32 *Upper plot* normalized power as a function of Ω for $\xi_L = 0.005$ ($Q_M = 100$) and different values of k_m^2 (*plain line* FTSECE, dashed line: simple rectifier). *Middle plot* optimal ϕ as a function of Ω. *Lower plot* optimal β as a function of Ω.

The normalized power as well as the optimal ϕ and β are plotted in Fig. 10.32 as a function of the operating frequency, for different values of k_m^2. High coupling coefficients have been considered because the FTSECE approach is especially promising in this case. Such coupling coefficients can be practically obtained using single crystals piezoelectric materials. For instance, a PEH for which k^2 equals 53 % ($k_m^2 = 112\,\%$) was presented in [42].

From the evolution of ϕ and β, it can be seen that the extraction times coincide with the voltage extrema (ϕ=0) only at the open-circuit resonance frequency of the PEH. In this case, if the coupling coefficient is high (typically larger than 5 %), only a small amount of the generated electrical charges are extracted (β tends to 1) to prevent too large damping of the PEH. When the vibration frequency is slightly shifted away from the open-circuit resonance frequency, the switching phase shift ϕ is adjusted to

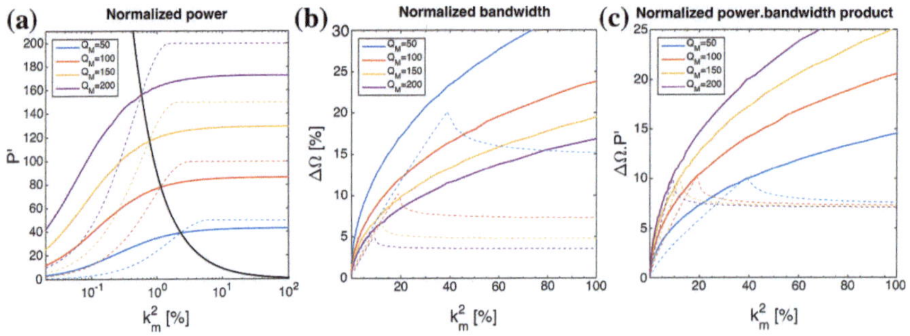

Fig. 10.33 **a** Normalized power versus k_m^2, **b** normalized bandwidth versus k_m^2, **c** FoM versus k_m^2 *Plain lines* FTSECE (Q_I=5, m=1), *dotted lines* simple rectifier

tune the resonance frequency of the PEH. This phase shift induces a decrease of the generated electrical charges, which is compensated by extracting a larger percentage of them. If the vibration frequency is shifted further, β becomes negative, which means that a portion of the generated electrical charges is injected back to the PEH with a reversed polarity. This effect increases the piezoelectric voltage amplitude and enhances the energy extraction.

The normalized power, bandwidth, and the figure of merit are plotted in Fig. 10.33 as a function of k_m^2 for different values of Q_M. Results using the FTSECE circuit are plotted as plain lines, and can be compared with results from the classical approach plotted as dashed lines.

Figure 10.33a shows that comparing to the classical rectifier approach the power using the FTSECE approach is larger when $k_m^2 Q_M < 0.9$ and slightly lower when $k_m^2 Q_M > 0.9$ (the black curve corresponds to $k_m^2 Q_M = 0.9$). Figure 10.33b confirms the bandwidth enhancement induced by the FTSECE circuit. Finally, Fig. 10.33c shows that the FTSECE approach gives higher figure of merit whatever the value of k_m^2.

The FTSECE circuit has not been practically implemented yet. The real-time tuning of two parameters (t_{on} and t_{lag}) requires the development of a dedicated switch control circuit, whose complexity may hinder the overall performances. Moreover, the effect of the piezoelectric voltage switching on highly coupled PEH may generate non-sinusoidal displacement of the dynamic mass, in which case the theoretical calculation detailed above become inaccurate. Because of the lack of experimental demonstration, this last section should then be considered with caution. It however suggests that resonant piezoelectric structures with high electromechanical coupling coefficient combined with dedicated nonlinear energy extraction circuit could lead to large bandwidth PEH (up to 20 % of the resonance frequency could reasonably be obtained).

To compare the FTSECE approach and the ideal impedance matching strategy (cf. Sect. 10.3), it is assumed that $\xi_E = 0.005$. Taking $k_m^2 = 100\,\%$ and $\xi_M = 0.005$ ($Q_M = 100$) leads to χ. The *FoM* using the impedance matching strategy would then be $\sqrt{\chi} = 200$, whereas it is only 20 using FTSECE (see red curve at $k_m^2 = 100\,\%$ in Fig. 10.33c). This result confirms that the ideal impedance matching gives an upper limit for a PVEH performance. It also suggests that some advanced energy harvesting circuits with enhanced bandwidth capability might be developed by further research.

10.8 Conclusion

This chapter presents a broad analysis of the existing power conditioning techniques for piezoelectric energy harvesting devices and the related circuits enabling practical implementation.

The classical impedance matching technique would enable in theory to get the best performances in terms of power and frequency bandwidth. However, practical implementation of this technique would require complicated, power consuming control algorithms. To date, very few experimental results can be found about this technique [8], confirming difficult implementation.

Overall, the so-called nonlinear interfaces bring several advantages. One of their most remarkable properties is the drastic power improvement of PVEH exhibiting low $k^2 Q_m$. For PVEH with higher $k^2 Q_m$, they give enhanced performance for pulsed excitation or out of resonance excitation.

Another advantage is that some of these nonlinear interfaces, such as SECE and OSECE, tend to minimize the effect of the electrical load on the energy conversion, which is a very interesting property when supplying a device with time varying electrical characteristics.

Finally, nonlinear interfaces, such as the FTCECE, may be used to electronically tune the PVEH resonant frequency, which could be used to drastically enhance their bandwidth.

Current trends in the domain of power conditioning circuits for piezoelectric vibration energy harvesting tend to push the limits towards "high" voltages (above 50 V) and "low" voltages (below 1 V) with ultra-low harvested power, typically in the range of 10 nW to 10 μW. Such developments raise new challenges in the domain of ultra-low power AC–DC and DC–DC converters, and in the design of ultra-low power ASICs.

References

1. Williams, C. B., & Yates, R. B. (1996). Analysis of a micro-electric generator for microsystems. *Sensors and Actuators A: Physical*, *52*(1), 8–11.
2. Arroyo, E., Badel, A., Formosa, F., Wu, Y. P., & Qiu, J. (2012). Comparison of electromagnetic and piezoelectric vibration energy harvesters: Model and experiments. *Sensors and Actuators A: Physical*, *183*, 148–156.
3. Krimholtz, R., Leedom, D. A., & Matthaei, G. L. (1970). New equivalent circuits for elementary piezoelectric transducers. *Electronics Letters*, *6*(13), 398–399.

4. Tilmans, H. A. C. (1996). Equivalent circuit representation of electromechanical transducers: I. Lumped-parameter systems. *Journal of Micromechanics and Microengineering*, *6*(1), 157–176.
5. Yang, Y., & Tang, L. (2009). Equivalent circuit modeling of piezoelectric energy harvesters. *JIMSS*, *20*(18), 2223–2235.
6. Renno, J. M., Daqaq, M. F., & Inman, D. J. (2009). On the optimal energy harvesting from a vibration source. *JSV*, *320*(1), 386–405.
7. Guyomar, D., Badel, A., Lefeuvre, E., & Richard, C. (2005). Toward energy harvesting using active materials and conversion improvement by nonlinear processing. *IEEE Transactions on Ultrasonics, Ferroelectrics, and Frequency Control*, *52*(4), 584–595.
8. Li, Y., Richard, C. (2014). Piezogenerator impedance matching using Mason equivalent circuit for harvester identification. In *Presented at the SPIE Smart Structures and Materials + Nondestructive Evaluation and Health Monitoring* (Vol. 9057, p. 90572I).
9. Erickson, R. W., & Maksimovic, D. (2001). *Fundamentals of power electronics*. US: Springer.
10. Lefeuvre, E., Sebald, G., Guyomar, D., & Lallart, M. (2009). Materials, structures and power interfaces for efficient piezoelectric energy harvesting. *Journal of* ....
11. Shu, Y. C., & Lien, I. C. (2006). Analysis of power output for piezoelectric energy harvesting systems. *Smart Materials and Structures*, *15*(6), 1499–1512.
12. Ottman, G. K., Hofmann, H. F., Bhatt, A. C., & Lesieutre, G. A. (2002). Adaptive piezoelectric energy harvesting circuit for wireless remote power supply. *IEEE Transactions on Power Electronics*, *17*(5), 669–676.
13. Ottman, G. K., Hofmann, H. F., & Lesieutre, G. A. (2003). Optimized piezoelectric energy harvesting circuit using step-down converter in discontinuous conduction mode. *IEEE Transactions on Power Electronics*, *18*(2), 696–703.
14. Lefeuvre, E., Audigier, D., Richard, C., & Guyomar, D. (2007). Buck-boost converter for sensorless power optimization of piezoelectric energy harvester. *IEEE Transactions on Power Electronics*, *22*(5), 2018–2025.
15. Yi, J., Su, F., Lam, Y.-H., Ki, W.-H., & Tsui, C.-Y. (2008). An energy-adaptive MPPT power management unit for micro-power vibration energy harvesting. In *Presented at the 2008 IEEE International Symposium on Circuits and Systems—ISCAS 2008* (pp. 2570–2573).
16. Kong, N., & Ha, D. S. (2012). Low-power design of a self-powered piezoelectric energy harvesting system with maximum power point tracking. *IEEE Transactions on Power Electronics*, *27*(5), 2298–2308.
17. Shim, M., Kim, J., Jung, J., & Kim, C. (2014). Self-powered 30 μW-to-10 mW piezoelectric energy-harvesting system with 9.09ms/V maximum power point tracking time. In *2014 IEEE International Solid- State Circuits Conference (ISSCC)* (pp. 406–407).
18. Bandyopadhyay, S., Mercier, P. P., Lysaght, A. C., Stankovic, K. M., & Chandrakasan, A. P. (2014). A 1.1 nW energy-harvesting system with 544 pW quiescent power for next-generation implants. *IEEE Journal of Solid-state Circuits*, *49*(12), 2812–2824.
19. Richard, C., Guyomar, D., Audigier, D., & Ching, G. (1999). Semi-passive damping using continuous switching of a piezoelectric device. In *Presented at the 1999 Symposium on Smart Structures and Materials*, *3672*, 104–111.
20. Badel, A. (2005). Efficiency enhancement of a piezoelectric energy harvesting device in pulsed operation by synchronous charge inversion. *JIMSS*, *16*(10), 889–901.
21. Lefeuvre, E., Badel, A., Richard, C., & Guyomar, D. (2004). High-performance piezoelectric vibration energy reclamation. *Smart Structures and Materials*, *5390*, 379–387.
22. Shu, Y. C., Lien, I. C., & Wu, W. J. (2007). An improved analysis of the SSHI interface in piezoelectric energy harvesting. *Smart Materials and Structures*, *16*(6), 2253–2264.
23. Lefeuvre, E., Badel, A., Richard, C., Petit, L., & Guyomar, D. (2006). A comparison between several vibration-powered piezoelectric generators for standalone systems. *Sensors and Actuators A: Physical*, *126*(2), 405–416.
24. Taylor, G. W., Burns, J. R., Kammann, S. M., Powers, W. B., & Wel, T. R. (2001). The energy harvesting Eel: A small subsurface ocean/river power generator. *IEEE Journal of Oceanic Engineering*, *26*(4), 539–547.

25. Badel, A., Benayad, A., Lefeuvre, E., Lebrun, L., Richard, C., & Guyomar, D. (2006). Single crystals and nonlinear process for outstanding vibration-powered electrical generators. *IEEE Transactions on Ultrasonics, Ferroelectrics and Frequency Control*, *53*(4), 673–684.
26. Lefeuvre, E. (2005). Piezoelectric energy harvesting device optimization by synchronous electric charge extraction. *JIMSS*, *16*(10), 865–876.
27. Makihara, K., Onoda, J., & Miyakawa, T. (2006). Low energy dissipation electric circuit for energy harvesting. *Smart Materials and Structures*, *15*(5), 1493–1498.
28. Lallart, M., & Guyomar, D. (May 2008). An optimized self-powered switching circuit for non-linear energy harvesting with low voltage output. *Smart Materials andStructures*, *17*(3), 035030–9.
29. Garbuio, L., Lallart, M., Guyomar, D., Richard, C., & Audigier, D. (2009). Mechanical energy harvester with ultralow threshold rectification based on SSHI nonlinear technique. *IEEE Transactions on Industrial Electronics*, *56*(4), 1048–1056.
30. Elliott, A. D. T., & Mitcheson, P. D. (2012). Implementation of a single supply pre-biasing circuit for piezoelectric energy harvesters. *Procedia Engineering*, *47*, 1311–1314.
31. Lallart, M., Garbuio, L., Petit, L., Richard, C., & Guyomar, D. (2008). Double synchronized switch harvesting (DSSH): A new energy harvesting scheme for efficient energy extraction. *IEEE Transactions on Ultrasonics, Ferroelectrics, and Frequency Control*, *55*(10), 2119–2130.
32. Shen, H., Qiu, J., Ji, H., Zhu, K., & Balsi, M. (2010). Enhanced synchronized switch harvesting: a new energy harvesting scheme for efficient energy extraction. *Smart Materials and Structures*, *19*(11), 115017.
33. Wu, W. J., Wickenheiser, A. M., Reissman, T., & Garcia, E. (2009). Modeling and experimental verification of synchronized discharging techniques for boosting power harvesting from piezoelectric transducers. *Smart Materials and Structures*, *18*(5), 055012–15.
34. Dicken, J., Mitcheson, P. D., & Stoianov, I. (2009). Increased power output from piezoelectric energy harvesters by pre-biasing. In *Presented at the PowerMEMS 2011*.
35. Liu, Y., Tian, G., Wang, Y., Lin, J., Zhang, Q. H. F., & Hofmann (2009). Active piezoelectric energy harvesting: general principle and experimental demonstration. *JIMSS*, *20*(5), 575–585.
36. Deterre, M., Lefeuvre, E., & Dufour-Gergam, E. (2012). An active piezoelectric energy extraction method for pressure energy harvesting. *Smart Materials and Structures*, *21*(8), 085004.
37. Wu, Y. P., Badel, A., Formosa, F., Liu, W. Q., & Agbossou, A. E. (2013). Piezoelectric vibration energy harvesting by optimized synchronous electric charge extraction. *JIMSS*, *24*(12), 1445–1458.
38. Wu, Y. P., Badel, A., Formosa, F., Liu, W. Q., & Agbossou, A. (2014). Self-powered optimized synchronous electric charge extraction circuit for piezoelectric energy harvesting. *JIMSS*, *25*(17), 2165–2176.
39. Gammaitoni, L., Neri, I., & Vocca, H. (2009). Nonlinear oscillators for vibration energy harvesting. *Applied Physics Letters*, *94*(16), 164102.
40. Eichhorn, C., Tchagsim, R., Wilhelm, N., & Woias, P. (2011). A smart and self-sufficient frequency tunable vibration energy harvester. *Journal of Micromechanics and Microengineering*, *21*(10), 104003.
41. Ahmed Seddik, B., Despesse, & Defay, E. (2012). Autonomous wideband mechanical energy harvester. In *Presented at the IEEE International Symposium on Industrial Electronics ISIE*, Hangzhou, China.
42. Badel, A., & Lefeuvre, E. (2014). Wideband piezoelectric energy harvester tuned through its electronic interface circuit. *Journal of Physics: Conference Series*, *557*, 012115.

MIX
Papier aus verantwortungsvollen Quellen
Paper from responsible sources
FSC® C105338

If you have any concerns about our products,
you can contact us on
ProductSafety@springernature.com

In case Publisher is established outside the EU,
the EU authorized representative is:
Springer Nature Customer Service Center GmbH
Europaplatz 3, 69115 Heidelberg, Germany

Printed by Libri Plureos GmbH
in Hamburg, Germany